MARINE ELECTROCHEMISTRY

Edited by

Joan B. Berkowitz
A. D. Little, Inc.
Cambridge, Massachusetts

Ralph Horne
A. D. Little, Inc.
Cambridge, Massachusetts

Mario Banus
B. U. Marine Program
Woods Hole, Massachusetts

P. L. Howard
P. L. Howard Associates, Inc.
Millington, Maryland

M. J. Pryor
Olin Corporation
New Haven, Connecticut

G. C. Whitnack
Naval Weapons Center
China Lake, California

H. V. Weiss
Naval Undersea Research
and Development Center
San Diego, California

THE ELECTROCHEMICAL SOCIETY, INC., Post Office Box 2071, Princeton, New Jersey 08540

Copyright 1973

by

The Electrochemical Society, Incorporated

Papers contained herein may not be reprinted and may not be digested by publications other than those of The Electrochemical Society in excess of 1/6 of the material presented.

Library of Congress Catalog Card Number: 73-75170

Printed in the United States of America

PREFACE

A great deal of electrochemical science and technology has been used in developing an understanding of the complex chemistry of the oceans, and in exploring the practical potential of the oceans as a natural resource. Inspite of the overlap of technical interests, few electrochemists are conversant with the general field of oceanography, and few oceanographers are fully aware of the kinds of problems that electrochemists might be able to solve. The purpose of this Symposium Volume is to place in perspective the present and projected role of electrochemistry in the solution of problems relevant to the marine environment.

The Marine Electrochemistry Symposium, upon which this volume is based, was sponsored by the New Technology Committee of the Electrochemical Society, in cooperation with the Battery, Corrosion, Electroorganic and Electrothermics and Metallurgy Divisions. The Symposium was part of the 142nd Meeting of the Society, held in Miami Beach, Florida, October 8-13, 1972. Forty papers were presented, of which 25 are included in this volume.

The keynote address by Dr. Dayton E. Carritt presents an overview of the theme of the Symposium. The following two papers by Dr. Ralph Horne and Dr. Paul Mangelsdorf respectively, discuss the oceans as a chemical system and as an electromagnetic system. Technical sessions were organized in the following areas:

(1) High pressure electrochemistry of aqueous systems - theory and practice.

(2) Batteries for deep sea submergence.

(3) Marine corrosion.

(4) The use of electrochemical devices in oceanography.

(5) Electrochemical monitoring and control of marine pollution.

The session Chairmen and Vice-Chairmen, the authors, and the participants, are due grateful recognition for their significant contributions to the success of the Symposium. Parts of the discussion are reprinted to give some indication of the lively interest generated by the papers presented.

On behalf of all who have had an interest in this undertaking, it is a pleasure to acknowledge the vital contribution of the Electrochemical

Society's Headquarters staff in handling the arrangements for the Symposium, and in guiding the publication of this Symposium Volume.

                                       Joan B. Berkowitz
                                       General Chairman

TABLE OF CONTENTS

PREFACE ... iii

INTRODUCTION TO MARINE ELECTROCHEMISTRY

Chairman - R. Horne   Vice-Chairman - J. Berkowitz

1. MARINE ELECTROCHEMISTRY: AN OVERVIEW

    Dayton E. Carritt 3

2. THE OCEANS AS A CHEMICAL SYSTEM

    R. A. Horne 22

    Discussion 29

3. THE OCEAN AS AN ELECTROMAGNETIC SYSTEM

    Paul C. Mangelsdorf, Jr. 31

THE EFFECTS OF DEEP SEA ENVIRONMENT ON ELECTROCHEMICAL SYSTEMS

Chairman - M. Banus   Vice-Chairman - M. Hotz

1. SURVEY OF TECHNIQUES TO STUDY AQUEOUS SYSTEMS AT PRESSURES OF 1-3 KBAR

    W. A. Adams and A. R. Davis 53

2. VOLUMES OF ACTIVATION FOR THE CONDUCTANCE OF MONOVALENT IONS IN WATER

    S. B. Brummer and A. B. Gancy 76

    Discussion 97

3. DESIGN AND APPLICATION OF A DEEP SEA pH SENSOR

    S. Ben-Yaakov and I. R. Kaplan 98

    Discussion 109

4. THE INCREMENTAL CONCENTRATION CELL AND ITS APPLICATION FOR STUDYING IONIC DIFFUSION IN SEAWATER

    S. Ben Yaakov 111

    Discussion 123

5. COMPUTER MODELING OF INORGANIC EQUILIBRIA IN SEAWATER

    Gordon Atkinson, M. O. Dayhoff, and David W. Ebdon 124

BATTERIES FOR DEEP SUBMERGENCE USE

1. HIGH CAPACITY SALT WATER BATTERIES UTILIZING FUSION-CAST LEAD AND CUPROUS CHLORIDE CATHODES

    T. J. Gray and J. Wojtowicz 141

    Discussion 165

2. CHARACTERISTICS OF LEAD ACID AND SILVER-ZINC BATTERIES IN THE DEEP OCEAN ENVIRONMENT

    K. K. Berju, E. L. Daniels, and E. M. Strohlein 166

3. SILVER-ZINC BATTERY POWER FOR DEEP OCEAN VEHICLES

    Guy W. Work 179

    Discussion 186

4. ENERGY DENSITY, SERVICE LIFE, AND COST OF DEEP-OCEAN VEHICLE POWER

    Donald O. Newton 187

5. A FUTURE LOOK AT SMALL MANNED SUBMERSIBLE ENERGY STORAGE SYSTEMS

    A. W. Petrocelli, E. S. Dennison, and A. S. Berchielli 196

MARINE CORROSION

    Chairman - M. J. Pryor   Vice-Chairmen - Z. A. Foroulis
                                                    J. A. Ford

1. CORROSION IN THE OCEAN

    F. L. LaQue 219

2. THE INFLUENCE OF WATER CHEMISTRY ON THE BEHAVIOR OF METALS IN HIGH-TEMPERATURE SEAWATER

    Oliver Osborn and Alan L. Whitted 242

3. SOME IDEAS ON THE MOLECULAR STRUCTURE OF SEAWATER BASED ON COMPRESSIBILITY MEASUREMENTS

    Iver W. Duedall 256

4. STRESS CORROSION CRACKING OF Ti-8Al-1Mo-1V ALLOY IN NATURAL SEAWATER

   W. R. Cares and M. H. Peterson 269

5. PRODUCTION OF FRESH WATER FROM THE SEA BY ELECTRODIALYSIS

   Frank B. Leitz and Mauro A. Accomazzo 278

6. PURITY OF DISTILLATE IN MULTI-STAGE FLASH DESALINATION PLANT

   Kenkichi Izumi 296

   ### THE USE OF ELECTROCHEMICAL DEVICES IN CHEMICAL OCEANOGRAPHY

   Chairman - G. C. Whitnack   Vice-Chairman - H. V. Weiss

1. ANODIC STRIPPING VOLTAMMETRY OF TRACE METALS IN SEAWATER

   Alberto Zirino, Stephen H. Lieberman, and Michael L. Healy .. 319

2. A TECHNIQUE FOR MEASURING CARBON DIOXIDE HYDRATION KINETICS IN SEAWATER

   James H. Mathewson 333

3. RECENT APPLICATIONS OF SINGLE-SWEEP POLAROGRAPHY, ABOARD SHIP AND IN THE LABORATORY, TO THE ANALYSIS OF TRACE ELEMENTS IN SEAWATER

   Gerald C. Whitnack 342

   ### ELECTROCHEMICAL MONITORING AND CONTROL OF MARINE POLLUTION

   Chairman - H. V. Weiss   Vice-Chairman - G. C. Whitnack

1. OIL POLLUTANTS IN THE MARINE ENVIRONMENT

   Sachio Yamamoto 355

2. BIOELECTRIC POTENTIAL MEASUREMENTS OF LIVING BIOLOGICAL MEMBRANES AS POLLUTION INDICATORS

   Anitra Thorhaug and Marcella Fernandez 368

3. A REVIEW OF LEAD, SULFUR, SELENIUM AND MERCURY IN PERMANENT SNOWFIELDS

   H. V. Weiss 385

INTRODUCTION TO

MARINE ELECTROCHEMISTRY

Chairmen:
Ralph Horne
Joan Berkowitz

MARINE ELECTROCHEMISTRY: AN OVERVIEW

Dayton E. Carritt
Professor of Marine Sciences
University of Massachusetts
Amherst, Massachusetts 01002

ABSTRACT

   The study of the open, deep ocean continues, but there is renewed interest in coastal processes and in the development and use of marine resources. Interinstitutional and programs involving international cooperation are growing.

   Electrochemistry has contributed to our present state of understanding of oceanic processes and use of marine resources largely through the development of instrumentation such as that for the precision measurement of the electrical conductance of solutions of electrolytes (sea water), stable electrodes as in the geomagnetic electrokinetograph, ion specific electrodes, and anodic stripping voltametry.

   Needs for the future appear to be associated with <u>in situ</u> measurements of chemical properties, continuous unattended measurements for days to months, and as yet unknown needs as coastal processes are studied. The latter involve a much compressed time and space scale, compared to the open ocean.

Mr. Chairman, members of the Electrochemical Society, ladies and gentlemen. I am indeed pleased and honored to be asked to be the key note speaker in your Symposium on Marine Electrochemistry. I should immediately acknowledge my appreciation for the help given to me by two of your members. First, to Dr. Ralph Horne for his suggestion that I might like to take part in this symposium. Secondly, to Dr. Joan Berkowitz for her continuing help in keeping me nearly on schedule in preparing for this meeting. Her ability to make a very respectable abstract from a few mumbled words on a long distance phone circuit is remarkable. My thanks to both.

My first thoughts on preparing an overview of marine electrochemistry were that although each field, oceanography and electrochemistry, by itself would have a rather massive literature and total effort of work in progress, the interaction between the two would be relatively small and that the entire subject could be wrapped up in one neat, small package rather easily. So I thought at first. I wasn't very far into the literature, however, before I discovered that I had been quite wrong.

There are several areas in which our present understanding of how the ocean behaves depends directly upon the inputs from electrochemical tools, techniques, and principles. Without them our present understanding of the sea would be lacking. A few of these I will discuss this morning. In addition, I have found a host of publications in the chemical and electrochemical literature, all of which are quite obviously related to significant developments in marine fields. In a few cases some of this peripheral but important work has been summarized in review papers and monographs. These I will refer to, but I have made no attempt to compile an exhaustive bibliography of all of these related works.

A glance at the scheduled sessions for this symposium indicates that several major areas of interaction between electrochemistry and the marine sciences will be covered in detail during the next few days.

It seems to me then that my primary function here this morning is to provide an oceanographic framework within which these specialized discussions of marine electrochemistry can take place. Since this is a meeting of the Electrochemical Society, I assume that the main contributions to your Symposium will be by and for electrochemists, and therefore it will be both unnecessary and inappropriate for me, an oceanographer but one time electrochemist, to undertake the "Coals to Newcastle" kind of task of developing a detailed review of electrochemistry as such. What I hope to do during the next few minutes is to help you, the electrochemists, to look at a few oceanographic problems, and to indicate how the use of electrochemical tools, techniques, and principles have helped to solve them and to provide a better understanding of what goes on in the ocean, and finally to do a little crystal ball gazing and to suggest some possible lines which might be followed in the near future. My function then is somewhat like that of a guide attempting to point out to you, the fishermen, where fish might be found. And to continue the analogy, I hope that in leading you to the fish I will be able to develop with you a feeling for the environment through which the many kinds of fish are distributed. And finally, perhaps together we can decide on the best kinds of bait and tackle to be used.

Let's first take a broad brush look at the field of oceanography. Put in the perspective of the past few decades, it seems to me that it is now fairly well into a period of a rather rapid change of emphasis on what is being done and why. A view of oceanography from the early 1900's through the 1960's shows several rather easily recognized trends. Until the late 30's the ocean was studied by very few small groups of scientists who in addition to being intrigued by the sea were for the most part either independently wealthy or were chiefs of state. As WW II developed, navy departments around the world saw possible advantages to having a better understanding of all aspects of the ocean than their potential enemies. The massive increase in federal funding for oceanography in this country, initially prompted in large part by the needs of our Navy, and which started slowly in the late 30's and early 40's, brought many new people, new instrumentation and equipment, and many new ideas into the field, with the result that not only has our Navy enjoyed the advantages of the inputs of modern scientific information, but also a much fuller understanding of oceanic processes, much of which has no direct connection with naval operations, has been developed.

The development during the past three or four decades has had one characteristic which is different from previous times and which now seems again to be changing. For the past thirty years or so, emphasis has been on the deep, open ocean with much less thought and support given to shallow, inshore, coastal regions. As a consequence we have developed a much better predictive capability for open ocean processes than for those occuring in our own front yard.

There is, however, a renewed interest in our coastal areas, an interest that recently was brought into focus by the 1969 Report of the Commission of Marine Sciences, Engineering, and Resources--the so-called Stratton Commission. This report, Our Nation and the Sea (1), while not neglecting the importance of the deep ocean as a national resource, did call attention to the poor state of understanding we have of our coastal waters and their resources and made specific recommendations that, if followed, would provide for the rational utilization of all of our inshore regions.

I mention this change in emphasis because, as we shall see, it may be impossible to obtain what now seems to be the necessary information about our coastal regions by doing in shallow waters what has proved to be effective and sufficient in the study of the deep ocean. Our neglect of the coastal waters has in one sense been a whistling in the dark, for it has been long recognized that not only are the inshore processes--that is, the physics, chemistry, biology, and geology--much more complicated there than similar processes occuring in the deep ocean, but also that coasts are regions where major social, economic, and political interactions occur--interactions that can for the most part be neglected in open ocean studies.

The increasing awareness that some of our resources are non-renewable, and also that many present multiple uses of coastal areas are incompatible, has provoked a renewed interest in coastal zone processes and the recognition of an obvious need to develop better management capabilities for our coastal resources.

An essential ingredient of any managerial system is a predictive capability that will permit the consequences of any recommended action to be known with reasonable certainty. Thus it is essential that we have an understanding of the basic processes that occur within any selected region, and have the capability of predicting both qualitatively and quantitatively how perturbations in one part of the system will be reflected in all other parts of the system. At the present time the reliability of many of our predictions concerning coastal processes is very low.

The point to be made here is that although we are continuing our efforts to develop a better understanding of deep ocean phenomena, we are now faced with the urgent need to develop an understanding of an entirely new set of problems, of which only a part are familiar to most of us who have been working in the physical, chemical, and biological studies of the deep ocean.

There are two other important aspects of the changing oceanographic scene that should be noted. The first of these is related to and may be considered a part of the return to coastal problems noted above. That is, there is an increasing emphasis on the use of oceans. The oceans, and especially coastal areas, are being considered more as resources than just the environment in which navies operate, and on which convenient transportation routes between continents exist. The commercial exploitation of marine resources has markedly increased during the past decade. And this is to be the subject of a part of this symposium. Interest in marine aquaculture is increasing in this country, suggesting that the farming point of view, as contrasted with the hunting, is increasing. This of course reflects the historical development of our terrestrial food sources. It is now clear that the ocean does not have "infinite capacity" as a dumping ground for many wastes and by-products of our expanding technologies. The finite capacity of the ocean is being considered as a resource that in many respects is a non-renewable resource. This may well be one facet of our increasing awareness of the need for rational management of all of our resources, especially our non-renewable ones.

The second of these features has to do with the way that the so-called "big problems" are being approached. Interinstitutional programs are increasing. These of course are not new, and perhaps to suggest they are increasing is to read noise as signal. Nevertheless, the Mid-Ocean Dynamic Experiment (MODE), the Geochemical Ocean Section Study (GEOSECS), the International Decade of Ocean Exploration (IDOE), and the International Biological Program (IBP) are programs through which both complex and long range studies are carried out by cooperation between several U.S. oceanographic institutions, and in the case of IDOE and IBP through international cooperation.

What are the scientific problems in oceanography into which electrochemistry has had an effective input in the past? What does the renewed emphasis on coastal problems mean in terms of continued use of the tools and techniques that have been effective in the past? There are three different but interrelated sets of problems that have engaged scientists in their study of the sea for many years. Electrochemistry has made significant contributions to our present state of understanding in each of them. These can be summarized in a rather uninformative way as: (a) ocean circulation, (b) chemical composition of the ocean, and (c) biological productivity of the sea.

The term ocean circulation can be taken to include a wide variety of theoretical and experimental interests. Much attention has been and is being given to questions such as, where has a given parcel of water come from? How fast is it moving, both horizontally and vertically? Where will it be at some time in the future? Although these are straight-forward questions, answers to them are not easily obtained. The difficulty here is that, except in well defined current systems such as the Gulf Stream and the Kuroshio Current, flow is extremely slow. Whereas the major permanent currents may flow with speed in excess of three knots (150 cm/sec), significant vertical motion may be as slow as 10^{-5} cm/sec, and horizontal motion in areas well away from the permanent currents rarely exceed 10 cm/sec. Thus, although relatively new techniques such as telemetering neutral density floats and moored verticle strings of recording current meters are providing direct measures of these slow horizontal velocities, by far the greatest part of what we think we know about ocean currents has been obtained by indirect methods. This may seem strange, for speeds of the order of 1 to 10 cm/sec while slow (1 mi/hr - 44 cm/sec) certainly can be measured with ease. This is so only if one has a stable platform on which to work and a firmly fixed reference from which distances can be measured. Neither of these conditions exists in most places where one wants to measure ocean currents. The role and pitch of a ship imparts a combination of horizontal and vertical motion to anything suspended from it, and the continual yawing on the end of several miles of anchor line makes the selection of a zero reference from which to measure centimeter intervals an academic exercise. Thus for these practical reasons most estimates we now have for motion of the ocean have been obtained by indirect methods. The techniques for obtaining these estimates, together with many of their difficulties and uncertainties, are well documented and need not be discussed in detail (2) (3) (4). There is, however, an essential part of the methodology that should be noted.

The indirect method of estimating ocean currents depends upon measurements that will describe the mass field in the ocean. Theory indicates that if we know the in situ density everywhere in the ocean, we will be able to calculate the speed and direction of water motion. A unique property of sea water has made the estimation of density a relatively simple task, at least simple in principle.

First, we should note that the density of sea water varies within rather narrow limits. In the open sea the property falls within the range 1.023 to 1.028, a range of some 5 parts per thousand. However, density or properties computed directly from it are being reported to ±0.000001, and significant decisions are made based upon changes of ±0.00001. These decisions thus assumed that density is known to within plus or minus one part per million! Achieving this accuracy or even precision in any kind of physical or chemical measurement is not a simple routine matter. However, having been achieved, the measurements which make it possible are now done routinely, often with all measured quantities being obtained with in situ transducers. This achievement is due largely to the development of precise, accurate, and reliable systems for measuring the electrical conductivity of aqueous solutions, for which the oceanographers have the electrochemists to thank.

Very briefly the situation is as follows. The density of an aqueous solution can be expressed as a function of temperature, chemical composition, and pressure. Pressure is included here because, despite the relative incompressibility of water and aqueous solutions, at the pressures encountered in the deep ocean (up to 1,000 atm) pressure coefficients must be considered if the required accuracy and precision is to be achieved. Classically, and still extensively used, the use of mercurial thermometers to provide estimates of in situ temperature as well as of pressure involves well documented procedures (2) (5).

The chemical composition of sea water, a density controlling factor, shows the unique property described as constant relative composition. This means that although the total dissolved salt content in samples of open ocean water may vary between approximately

33.0 to 38.0 parts per thousand, the ratios of the concentrations of
any two major constituents, of any one to the total and of any one to
the density, are remarkable constant. The main variable then is the
concentration of water. As a result, measured values of colligative
properties will be proportional to the total salt content and to density. Thus, measured values of electrical conductivity can be used,
when properly calibrated, to provide estimates of the composition factor in the density relations noted above. Two papers in this afternoon's session will examine the pressure-conductance problem.

The development of the instrumentation and techniques for measuring the electrical conductivity is certainly of primary interest here,
for it represents a significant contribution by electrochemists to the
development of oceanography. Like many significant advances in the
scientific fields, our National Bureau of Standards played a leading
role in the studies which lead to the routine use of conductance
bridges in oceanography (7). The first routine use of the system at
sea appears to have been by the U.S. Coast Guard in its International
Ice Patrol Service on the cutter Tampa during 1923. This equipment
and all of the modifications of it for many years had one common feature which prevented further modification into instruments capable of
in situ measurements. All of the early equipment contained specially
designed conductance cells in which the conducting path was defined by
two platinized electrodes in a geometry which minimized interactions
between electrode leads and other circuit components. The conductance
cells and critical circuit components were carefully thermostated.
I'm sure that one of the characteristics of this kind of equipment
would be readily predicted by most electrochemists today. The tendency of platinized platinum electrodes to change characteristics, to
poison, in even the cleanest of systems is now well known, so that it
probably is no surprise to find that frequent restandardization or redetermination of the "cell constant" was necessary.

The development of the so-called "electrode-less cell" in which
a well defined sea water path is made the connecting link between two
isolated transformer elements, completely solved the electrode problem
and made restandardization unnecessary even after many months of hard
use. See for example Pritchard in (5) for a summary of early work and
Reeburgh (6) for a more recent study. With this development it became
possible to move toward in situ instrumentation, with the result that
today several excellent commercially produced instruments are available and the electrode-less cell has also been incorporated into laboratory instruments for the analysis of individual samples.

Oceanographic instrumentation has developed to the state where
continuous traces of temperature and electrical conductivity from
sea surface to the bottom can be obtained with devices which incorporate resistance thermometers, electrical pressure transducers,
and "electrode-less" conductivity units. Most such instruments also

contain a simple computer system which converts these parameters and
gives a plot of temperature and salinity vs depth. In addition some
instruments store, in digital form, the unprocessed information.

The National Oceanographic Instrumentation Center, a part of the
National Oceanic and Atmospheric Administration (NOAA) maintains a
laboratory which tests and evaluates many kinds of oceanographic
instruments. They periodically publish an <u>Instrument Fact Sheet</u> (8)
reporting the results of tests of commercially available instruments.
A recent issue, June 1972, listed eight different Salinity/Conductiv-
ity Systems that have been or are being evaluated.

In addition to transport or advective processes of the kind
represented by ocean current studies, physical oceanographers are
concerned with diffusion or mixing processes: the so-called turbulent
diffusion processes as distinct from the more familiar molecular
diffusion.

Turbulent mixing processes are of special significance in
coastal areas for it is there that many pollutants are disposed of in
attempts to take advantage of what is incorrectly assumed to be the
infinite dilution capacity of the sea. These problems are rapidly
multiplying as new power plants, especially nuclear plants, are
frequently sited in coastal areas where large volumes of cooling
water are available.

There appears to be no unique electrochemical application in
turbulent diffusion studies, although possibly new tracer techniques
might be developed. Okubo (9) recently prepared an excellent review
of both the theory and some of the practical aspects of mixing
processes.

It may not be well known among electrochemists, but Mr. Faraday,
whose electrochemical expertise I'm sure is well known, was also in
some respects a practicing oceanographer. In the 1832 Bakerian
Lecture before the Royal Society, Faraday noted: "Theoretically,
it seems a necessary consequence that where water is flowing, there
electric currents should be formed; thus, if a line be imagined passing from Dover to Calais through the sea, then returning through the
land beneath the water to Dover, it traces out a circuit of conducting matter, one part of which, when the water moves up or down the
channel, is cutting the magnetic curves of the earth, while the other
is relatively at rest."

Although Faraday was unable to demonstrate the existence of such
an electric current produced by the flow of the Thames River--he
attempted measurement with bright copper electrodes--subsequent
investigators not only have shown its existence but have used the
phenomenon in practical ways. In 1950 Von Arx (10) described an
instrument that would provide nearly continuous estimates of surface

currents obtained from a ship while under way. The device measured the potential difference between two points in the surface current generated by its motion through the earth's magnetic field. He named the instrument the Geomagnetic Electrokinetograph or GEK. In principle the GEK is a simple electrochemical device. It consists of a pair of matched silver-silver chloride electrodes attached to two insulating cables and arranged so that they can be towed behind a ship with a predetermined interelectrode spacing. The potential difference between the electrodes was measured with a simple recording voltmeter. Figure I is an example of a GEK record and Figure II is an analysis of a Gulf Stream eddy constructed by combining GEK measurements with temperature measurements in the top 200 meters of water.

Mangelsdorf (11) described the measurement of coastal currents in a paper entitled The World's Longest Salt Bridge. Based upon the same general principle as the GEK, Mangelsdorf removed the electrodes from direct exposure to the environment at the points where the potential is to be measured by interposing salt bridges to a pair of electrodes contained in an isolated, shielded, and isothermal chamber. The salt bridges, which at times were 200 meters or more long, served to isolate the electrodes from the effects of changing temperature and salinity and made possible measurement of the vertical potential gradient, from which it should be possible to obtain estimates of the east-west transport at the point of measurement. In addition, the salt bridge technique was shown to be useful in measuring the current through narrow channels by placing the ends of the two salt bridges on opposite sides of the channel. Figure III shows the current record for one tidal cycle in the channel leading from Woods Hole to Long Island Sound.

Hughes (12) took advantage of three existing submarine telephone cables, total length 166 km, which together lie across all of the channels leading into a rather large segment of the northeastern Irish Sea, and using measurements of the electromagnetically induced emf, obtained estimates of the transport into and out of the region. Transport estimates from emf measurements agreed well with that computed from tidal heights given in Admiraly Co-tidal Charts, 85 km^3 as compared with 89 km^3.

I mention the GEK and the salt bridge techniques not only because they indicate how relatively simple electrochemical measurements provide information that would be very difficult to get otherwise but also because it seems to me that with the developing emphasis on inshore and estuarine processes, these techniques might provide the means of getting more detailed information of transport processes there.

The chemical compostition of sea water has been studied since pre-Challenger days. It was analyses by Dittmar of samples taken during the first round-the-world oceanographic cruise of HMS Challenger, 1872-76, that confirmed the notion of the constancy of relative proportion of the major constituents of sea water. Since that time chemical studies have lengthened the list of elements found in sea water and established the limits within which the constancy notion appears to be valid. Periodically new listings of what is reported to be the Chemical Composition of Sea Water appear in the literature. Each new listing includes the results of the latest analyses. Simple listing is unfortunately often misleading, not because the values given are incorrect but because it is possible to indicate in a simple list only a few of the important features of sea water composition.

Figure IV taken from Mac Intyre (13) shows graphically several important features of the dissolved constituents of sea water. Note the log scale showing concentrations varying over eleven orders of magnitude. Note that for all trace elements, substances present at less than micromolar concentrations, a wide range of concentrations is given. For these elements the constancy law does not apply. Some forty elements are shown in the figure, and we could add another thirty or more, all of them trace constituents. It is perfectly reasonable to believe that the sea contains as dissolved inorganic constituents all of the elements of the periodic table, including the permanent atmospheric gases, as well as a wide variety of biologically produced organic substances.

All of marine chemistry is not just the measurement of the percentage composition of sea water. Recent emphasis has been on the chemical species in sea water and the structures present there, together with attempts to learn how these properties change within the limits of the temperatures and pressures that occur in the ocean. Observation of anamolus sound attenution by sea water suggested the possibility of the presence of structures that could enter into a resonance absorption like reaction at well defined frequencies. It has been demonstrated since then that the major dissolved inorganic constituents are not present just as simple hydrated ions but do form anion-cation pairs with the production of species with neutral charge and species with reduced charge. Garrels and Thompson (14) examined the distribution of major anions and cations of sea water, and the ion pairs formed from them. Table I summarizes their results. It should be emphasized that these are computed values obtained by using values for ion pair stability constants which in turn were based upon thermodynamic data and estimated values for activity coefficients. Clearly, significant fractions of several of the major constituents appear to be tied up as ion pairs-notably, nearly 50 percent of the sulphate, two thirds of the bicarbonate, and nearly 15 percent of the magnesium and 10 percent of the calcium.

TABLE I

Distribution of major cations as ion-pairs with sulphate, carbonate and bicarbonate ions in sea water of chlorinity 19°/₀₀, pH 8·1 at 25°C and one atmosphere pressure. (from Garrels and Thompson, 1962)

Ion	Molality	Free Ion	Sulphate ion-pair (%)	Bicarbonate ion-pair (%)	Carbonate ion-pair (%)
Ca^{2+}	0·0104	91	8	1	0·2
Mg^{2+}	0·0540	87	11	1	0·3
Na^+	0·4752	99	1·2	0·01	---
K^+	0·0100	99	1	--	---

			Ca ion-pair (%)	Mg ion-pair (%)	Na ion-pair (%)	K ion-pair (%)
$SO_4^=$	0·0284	54	3	21·5	21	0·5
HCO_3^-	0·00238	69	4	19	8	---
$CO_3^=$	0·000269	9	7	67	17	---

More recently measurements of ion activity using ion specific electrodes, in both synthetic and real sea water, have confirmed the formation of ion pairs but have changed the quantitative picture. Kester and Pytkowicz (15) from measurement of magnesium sulphate association show that less than 40 percent of the sulphate is free and that significant quantities of the single negatively charged sodium sulphate and potassium sulphate are present. The same authors (16) studied the sodium, magnesium, and calcium sulphate ion pairs in sea water and in synthetic sea water. Their measured values for free sodium, free magnesium, and free calcium ions in sea water show excellent agreement with values calculated from constants obtained from measurements in artificial sea water.

The increased reliability and specificity of modern single ion electrodes is a welcome contribution of electrochemistry to oceanography and limnology. The fluoride specific lanthanum fluoride electrode, for example, has been used by several investigators to study the distribution and behavior of fluoride in the marine environment. Fluoride, as the fluoride ion, is often considered to be a major dissolved constituent of sea water (approximate concentration 1.3 mg/kg). Pre-electrode colorimetric analyses suggested a nearly constant F/Cl ratio of 6.7×10^{-5}. Greenhalgh and Riley (17) reporting on over 300 fluorine analyses showed that the F/Cl ratio in surface waters was remarkably constant but that deep ocean waters from some localities has unusually high fluorine content. Like Florida weather, this lead to the designation normal and anomalous. Brewer, Spencer, and Wilkniss (18) in an attempt to discover the cause of the anomalies, analyzed samples from 33 North Atlantic stations, two within the anomalous region, using colorimetric, fluoride ion specific electrode, and photon activation techniques. Their results confirmed the anomalous deep water values but only when analyses were by colorimetric or photon activation methods--electrode results indicated a constant F/Cl ratio in all samples. They concluded that since the electrode "sees" only free fluoride ion, and the colorimetric and activation methods the total fluorine, the anomalous values represented the presence of fluorine in some combined form. They demonstrated the effect of the formation of the single positively charged magnesium fluoride ion-pair on the electrode response, but ruled out that species as the cause of the apparent anomalies. They speculated that a collodal form of fluorine might be the cause. Warner (19) using only electrode measurements refined the F/Cl ratio in so-called normal sea water, giving the value $6.75 \pm 0.03 \times 10^{-5}$.

The presence of dissolved complex species in natural water is of practical as well as of theoretical interest. All of the chemical, the biological, and the geological reactions--the geobiochemical cycles--of substances in natural systems are controlled by the chemical species of the reactants. Just as the behavior of sulfur will be very different as sulfide compared with sulphate, so the behavior of **ion pairs,** of complexes, of chelates and of colloids will be different from the simple, free hydrated ions which form them. Specific-ion electrodes provide a relatively new tool with which some of these phenomena may be examined.

In additon to ion specific ion electrodes, polarographic techniques allow certain reaction processes and their products to be examined. The techniques of anodic stripping voltametry are especially appealing as the sensitivity is high, in some cases being useful at concentration of 10^{-10}M without pretreatment of the sample. This latter is important, for many techniques require manipulations of the sample which alter or destroy the natural system. Matson (20) combined the anodic stripping technique with a composite mercury graphite electrode which has the desirable characteristics of thin film electrodes but with added stability, ease of preparation, and desirable hydrogen overvoltage characteristics. By combining anodic stripping techniques with Sephadex elutions, the presence of high molecular weight copper and lead complexes in natural fresh waters was shown, together with estimates for the stability constants and kinetics of formation.

Fitzgerald (21) extended these techniques to sea water analyses and improved upon the elimination of mutual interferences in multi-element analysis.

Corrosion, a multi-billion-dollar phenomenon, is to be discussed in detail during this symposium. Let me make three comments. First, sea **water is considered,** at least by oceanographers, to be one of the most corrosive substances in the world. Secondly, research vessels, essential tools of oceanographers, are described as being holes in the water lined with rust into which one pours money. Thirdly, I can assure you that advances in the art and science of corrosion prevention will be very much welcomed by oceanographers, for that hole in the water could be made less rusty and the money poured into it could be used for more interesting purposes.

To summarize, it seems to me that the big problems in oceanography at present are ones that have been around for many years but with a new urgency given to some of them, plus a new set of problems, some of which we seem to be poorly prepared to handle. In physical oceanography turbulent diffusion problems, the mixing problems, sea-air interactions looking at transfer processes are not new problems but some of them take on new dimensions when attention is moved from the deep ocean to inshore regions. Chemists are still concerned with sea water composition but with emphasis on chemical species and the processes that form and destroy them, the kinetics of geobiochemical reactions. Interest is developing in mass balance kinds of experiments, in modeling a way that will provide predictive capabilities. The sea-air interface, the film that is found on all natural water surfaces, seems to have an important role in material transfer between air and water.

Clearly, there are several areas within the marine sciences where electrochemical inputs can provide the kind of information that appears to be needed.

The need for *in situ* measurements is more pressing as emphasis on coastal processes increases. It is here that significant rates of change of properties is measured in minutes and hours rather than months and years as in the deep oceans.

Here we would like information about the activity and concentration of a variety of chemical species, especially those that are the results of man's activities. Needed also are continuous records of long enough duration to provide good estimates of mean values from which trends can be noted.

REFERENCES

(1) Our Nation and the Sea. A Plan for National Action. Report of the Commission on Marine Science, Engineering, and Resources. G.P.O. Washington, D.C. January 1969.

(2) Sverdrup, H. U., M. W. Johnson, and R. H. Fleming. The Oceans Chaps. X thru XV. Prentice-Hall. N.Y. (1946).

(3) Hill, M. N. (ed.) The Sea: Ideas and Observations on Programs in the Study of the Sea. Interscience. N.Y. (1962).

(4) Dietrick, G. General Oceanography, Section III. John Wiley N.Y. (1963).

(5) Physical and Chemical Properties of Sea Water. Publication 600, National Academy of Sciences, National Research Council, Washington, D.C. 1959.

(6) Reeburgh, W. S. Measurements of the Electrical Conductivity of Sea Water. J. Mar. Res. 23 #3, 187-99 (1965).

(7) Wenner, F., E. H. Smith, and F. M. Soule. Apparatus for the determination aboard ship of the salinity of sea water by the electrical conductivity method. Nat. Bur. Stds. J. Res. 5 711-732. R.P. 223 (1930).

(8) Instrument Fact Sheet. U.S. Department of Commerce, National Ocean Survey, National Oceanographic Instrumentation Center. Washington, D.C.

(9) Okubo, A. Chapter 4 in Impingement of Man on the Oceans. D. W. Hood (ed.). Wiley-Intersciences N.Y. (1971).

(10) von Arx, W. S. An electromagnetic method for measuring the velocity of ocean currents from a ship under way. Paper in Physical Oceanography and Meteorology. WHOI Woods Hole, Mass. 1950.

(11) Mangelsdorf, P. C., Jr. The World's Longest Salt Bridge. Marine Sciences Instrumentation Vol. 1. Plenum Press. N.Y. (1962).

(12) Hughes, P. Submarine cable measurements of tidal currents in the Irish Sea. Limrol. Oceanogr. 14 #2, 269-78 (1969).

(13) Mac Intyre, F. Why the sea is salt. Scientific American. November 1970.

(14) Garrels, R. M. and M. E. Thompson, Amer. J. Sci. 260 57 (1962).

(15) Kester, D. R. and R. M. Pytkowicz. Magnesium sulfat association at 25C in synthetic sea water. Limrol. Oceanogr. 13 670-674 (1968).

(16) Kester, D. R. and R. M. Pytkowicz. **Sodium, m**agnesium, and calcium sulfate ion-pairs in sea water at 25C. Limrol. Oceanogr. 14 686-692 (1969).

(17) Greenhalgh, R. and J. P. Riley. Occurence of abnormally high fluoride concentrations at depth in the oceans. Nature, London 197, 371-2 (1963).

(18) Brewer, P. G., D. W. Spencer, and P. E. Wilkniss. Anomalous fluoride concentrations in the North Atlantic. Deep Sea Res. 17, 1-7 (1970).

(19) Warner, T. B. Normal fluoride content of sea water. Deep Sea Res. 18 1255-1263 (1971).

(20) Matson, W. R. Trace metals, equilibrium and kinetics of trace metal complexes in natural media. Ph.D. thesis. Department of Chemistry, M.I.T., Cambridge, Mass. 1968.

(21) Fitzgerald, W. F. A study of certain trace metals in sea water using anodic stripping voltammetry. Ph.D. thesis, Dept. of Earth and Planetary Science, M.I.T. and Dept. of Chemistry WHOI, (1970).

Figure I

A 45 minute segment of a geomagnetic electrokinetograph (GEK) record showing two estimates of true current. The "noise" provides data on wave periods.

[From von Arx (10)]

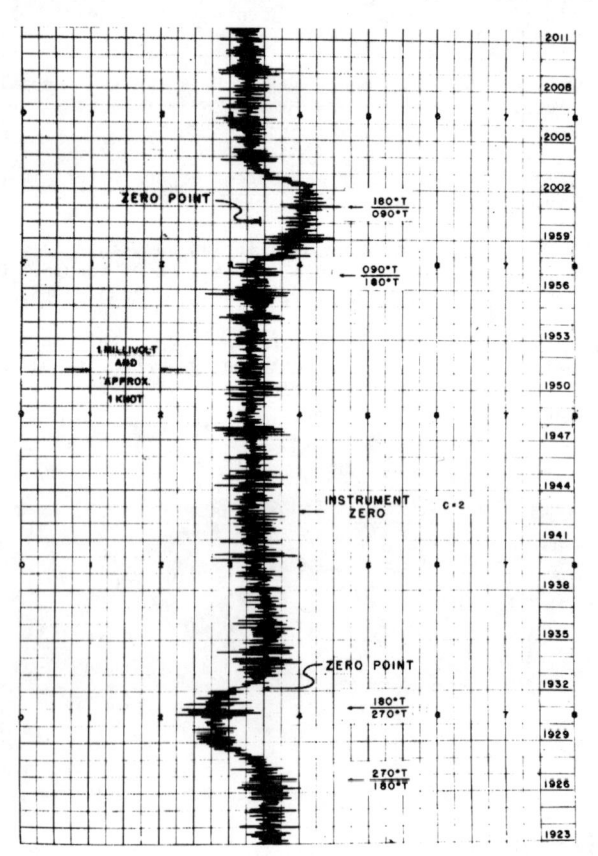

Figure II

An eddy of the Gulf Stream. Contours are temperature °C. The arrows show current direction from GEK estimates.

[From Fuglister and Worthington, quoted in Dietrick (4)]

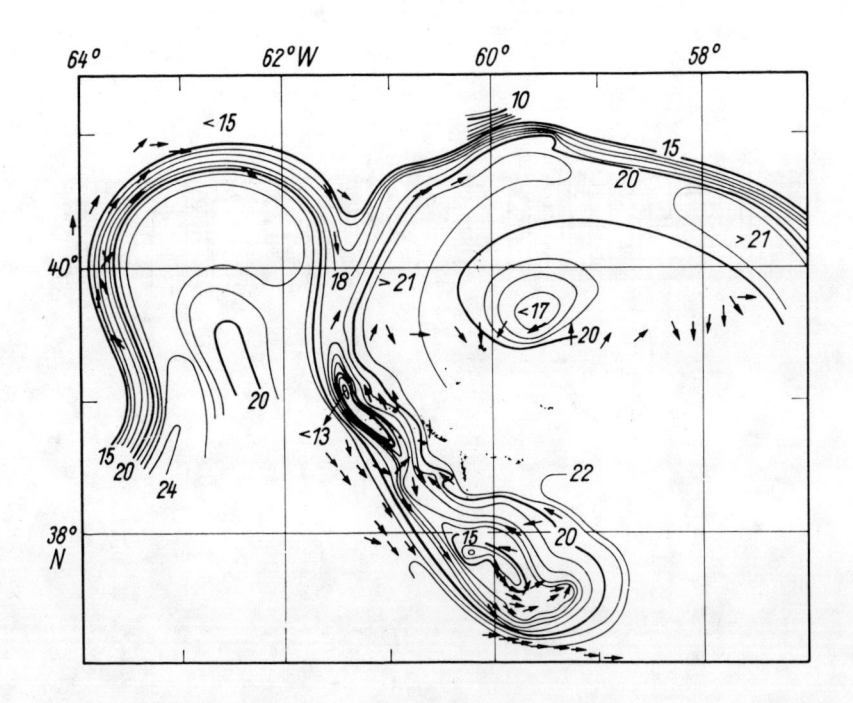

Figure III

An 18 hour record of the geomagnetic potential induced across the channel at Woods Hole, Massachusetts.

[From Mangelsdorf (11)]

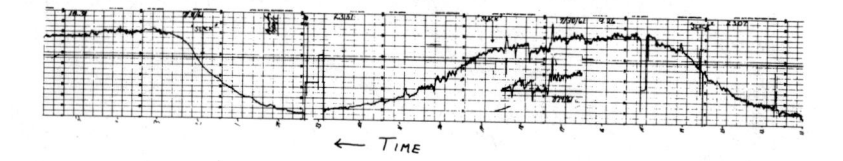

Figure IV

Chemical composition of sea water (partial).

[From Mac Intyre (13)]

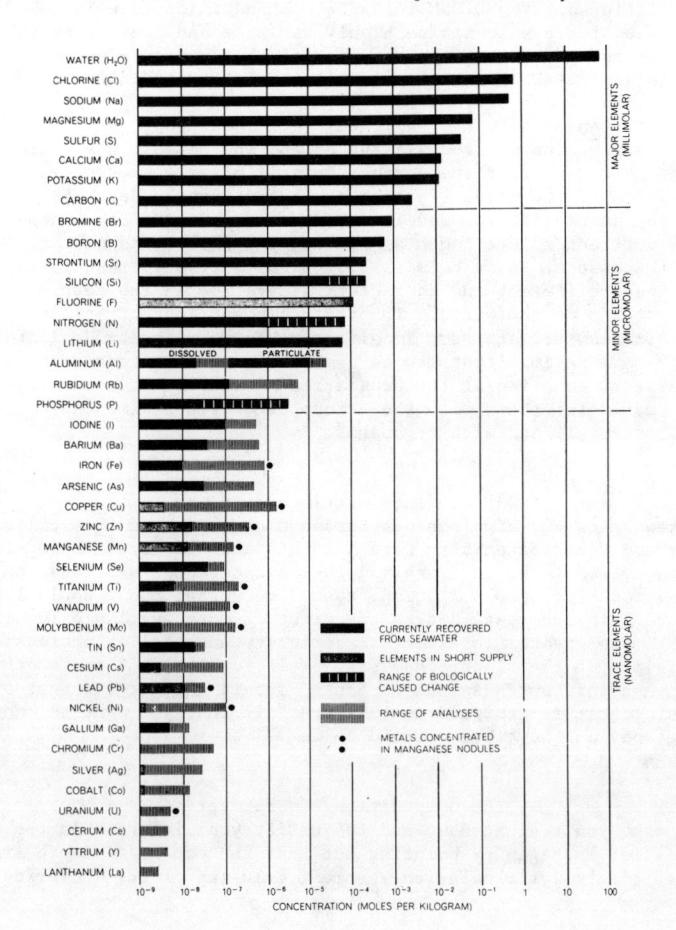

The Oceans as a Chemical System

R. A. Horne

Arthur D. Little, Inc.

Acorn Park, Cambridge, Massachusetts 02170

Abstract

Seawater is an approximately 0.5M NaCl aqueous electrolytic solution 0.05M in $MgSO_4$. In addition to solutes seawater also contains highly variable and small amounts of suspended particulate material, both organic and inorganic. While the total salt content or salinity of seawater can be variable, the ratios of the major constituents, Na^+, K^+, Mg^{++}, Ca^{++}, Cl^-, and $SO_4^=$, are nearly constant for all the oceans and seas of the world indicating that the oceans are well mixed. The speciation of the major elements can be adequately described by an equilibrium model for seawater. On the other hand nutrient solutes such as silica, phosphate, and nitrate, because of their role in biological activity, tend to have different and characteristic ratios in the several oceans of the world. The trace element concentrations in seawater are very imperfectly known. A great deal of the more important marine chemistry is interfacial chemistry occuring at the boundaries of the oceans where seawater interacts with the atmosphere, with sedimentary material, and with organisms.

1
Introduction

A few years ago simultaneous Gordon Conferences on Chemical Oceanography and Electrochemistry were held at Santa Barbara. More electrochemists seemed to be infiltrating the oceanographic sessions, especially those touching upon pollution, than vice versa, and I would like to attribute this interest, not to the scent of more available funding, but rather to the greater intellectual curiosity and smaller professional parochialism of the electrochemists. Well for those of you who might gave peeked into our sessions at Santa Barbara and for the rest of you today's program is designed to tell you "all that you want to know about oceanography but have been afraid to ask."

2
The Composition of Seawater

To make you feel at home and to justify your interest in the marine sciences let me begin by pointing out that the oceans of Earth are an aqueous, electrolytic solution - some 1.4×10^{21}kg of it - covering about

71% of this planet's surface. Seawater is a roughly 0.5M NaCl solution, 0.05M in $MgSO_4$ and containing a trace of just about everything else imaginable. For the overwhelming majority of engineering calculations to take seawater as an 0.5M NaCl solution is an adequate approximation. Now those of you who are solution electrochemists will quickly recognize that 0.5M is just about the messiest concentration range imaginable - too concentrated for the many theoretical treatments that have been developed for "ideal solutions" to be applicable, yet more dilute than the concentrated solutions which are now finally receiving some much needed attention (largely as a result of their importance as electrolytes in electrochemical power sources). To make matters worse it is also a mixed electrolyte solution. With respect to solvent properties seawater is still fairly ideal (0.4% vapor pressure deviation from Raoult's Law), but with respect to solute properties it is far from ideal (26% deviation from ideality in the equivalent electrical conductance) (Horne, 1969, ch. 2).

The single most important chemical generalization characterizing seawater was discovered in 1819 by Marcet and was restated by Maurey in 1855 as follows..."the constituents of seawater are as constant in their proportions as are the components of the atmosphere." That is to say, although the total salt content (or "salinity" - see Horne, 1969, pp. 146-147 for the definition of this term) can be highly variable in space and time, the ratios of the major constituents (see below) of seawater are remarkably constant throughout all the oceans of Earth, indicating that the marine hydrosphere, like the atmosphere, is well mixed. Many factors affect the salinity of seawater (see Horne, 1969, pp. 156-161), for examples, river runoff and the salinity of surface waters in the open ocean are highly dependent on the balance between evaporation and precipitation in any geographic area.

Like the salinity the temperature of the water column in the ocean tends to be variable near the surface, but the deep ocean is a very stable thermostat with a constant salinity near 35 0/00 and temperature near $+4^oC$. The greatest depths of the ocean correspond to hydrostatic pressures of about 1000 atm. In terms of chemical equilibria among condensed phase species this is a fairly modest pressure, and, generally speaking, temperature and salinity effects are much more important in marine chemistry than pressure effects. For equilibria which do not involve gases the pressure effect rarely amounts to more than a few per cent, but subsequently I will mention one important exception, $MgSO_4$ ion-pair formation. For those of you who might be interested I have extensively reviewed the effects of high pressure on aqueous electrolytic solutions elsewhere (Horne, 1969a, ch. 3; Horne, 1969b) and the subject has been treated more recently by Brummer and Gancy (1972).

In addition to solutes and colloids seawater also contains a great variety of suspended material (Horne, 1969a, ch. 8). An Atlantic sample, for example, taken at a depth of 4030m contained a relatively "large" concentration of solid material - 0.003g. But when one compares this with the 30-40g of dissolved salts in seawater it is a rather insignificant amount. Nevertheless while small in quantity, the suspended material in seawater probably plays a disproportionately large role in marine chemistry. The suspended material in seawater can be both inorganic and organic. With respect to the latter, the organic matter in the euphotic zone (surface waters penetrated by light) is about 89% dissolved, 9% particulate detritus, and the remaining 2% organic ranges from plankton to whales. Unlike the dissolved salts, the level of suspended material in seawater is very highly variable, the amounts of both organic and inorganic material being relatively high and especially variable, just as one would suspect, in coastal waters.

For our purposes today we can classify the dissolved materials in seawater into four chemical categories:
> major constituents,
> trace elements,
> nutrients, and
> organic materials.

Five ionic species in seawater account for more than 98% by weight of its dissolved salts. These <u>major constituents</u> are Cl^-, Na^+, $SO_4^=$, Mg^{++}, and Ca^{++} and their amounts in 35 o/ooS seawater are 19.4, 10.8, 2.7, 1.3, and 0.4 g/kg respectively. As we have said their ratios in the oceans of the world are highly invariable, thus, for example, the value of $(Na^+)/$ o/ooCl is 0.5544 to 0.5567 in the Atlantic Ocean, 0.5497 to 0.5561 in the Pacific Ocean, 0.5310 to 0.5528 in the Mediterranean Sea, 0.5536 in the Baltic Sea, 0.5518 in the Black Sea, and 0.5573 in the Irish Sea. The species K^+, HCO_3^-, Br^-, and Sr^{++} account for about another percent. However, most of the natural radioactivity of seawater is due to the decay of K^{40} and HCO_3^- is extremely important, along with silicate minerals, in the narrow control of the pH of seawater (pH = 7.5 to 8.4 in the oceans or slightly alkaline). The chemistry of the major constituents is for the most part fast ionic metathases. On the basis of known equilibrium constants Garrels and Thompson (1962) have constructed a chemical model for sea water and they have calculated the speciation given in Table 1 at 25°C and 1 atm. The

Table 1
Major Species in Seawater

Ion	Molality	Speciation		Ion	Molality	Speciation	
Na^+	0.4752	Na^+	99%	$SO_4^=$	0.0284	$SO_4^=$	54%
		$NaSO_4^-$	1.2			$MgSO_4$	21.5
		$NaHCO_3$	0.01			$NaSO_4^-$	21
Mg^{++}	0.0540	Mg^{++}	87			$CaSO_4$	3
		$MgSO_4$	11	HCO_3^-	0.00238	HCO_3^-	69
		$MgHCO_3^+$	1			$MgHCO_3^+$	21.5
		$MgCO_3$	0.3			$NaHCO_3$	21
Ca^{++}	0.0104	Ca^{++}	91			$CaHCO_3^+$	3
		$CaSO_4$	8	$CO_3^=$	0.000269	$MgCO_3$	67
		$CaHCO_3^+$	1			$NaCO_3^-$	17
		$CaCO_3$	0.2			$CO_3^=$	9
						$CaCO_3$	7

temperature and even more so the pressure effects are modest. We
(Courant, Horne, and Kester, 1972) have used this model to calculate
the ionic strength of seawater ($\mu = 0.0054 + 0.0184$ So/oo + 1.78×10^{-5}
(So/oo)^2 - $3.0 \times 10^{-4}(25 - T^oC) + 7.6 \times 10^{-6}$(P atm - 1)) and the electrical conductivity and colligative properties and in our opinion it
is an entirely satisfactory description of seawater. As you can see
there is very little interesting chemistry going on in a representative
element of seawater in the open ocean. Most of the important marine
chemistry is the chemistry of the interaction of the sea with its atmospheric, lithospheric, and biospheric boundaries. An exception is
the equilbrium

$$Mg^{++} + SO_4^= \rightleftharpoons MgSO_4$$

Due to electrostriction the hydrated $MgSO_4$ ion-pair is larger than the
sum of the volumes of the hydrated Mg^{++} and $SO_4^=$ ions, and as a consequence hydrostatic pressure favors the left hand side of this equilibrium. This pressure-sensitive equilibrium is responsible for the
"anomalous" absorption of acoustic energy in seawater and has been studied in detail by Fisher (1962, 1965).

All of the naturally occurring chemical elements are present in
seawater at least in trace amounts (see Horne, 1969a, ch. 5, table 5.3),
but whereas our understanding of the major constituents is in good shape
our knowledge of these trace elements in in a mess. Since enormous
quantities of public funds have been and continue to be expended on
trace element analyses, I might even venture to say a scandalous mess.
The results of an interlaboratory check reported at the Gordon Conference mentioned earlier point in the direction that there is not a single trace element, not even iron, whose concentration in seawater is
reliably known. The level of professional competence in chemical oceanography is not distinguished, and it is clear that some people have been
attempting extremely difficult analytical tasks for which they are inadequately trained. The question has been raised before, what can professional analytical chemists do to help resolve this mess, and the
question might be raised here of what assistance electrochemists might
offer. But before you plunge in let me warn you that there are some
very real and formidable difficulties. The levels of many of the elements are exceedingly low and tax even the most sophisticated techniques,
the contamination problems are horrendous, there is both dissolved and
suspended material, and a seawater sample is not a sterile solution
but rather a biological broth in which many complex organic reactions
are occurring and will continue to occur long after the sample is
taken. But perhaps the greatest difficulty may arise from the fact that
a specimen taken from the deep open ocean is very peculiar in the sense
that it has not been exposed to a surface for a very long time. Thus
a seawater sample is highly unstable with respect to surface chemistry
and acidification is probably only partially effective in preventing
surface reactions of dissolved trace metal species.

The so-called <u>nutrients</u> are fascinating for, unlike the major species, their ratios in seawater are not only highly variable but even tend to have characteristic values in the several great world oceans (Table 2). These species are consumed by organisms and their ratios, in addition to providing a measure of biological activity, also provide a sort of biological clock that can indicate the age of water masses. The young well-flushed waters of the Atlantic have quite different ratios

Table 2
Inorganic Anion (Nutrient) Ratios in Deep Waters
of the Major Oceans (From T. J. Chow and A. W.
Mantyla, <u>Nature</u>, <u>206</u>, 383 (1965)

Ocean	Silicate/Phosphate	Silicate/Nitrate	Nitrate/Phosphate
S.E. Pacific	55-65	3-5	13-14
Equat. Indian	40-50	3	15
N. Atlantic	20-40	1-2	12-16

from the old, tired waters of the Pacific (notice that the northern Pacific is nearly a cul-de-sac.)

I do not want to say anything further here about the organic material in seawater except to add that it provides a great variety of complexing ligands for metallic cations and that this variety is being increased by human pollution. Since I have mentioned the particulate material in seawater I suppose I must also mention the dissolved gases and I will have more to say subsequently about the carbon dioxide-carbonate system. The dissolved gases are discussed in detail in Chapter 7 of my book (Horne, 1969a) and the biochemistry of the oceans in Chapter 9.

3
The Oceans as Part of the World Chemical System

As I have said here and elsewhere most of the interesting marine chemistry occurs not in the seawater itself but in the interaction of the oceans with their boundaries, of the hydrosphere with atmosphere, lithosphere and biosphere.

<u>Air-Sea Interactions</u>. The supreme importance of the evaporation-precipitation balance of the oceans in the enormous solar-powered hydrologic cycle is well known, but perhaps you are not so familiar with the great variety and quantities of chemical substances other than water that are transported across the air/sea interface. The elemental ratios in the marine aerosol differ appreciably from the corresponding values in seawater and these materials now appear to fall into three main groups: human pollutants, chemicals for which the sea is a sink (such

as air carried dust blown from the continental land masses), and chemicals for which the oceans are a source but which may be fractionated by physical-chemical processes upon their transport through the air/sea interface (see Horne, 1969a, ch. 8 and 11). Atmospheric gases dissolve in seawater. Nitrogen, because of its relative inertness, serves as a sort of chemical blank, while the oxygen, both dissolved from the atmosphere and produced by photosynthesis is essential to and utilized by the decomposition of organic matter and animal life. Certain ocean areas, such as the Black Sea depths, are depleted in oxygen; these anoxic waters are literally "dead seas" with little life, quantities of undecomposed organic matter, high H_2S levels, and a chemistry, based on the sulfur rather than the oxygen system, quite different from "normal" waters. Of particular interest is the dissolution of atmospheric CO_2 in the oceans (Horne, 1969a, ch. 7). The system $CO_2^-HCO_3^-CaCO_3$ (along with silicate minerals) buffers seawater and maintains the slight alkalinity of the oceans. The crucial rate-determining step appears to be the dissolution of gaseous CO_2 in seawater, and the fate of life on this planet may hinge on the question of whether this rate is fast enough to remove the enormous quantities of CO_2 now being produced by the combustion of fossil fuels.

Hydrosphere-Lithosphere Interactions. If we compare data on the quantities of chemical elements in seawater with what we would expect from river input due to the weathering of crustal rocks we find that they tend to fall into three groups: those like Na, with relatively little chemistry, present in the "expected" amount; those in excess, such as Cl, S, and Br; and those, representing by far the greatest group, which are depleted (Horne, 1969a, ch. 12). The so-called "excess volatiles" are now believed to have come from volcanic and other such emissions, in particular the oceans throughout geological time have been a sink for chloride ion. As for the depleted elements a few of them such as Si and Ca are removed by incorporation into the life cycle and subsequent precipitation as SiO_2 and $CaCO_3$ which sinks out of the water column and is added to the sedimentary accumulation. Many more of them, especially trace metal cations such as Fe, Mn, Ni, V, Hg, etc., are presumably absorbed by solid surfaces and thus ultimately incorporated into marine sediments. Recent developments in our knowledge of continental drift and ocean-floor spreading have opened up enormous new possibilities of hydrosphere-lithosphere interactions. Not only may solutes be leached into seawater from freshly exposed lithosphere at the mid-ocean ridges, but the sedimentary burden is slowly carried by ocean-floor spreading to the continental margins where it sinks down beneath the continental mass and under the influence of temperature and pressure may be re-incorporated into the magma, carrying a fraction of the pore fluids and water of hydration along with it, and thus linking the hydrological and geochemical (or rock) cycles. I might mention in passing that cold earth hypotheses are currently somewhat more popular than hot earth models so that the water of the earth's hydrosphere is now believed to have been released from the magma rather than have condensed from a primitive atmosphere (Horne, 1969a, ch. 12).

Hydrosphere-Biosphere Interactions. This is hardly the place for a detailed discussion of the biochemistry of the oceans and of the origin of life in the oceans of Earth. Most of this planet's biomass is in its oceans. I would like to conclude, however, by stressing a single point. Man oftentimes excuses his abuse of the oceans as a dump with the argument that his activity represents an insignificant perturbation. In its geological history the world-ocean has undergone one profound chemical change and that change was brought about by the activity of organisms far smaller and more insignificant than man. The primitive ocean, like the primitive atmosphere, was reducing. The present oxygen content of the atmosphere is the product of photosynthesis, and with the advent of photosynthesis the oceans of earth went from a reducing to an oxidizing environment. The Fe(II) in seawater was oxidized up to Fe(III) which then precipitated down in the form of hydrous oxides, and laying down the vast ore deposits which form the basis of our Age of Iron. Stand on the rim of Grand Canyon - that is how all that sandstone got painted red!

References

R. A. Horne, Marine Chemistry, Wiley-Interscience, New York, N.Y., 1969a.
R. A. Horne, Adv. High Pressure Res., 2, 169 (1969b).
S. B. Brummer and A. B. Gancy, in R. A. Horne (ed.) Water and Aqueous Solutions, Wiley-Interscience, New York, N.Y., 1972.
R. M. Garrels and M. E. Thompson, Amer. J. Sci., 260, 57 (1962).
R. A. Courant, R. A. Horne, and D. R. Kester, Limnol. Oceanogr., in press, 1972.
F. H. Fisher, J. Phys. Chem., 66, 1607 (1962); 69, 695 (1965).

Additional Reading

G. Dietrich, General Oceanography, Interscience, New York, N.Y., 1963.
M. N. Hill (ed.), The Sea, Interscience, New York, N.Y., 1963.
R. A. Horne, Water Resources Res.; 1, 263 (1965).
R. A. Horne, Adv. Hydrosci., 6, 107 (1970).
F. MacIntyre, Sci. Amer., 227, No. 5, 104 (Nov., 1970).
K. Park, in R. A. Horne (ed.), Water and Aqueous Solutions, Wiley-Interscience, New York, N.Y., 1972.
J. P. Riley and G. Skirrow (eds.), Chemical Oceanography, Academic Press, London, 1965.
M. Sears (ed.), Oceanography, Amer. Assoc. Adv. Sci., Pub. No. 67, Washington, D.C., 1963.

DISCUSSION

H.S. Spacil, G.E. R&D Center, Schenectady, New York: To what extent does the ocean surface show equilibrium with the atmosphere?

Dr. Horne: Chemical models of the ocean assume that each element of seawater is at some previous point in its career at equilibrium with atmospheric gases. Yet I get the distinct impression that the atmosphere and the sea are hardly ever at equilibrium. If you look at charts of various atmospheric gases such as CO_2 in the surface waters of the world's oceans, you see such and such a percentage undersaturated, such and such a percentage supersaturated, but rarely saturated. Not only is gas solubility dependent on temperature, pressure, and salinity, all variables in the ocean, but biological activity influences the levels of gases such as Co_2 and O_2 in the waters.

Robert N. O'Brien, University of Victoria, Canada: Have you read a recent paper in Science reporting satellite observation above the ionosphere in which it is suggested that the main source of terrestrial O_2 is cosmic ray disassociation of H_2O, the H_2 going into a deep outside layer and O_2 at M.W. 32 being held in and mixed in the atmosphere. This is suggested as an alternative to the photosynthesis theory of the production of O_2 on earth.

Dr. Horne: I saw the title and I filed the abstract, but I haven't read the paper yet. The photosynthetic theory of the origin of the atmosphere's oxygen is in such good shape and is supported by very formidable evidence from many directions that I would say that any alternative theory would have to document some very convincing arguments to be competitive. Remember that the universe as a whole is a strongly reducing environment. The Earth's oxidizing environment is thus a very noteworthy feature and presumably could have been the result of only a very noteworthy phenomenon - life.

Walter J. Hawer, 3028 Dogwood St., N.W., Washington, D.C., 20015: Is the total volume of the oceans decreasing?

Dr. Horne: The volume of the oceans is subject to two types of changes - short range and long range. The former are largely the result of climatic changes and the advance and retreat of the polar icecaps. The water in the oceans comes not from the condensation out of a primitive atmosphere, but from the Earth's mantle. Some authorities see the volume of water as relatively constant throughout geological time, some increasing rapidly in the remote past, others increasing rapidly in the more recent past, and others increasing gradually throughout geological time. For no good reason, the gradual accumulation, with perhaps somewhat faster increase when the crust was younger and less stable, seems the most reasonable hypothesis to me.

Gerald D. Mitchell, The Applied Physics Laboratory, 8621 Georgia Ave., Silver Spring, Maryland, 20910: With the complexity of interfacial processes in the oceans, do you see in the future a generalized physical and chemical model for the ocean?

Dr. Horne: It depends on whether you are interested in big effects or little ones. In the case of the larger phenomena, the relative importance of more subtle surface phenomena tends to be minor. Thus, we have very good models at the present time for ocean mixing processes and for the chemical reactions among the major constituents of seawater. There are intermediate phenomena, such as the mechanical coupling between wind and waves where again good models are available, but they have a provision in them for, say, the effects of surface slicks. Finally, there are those phenomena which are never going to be really understood unless we have some very detailed knowledge of interfacial properties - these include the chemical partitioning between seawater and the marine aerosol and the chemistry of trace elements in the marine environment. This is most important because it includes the partitioning and chemistry of pollutants in the marine environment. Finally, we must remember that superimposed on the whole is an extremely complex interfacial system - the biosphere. Now this doesn't effect currents and tides very much, but it effects the oxygen content of the atmosphere, the chemistry of nutrients and trace elements in the sea, and sedimentation rates.

THE OCEAN AS AN ELECTROMAGNETIC SYSTEM

Paul C. Mangelsdorf, Jr.
Department of Chemistry
Woods Hole Oceanographic Institution
Woods Hole, Mass. 02543

(also)
Department of Physics
Swarthmore College
Swarthmore, Pa. 19081

Abstract

The motion of sea water flowing in the earth's magnetic field induces large scale EMFs and electric currents. These can be measured with submarine cables, with towed electrodes (GEK), or with a new free-fall instrument, to provide useful information about the water movement. Other electromagnetic effects in the ocean are briefly discussed.

For electrochemists, the ocean is the grandest electrolyte solution there is, with all of its 10^{21} liters of mixed salt solution inviting us to dunk in our electrodes and start measuring voltages. What kinds of voltages can we expect, and how are they generated?

There are, of course, two basically different kinds of electrode measurements that one can make in a large heterogeneous system like the ocean. One could compare two different electrodes at the same location and thereby learn something about the local chemistry of the ocean, but that properly belongs to the electroanalytical chemistry of sea water. It could be done just as well on a sea water sample in the laboratory as in the ocean. Moreover, that kind of measurement will be discussed in detail by other contributors to this symposium.

The other kind of electrode measurement one can make is to determine the potential difference between two matching electrodes located at different places in the ocean. What happens if I put an electrode in the ocean here at Miami and put another one in at Bimini 90 kilometers across the Florida Straits to the east? This measurement has been made, indirectly, by Sanford and Schmitz[1]. Their result is that Miami is about 2.2 volts positive with respect to Bimini. What is happening is that sea water is flowing northward across the Miami-Bimini line at speeds ranging up to 4 knots at the surface, with a total flow amounting to 33 million cubic meters per second. (A cubic meter of sea water weighs about 1.025 metric tons and contains about 35 kilograms of dissolved salts.) This particular ocean current is popularly known here as the Gulf Stream, but oceanographers prefer to call it the Florida Current, reserving the title of Gulf Stream for the much larger current system -- the Florida Current included -- which characterizes the westward edge of the North Atlantic circulation.[2] As the Florida Current moves northward across the vertical component of the earth's magnetic field, this motion generates a transverse EMF, just as in any other conductor cutting through a magnetic field.

There is nothing very new about this. It was all perfectly clear to Michael Faraday 140 years ago. In his Bakerian lecture before the Royal Society,[3] he mentioned an unsuccessful attempt to measure the flow in the River Thames by observing the potential difference between two metal plates lowered into the water at opposite ends of Waterloo Bridge. He then went on to say:

"Theoretically, it seems a necessary consequence that where water is flowing, there electric currents should be formed: thus, if a line be imagined passing from Dover to Calais through the sea, and returning through the land beneath the water to Dover, it traces out a circuit of conducting matter, one part of which, when the water moves up or down the channel, is cutting the magnetic curves of the earth whilst the other is relatively at rest. This is a repetition of the wire experiment, but with worse conductors. Still there is every reason to believe that electric currents do run in the general direction of the circuit described, either one way or the other, according as the passage of the waters is up or down the channel. Where the lateral extent of the moving water is enormously increased, it does not seem improbable that the effect should become sensible; and the gulf stream may thus, perhaps, from electric currents moving across it, by magneto-electric induction from the earth, exert a sensible influence upon the forms of the lines of magnetic variation."

Using Faraday's Law of Induction for a conductor moving in a magnetic field, it is easy to calculate the magnitude of EMF one should expect.

$$\Delta \Phi = \left(\vec{V} \times \vec{B} \right) \cdot \vec{\Delta \ell} \tag{1}$$

where $\Delta \Phi$ is the potential difference across the directed length interval $\vec{\Delta \ell}$, $\vec{V}$ is the velocity vector, and $\vec{B}$ is the magnetic field.

If we realistically limit our calculation to horizontal motion in the sea and, for the moment, to horizontal EMFs, then we are only concerned with the vertical component of the earth's magnetic field. This component is typically of the order of 0.5 gauss, directed <u>downward</u> in the Northern magnetic hemisphere, <u>upward</u> in the Southern hemisphere, vanishing at the magnetic equator.

For simplicity in the subsequent discussion, I will use coordinates such that the velocity is along the y axis, the vertical component of the magnetic field is along the z axis, and the cross-stream length interval is in the x direction.

Thus

$$\Delta \Phi / \Delta \ell_x = V_y B_z \tag{1a}$$

Here at Miami, the earth's vertical field is about .44 gauss (.44X10^{-4} webers/meter^2) downward. In this field, a northward flow of 1 knot (.51 meter/sec) generates a westward EMF of 22 microvolts/meter.

For the simplest possible case, that of uniform flow in a perfectly uniform straight channel with the same velocity everywhere, the cross-stream potential gradient would be exactly equal to this induced EMF, and a voltage measurement would be easy to interpret (Figure 1). In the real world, things are not so simple: a velocity cross section of the Florida Current would reveal a fast-moving core at the surface near the middle with velocities tending to vanish near the bottom and near the shores on either side.[5] The EMF produced in a parcel of fast-moving water will tend to be "shorted out" by the electrical currents it will produce in slower-moving water around it, as indicated schematically in Figure 2. The measured potential drop within the parcel will be diminished by the IR drop produced by such currents:

$$\Delta \Phi / \Delta \ell_x = V_y B_z - J_x / \sigma \qquad (2)$$

Here J_x is the cross-stream component of the local electrical current vector and σ is the local electrical conductance.

Although further contributions to the EMF can be generated in transient cases where either the electric current or the magnetic field is changing with time, for the purpose of DC measurements this equation is complete and exact. It should be noted, however, that in the deep ocean even the tidal components of the flow change fast enough at two cycles per day to cause appreciable AC inductive components in the electrical impedance of the sea.[6]

To use equation (2) to describe the results of a point-to-point measurement across the ocean, across the Gulf Stream, or even across a confined channel such as the Florida Straits, we must do some vertical averaging and some horizontal integration. Further assumptions are needed.

As a general rule, the horizontal scale of any oceanic property is much greater than the vertical scale. The typical ocean basin is a few thousand kilometers broad, while the average depth of the ocean is about 4 kilometers. The Gulf Stream north of Cape Hatteras is about 100 kilometers wide whereas the continental rise along which it runs is only about 2 or 3 kilometers deep. The Florida Straits here off Miami is about 80 kilometers wide but less than a kilometer in depth.

This tremendous disparity between horizontal breadth and vertical shallowness is further enhanced by the marked density stratification which the ocean exhibits throughout all the tropical and temperate regions of the globe. Most of the density variation is due to temperature differences -- deep ocean water is always ice cold even at the equator -- water cooler than 6°C can be found near the bottom out here in the relatively shallow Florida Straits. Occasionally the dominant density variation is due to salinity differences, especially along coastlines where fresh water runs off the continents into the sea. In either case the resulting density stratification dominates the dynamics of the water movement so that steady vertical motions are negligible. The different layers of water can be thought of as broad thin sheets horizontally uniform in velocity, density, and conductivity over expanses which are very large compared to the depth. Under these circumstances, the electric current vector $\vec{J}$ is almost entirely horizontal at every depth, and the horizontal potential gradient $\Delta\Phi/\Delta\ell_x$ is essentially independent of depth.

In the case of flow confined to an insulating channel, such as would be a model for our Florida Current, all of the cross-stream electric current is generated within the channel and the net cross-stream current integrated from top to bottom has to vanish:

$$0 = \int_{BOTTOM}^{SURFACE} J_x \, dz = B_z \int_{BOTTOM}^{SURFACE} \sigma V_y \, dz - \left(\frac{\Delta\Phi}{\Delta\ell_x}\right) \int_{BOTTOM}^{SURFACE} \sigma \, dz \quad (3)$$

In this case we can recapture Faraday's Law in its simplest form

$$\frac{\Delta\Phi}{\Delta\ell_x} = \hat{V}_y^* B_z \quad (4)$$

where $\hat{V}^*$ is now the conductivity-weighted average velocity, defined by

$$\hat{V}_y^* \equiv \int_{BOTTOM}^{SURFACE} V_y \, \sigma \, dz \bigg/ \int_{BOTTOM}^{SURFACE} \sigma \, dz \quad (5)$$

In the more general case where the channel cross-section varies along the stream, or where there is no insulating channel, it is possible to have a non-vanishing net electric current arising from distant sources. Such a non-local net current will augment the potential gradient by an amount $\left(\Delta\Phi/\Delta\ell_x\right)_{REGIONAL}$ which will be independent of depth but may vary slowly across the stream.

The peculiar conductivity-weighted average $\hat{V}_y^*$ is not a quantity of great intrinsic interest. For tidal flows in shallow waters where neither the flow speed nor the conductivity change much over the narrow range of depth, $\hat{V}_y^*$ might be expected to approximate the true speed of the water column. In such shallow waters, however, the conductance of a sediment layer under the water is apt to be large enough to seriously diminish $\hat{V}^*$. This effect of bottom conductance is readily included in equation (5) by treating the bottom as a further static layer of sea water, augmenting the integral in the denominator without affecting the numerator.

Using observations from underwater telephone cables, Longuet-Higgins[7] found that bottom conductance diminished the electrical potential across the English Channel by as much as 85%. Sanford and Schmitz[1] estimate that the effect of bottom conductance in the Florida Straits diminishes V^* here by about 10%. My own work on the Gulf Stream southeast of Cape Cod suggests a 30% reduction there.[8]

Although this effect of bottom conductance is less in deeper waters, the effect on $\hat{V}^*$ of vertical variation in water conductivity becomes more marked. Hot salty surface water can have a conductivity approaching .06 mhos/cm (6 mhos/meter), while cold fresher bottom water can run as low as .03 mhos/cm.[9] Thus the conductivity-weighted average tends to be overweighted in favor of the surface where the faster currents are usually found, so that $\hat{V}^*$ is neither fish nor fowl, being greater than a true average velocity but less than the surface velocity. Sanford and Schmitz[1] estimate this velocity-conductivity correlation effect to add about 10% to $\hat{V}^*$ in the Florida Straits, just cancelling the effect of bottom conductance. In my own Gulf Stream studies farther north,[8] this correlation effect was less than the effect of bottom conductance.

The big drawback of $\hat{V}^*$ as a velocity measurement is that it is an indication of velocity and not of total flow. To measure the total flow through the Florida Straits, one would integrate the true average water column velocity V, against the depth D, across the channel:

$$\text{TOTAL FLOW} = \int_{\text{WEST}}^{\text{EAST}} \hat{V}_y \, D \, dx \tag{6}$$

On the other hand, if one measures the total electrical potential difference across the Straits, one obtains

$$\Delta \Phi_{\text{TOTAL}} = B_z \int_{\text{WEST}}^{\text{EAST}} \hat{V}_y^* \, dx \tag{7}$$

in which the depth does not enter. Thus a given average speed $\hat{V}^*$ in a shallow part of the channel contributes as much to the cross-stream potential difference as does the same average speed in a deep part of the channel. Or, to put it the other way, a given total flow in shallow water contributes much more to the cross-stream potential than does the same flow in deep water.

Such an effect has been invoked by Sanford and Schmitz[1] to account for the long-period fluctuations in the potential difference between Key West and Havana found by Wertheim[10]. The channel there is distinctly assymmetrical, being much deeper near Cuba than it is near Key West. The core of the current is known to shift back and forth across a channel[10], and these excursions would probably be enough to account for the slow fluctuations in potential difference.

The same effect is bound to occur with the Gulf Stream north of Cape Hatteras after it leaves the relatively shallow Blake Plateau. The path of the Stream meanders freely back and forth by several hundreds of miles,[2] so that the water depth immediately under the Stream may vary from place to place, and from time to time, by as much as 50%. If we think of waters along the east coast as tending to be at one average potential and the Sargasso Sea beyond the Gulf Stream as being at another fairly uniform potential, we must imagine that the extra potential difference generated locally by shallower coastward meanders will create widespread regional electric current patterns of the sort depicted in Figure 3. This kind of problem has been considered by Sanford[12], who made a model calculation based on a stream meandering sinusoidally over a sloping bottom.

A similar problem can arise with tidal flows in shallow irregular passages and estuaries where flow can be concentrated at one part of the channel on the flood tide and another part on the ebb. Such an effect may account for some of the marked disparity observed at Woods Hole passage[13], where it appeared that more than 2½ times as much water came through to the eastward on the flood tide as was measured on the preceding ebb.

Despite these difficulties, measurements of potential difference across channels are still valuable because they can be made continuously and inexpensively at shore stations, without interruption from shipping traffic or from bad weather, and without mechanical failure due to marine fouling. Such measurements have been used for time studies of flow fluctuations in the Straits of Dover[14], the Irish Sea[15,16], the Kuroshio current off Japan[17], the Gulf of Naples[18], in tidal channels on the Pacific northwest coast[19], between widely spaced points in the North Atlantic[20], and at several deep water locations off the coast of California[6,20,21]. The theory of such measurements has been extensively developed for both quasi-static flows[7,22,24] and for fluctuating flows with self-inductive effects[6,25].

Measurements with Towed Electrodes

Because the distribution of submarine cables is determined by communications needs rather than by the research interests of oceanographers, the number of places where useful cable measurements can be made is necessarily limited. This limitation was successfully overcome by von Arx[26] who described in 1950 a method for towing behind a ship a length of cable carrying a pair of silver - silver chloride electrodes spaced 100 meters apart. The von Arx instrument, known as the G.E.K. (for Geomagnetic Electro-Kinetograph), gained rapid acceptance as a standard oceanographic tool.

To a very good approximation, the average motion of such a cable dragging through the water is entirely parallel to the cable direction so that no large scale EMF is generated by the forward motion of the cable along the ship's track. (Small corrections may be necessary for cable droop[28], or for systematic transverse deflection of the cable by the ship's wake[8], but these are usually constant.) On the other hand, the very fact that the cable has no transverse motion relative to the surrounding water assures that the cable shares fully whatever transverse displacement the local surface water is undergoing. Any EMF generated by the transverse motion of the surface water is generated equally in the G.E.K. line and cannot be detected by the G.E.K. Therefore, the measured potential difference between the G.E.K. electrodes (usually a few millivolts in magnitude) is solely due to the IR drop produced by whatever current is flowing in the sea:

$$\left(\frac{\Delta\Phi}{\Delta\ell_x}\right)_{GEK} = -\left(\frac{J_x}{\sigma}\right)_{SURFACE} \qquad (8)$$

This expression can be more usefully written as

$$\left(\frac{\Delta\Phi}{\Delta\ell_x}\right)_{GEK} = B_z\left(\hat{V}_y^* - V_{y,SURFACE}\right) \qquad (9)$$

using equations (2) and (4). From this, it appears that the G.E.K. signal is a measure of the difference between the transverse water velocity at the surface where the cable is being towed, and the conductivity-weighted average $\hat{V}_y^*$.

In deep ocean currents, $\hat{V}^*$ is generally much less than the surface velocity so the G.E.K. signal is usually dominated by the surface motions. The greatest use of the G.E.K. has been as a surface-current indicator, and numerous surface-current studies have been made using the instrument in this way. To allow for the effect of $\hat{V}^*$, von Arx introduced a "K factor", usually between 1.0 and 1.1 in the deep ocean, by which the G.E.K.-measured velocity difference could be multiplied to estimate the surface velocity.

A few GEK studies have been made in which the surface speed $V_{surface}$ has been independently determined by precise navigational measurements [1,8,29], so that the GEK signal could be used to recover an estimated value of $\hat{V}^*$ such as would be measured with a fixed cable. The possible advantages in thus measuring total transverse flows directly from a moving vessel have been largely offset in the past by the difficulty of obtaining navigation measurements of the necessary precision.

The Free-Fall Electromagnetic Current Meter

The most remarkable and most recent advance in the study of the motional electrical signals in the sea was made by Drever and Sanford[30] who described in 1970 the use of a free-falling instrument which, like the GEK, measures the velocity difference ($\hat{V}^* - V$) but does so at all depths, not just at the surface.

The free-fall instrument, shown in Figure 5, has pitched vanes so that it rotates once in every eight meters of fall. Horizontal electric field sensors at the ends of the vanes are connected to the electrodes by salt bridges, as shown in Figure 6. Although the potentials being measured are exceptionally small (a few tens of microvolts), the AC signal produced by the rotation can be processed well enough to yield velocity data with noise of only about ± 1 cm/second.

The instrument provides a check on its own performance: when it reaches the bottom, it releases a weight and comes back up (Figure 7) making the same measurements all over again. Current profiles obtained with the free-fall instrument, such as shown in Figure 8, give far more detail and more precise detail about the vertical structure of ocean currents than has been heretofore available. At the present time, the constant value of $\hat{V}^*$ is not determined by this instrument so that absolute velocities are not known unless an independent determination of $\hat{V}^*$, or of the surface current, is made at the time of the drop. This is not, however, a serious drawback, because there is a natural theoretical separation between the vertical velocity structure, which is related to the density structure of the water ("baroclinic flow") and is relatively stable, and the average velocity, which is subject to tidal fluctuations and storm surges ("barotropic flow").[27]

Vertical Potential Differences

So far we have only considered the effects of horizontal motions across the vertical component B_z of the earth's magnetic field. There is also a horizontal component well-known to be directed toward the North magnetic pole. This north-south component can induce an EMF in water moving vertically, but such motions in the ocean are either transitory or minute. It can also induce a vertical EMF in water moving in a magnetic east-west direction. (It might be noted here that whatever may be the potential gradients and currents thus induced by the north-south magnetic field, they are quite independent of those induced by the vertical component so that any resultant gradients and currents are obtained as a linear superposition of the two. Moreover, the horizontal effects of vertical EMF are very small and can usually be ignored in horizontal potential measurements[25].)

Because the surface of the ocean is an insulator, the effect of short-circuiting on such vertical EMFs is much reduced. Malkus and Stern[23] suggested that measurements of the vertical potential gradient be used directly to measure east-west transport. Unfortunately this simple physical conception leads the way into an electrochemical nightmare.

To a physicist an electrode is an electrode, and anything that makes electrical contact will do. Many of the potential measurements on submarine cables were made using the metallic cable shielding as an electrode surface. So long as the potentials are the order of a volt in magnitude, this is acceptable, especially if tidal fluctuations are the main subject of interest. For the much smaller signals obtained with the GEK, much better electrodes were necessary, and von Arx took considerable pains to use massive silver/silver chloride electrodes which were well-matched for temperature coefficient and which were somewhat protected from direct contact with the sea.[25] Even so, the problems of electrode zero signal and electrode drift were serious, so that periodic course reversals were recommended for monitoring the zero. The problem is greater than one might expect because Ag/AgCl electrodes are much less reversible in sea water than in pure chloride solutions, presumably because the equilibrium bromide/chloride ratio in the silver halide is much greater than in the sea salt mixture.

When it comes to measuring signals at the submillivolt level in sea water, any electrode is very nearly a universal transducer, being sensitive to temperature, pressure, salinity, oxygen level and almost anything else one can think of. For the purpose of horizontal GEK measurements, this sensitivity is not very critical, because as emphasized earlier, the ocean tends to be very uniform horizontally over wide areas. Not so in the vertical!

In addition to the intrinsic increase of pressure with depth, we have the typical decrease of temperature associated with the density stratification. Commonly there is also an oxygen variation. Then, in most places, the salinity decreases somewhat with depth because the warm surface water underwent some evaporation while it was being heated. Since the equilibrium salt distribution in an unstirred ocean would actually be one with a Boltzmann distribution increasing with depth at about 10 percent per kilometer[31,32], the ocean has far too much salt near the surface and presents us, in effect, with the world's most extensive liquid junction.

Vertical potential measurements have been attempted with electrodes[33-36] but it is difficult to know how to interpret the results. At Woods Hole we have attempted vertical measurements with long salt bridges consisting of polyethylene tubing filled with sea water[13]. The high impedance of such circuits (order of megohms) requires the use of electrometer amplifiers and special low-leakage components. Our most ambitious effort was an attempt in 1963 to measure the Atlantic Equatorial undercurrent, which was only marginally successful. Such velocity measurements as we obtained were all determined relative to the ship's drift which was not at the time easily estimated. Potentials near the surface were badly distorted by the cathodic protection devices on the ship's hull.

Because the salt-bridge technique permits a direct measurement of electrode zero by means of a solution shunt, it also has some advantages for horizontal GEK measurements where high precision is desired[8].

Other Electromagnetic Effects

Everything discussed thus far has related to the EMF and resultant current produced in a moving conductor by a magnetic field. We should also expect other typical electromagnetic effects[4]: a force on a current-carrying conductor in a transverse magnetic field; a circuit EMF induced by changes in the magnetic field; magnetic field components produced by currents, and even self-inductance and mutual inductance associated with fluctuating currents. These effects are all present, though none are terribly conspicuous.

The force per unit volume produced magnetically on a transverse current in a parcel of moving water is given[4]

by
$$F_y / \text{Volume} = -J_x B_z \tag{10}$$

From our previous discussion, it appears that the current density J_x, which is what the GEK measures, is equal to

$$-\sigma B_z (\hat{V}_y^* - V_y) \quad \text{(c.f equations 8 and 9) so that}$$

$$F_y / \text{Volume} = \sigma B_z^2 (\hat{V}_y^* - V_y) \tag{11}$$

If we relate this force to the acceleration it would produce, we find

$$F_y/V_{\text{volume}} = \rho \frac{dV_y}{dt} = \sigma B_z^2 \left(\hat{V}_y^* - V_y\right) \qquad (12)$$

where ρ is the density of the sea water. Thus the electrical drag on moving water in the sea is a body force trying to bring all the fluid exponentially to the conductivity-weighted local average $\hat{V}_y^*$. The exponential decay time is $\rho/\sigma B_z^2$, about 3000 years. (Momentum exchange by turbulent transfer is much more rapid.)[37] This is a typical eddy current damping, showing that, after all, the currents we have been discussing are really just very large eddy currents.

The magnetic fields produced by steady electric currents are probably barely detectable. Sanford[25] has estimated that the solenoidal currents produced by the Gulf Stream should generate an axial magnetic field along the Stream which would be in excess of 100γ (10^{-3} gauss) just under the core of the stream. Unfortunately, this small component is largely perpendicular to the 500-fold greater earth's field so that it would tend to produce a change in field direction rather than in the more readily measured field intensity.

Small fluctuating magnetic fields associated with tidal motions have been identified by Larsen[38] and by Malin[39]. These fields are quite small, generally a fraction of a γ in amplitude, detectable only by harmonic analysis of long time-series of observations, but they are in reasonable accord with theoretical predictions.[40] Larsen estimated that the transient magnetic field produced by a single <u>tsunami</u> (the so-called "tidal wave" produced by an earthquake, usually in the Pacific Ocean) would be too small to be detected.

To deal with effects associated with time-varying fields and currents, it is customary to bring up the full panoply of Maxwell's equations. For the case of variations produced by sources outside the ocean, we can treat the problem as a special instance of the general problem of propagation of electromagnetic radiation into a conducting medium. It turns out that the ocean, just like any other conductor, has a "skin depth" for the penetration of low frequency electromagnetic radiation:[41]

$$\delta = \left(\frac{10^7}{4\pi^2 f \sigma}\right)^{1/2} \qquad (13)$$

where f is the frequency and σ is the conductivity. For a typical conductivity of 4 mhos/meter, the product of skin depth by the square root of frequency is a constant.

$$\delta f^{1/2} \simeq 250 \text{ meters/sec}^{1/2} \qquad (14)$$

Thus even at periods of 1 second, the effective signal penetration is only a quarter of a kilometer. To get a skin depth equal to the average depth of the ocean, which would allow e^{-1} of the incident radiation to reach the bottom, one would have to employ a signal with a period of at least 4 minutes. This skin depth formula is not applicable, of course, at optical frequencies where ionic conductivity is insignificant. Otherwise the ocean would have a fine metallic lustre and no food value.

The skin depth limitation has always been an obstacle to radio communication with submarines even when VLF transmissions are used. It also precludes the use of radio-telemetry in the sea, and accounts for the extensive use of acoustic sounding, acoustic ranging, and acoustic telemetry in deep-sea research.

To geophysicists wishing to study the electrical conductivity of the earth's mantle at depths of several hundred kilometers, the conductivity of the ocean has seemed a special nuisance. Large-scale ionospheric disturbances cause regional fluctuations in the earth's magnetic field which, in turn, induce electrical "earth currents" in the earth's crust and mantle. By correlating horizontal potential gradients with the magnetic fluctuations that produce them, the geophysicist can, in principle, determine the underlying conductivity structure of the earth.

The various ways in which the ocean has interfered with such magneto-telluric measurements are fully discussed in extensive reviews by Cox, Filloux, and Larsen[21] and by Bullard and Parker.[37] Not only do the oceans effectively shield the underlying oceanic crust from the higher frequency fluctuations, they also strongly perturb the earth current distribution on adjacent land areas. In addition, of course, the potentials and currents produced by the motions of the sea have fluctuating components which further complicate the problem.

The earth currents perpendicular to coastlines are strongly intensified near the shore where the sharp edge of the conducting sea water wedge extends furthest over the poorly-conducting crust. In shallow coastal waters, potential measurements normal to the shore are often completely dominated by such fluctuating earth currents, although elsewhere in the ocean those currents contribute a very minor background to measured DC potentials.

It may be well to note here that potential measurements in waters near urban or industrial areas are also liable to interference from man-made earth currents produced by cathodic protection systems or by electrical trains and subway systems with ground returns.

Magnetic Anomalies in the Sea

Extensive detailed surveys of the magnetic field at sea have been made.[42] The standard instrument for this purpose is the proton resonance magnetometer which can be towed behind a ship, or even an airplane, measuring total magnetic intensity for a precision of ± 1 gamma ($1\gamma = 10^{-5}$ gauss = 10^{-9} webers/meter^2).

The local departures of the magnetic field from a smoothed regional field are called "magnetic anomalies," and are sometimes associated with structural features such as sea-mounts.[43] In recent years the overwhelming interest in these anomalies has focussed on regional patterns of magnetic lineations covering most of the ocean floor. Along the mid-ocean ridges these lineations run parallel to the ridges and typically form symmetric patterns around a central rift valley.

In 1963, Vine and Matthews[44] advanced the hypothesis that these magnetic anomaly lineations were caused by spreading of the sea floor at the mid-ocean ridges over geological time, the alternation of magnetization being due to long-term reversals of the earth's magnetic field[45,46] being recorded in the new sea floor as it formed and cooled. This hypothesis, which has been overwhelmingly confirmed by subsequent research, has opened up a new scientific revolution in geology and geophysics. It would probably not be too much of an exaggeration to say that the new "plate tectonics" represents the single most important scientific advance of the past decade. The spreading rates and plate motions deduced from magnetic anomaly measurements have been shown to be directly related to almost all major patterns of earthquakes, volcanic activity, and mountain-building throughout the globe.[47]

One consequence of these spectacular developments is that ways of studying the properties of the earth's mantle at depth of several hundred kilometers are more important now than ever before. What is going on down there that enables great thick slabs of the earth's surface to form, to slide around, and finally to disappear by ducking under other slabs? Magneto-telluric measurements made at the deep ocean floor, well away from coast lines and major oceanic currents, offer one of the very few reasonable possibilities for finding out. The next decade should certainly see a greatly intensified interest in all kinds of electrical and magnetic measurements made at sea.

Acknowledgments

The author wishes to thank Dr. Thomas Sanford for helpful suggestions and critical review of the manuscript, as well as for permission to reproduce some of the illustrations used here. Support of the author's own research on potential measurements at sea from NSF Grants G12178, GP3218, and GA1140 is gratefully acknowledged.

This is Contribution No. 2960 from the Woods Hole Oceanographic Institution.

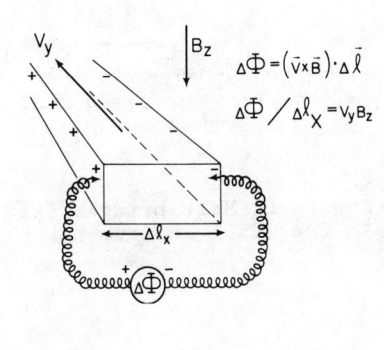

Figure 1. EMF induced in moving water according to Faraday's Law.

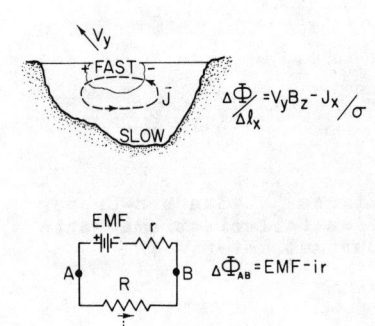

Figure 2. Observed potential differs from EMF because of electric current flow.

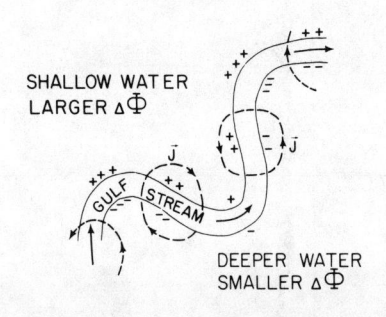

Figure 3. Regional currents produced by meandering Gulf Stream over sloping bottom (after Sanford).

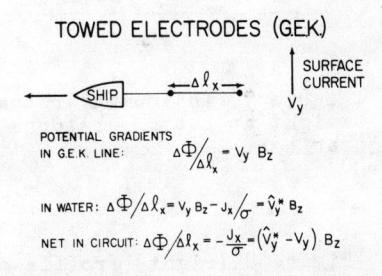

Figure 4. The von Arx GEK measures velocity difference from local conductivity-weighted average.

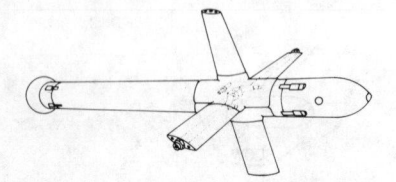

Figure 5. Sanford-Drever free-fall electromagnetic current meter.

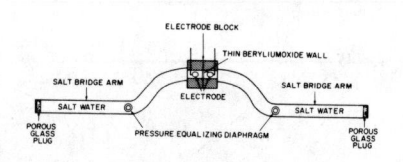

Figure 6. Salt bridges used in free-fall instrument.

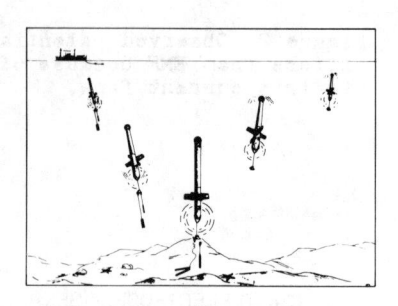

Figure 7. Instrument releases weight at bottom, returns to surface.

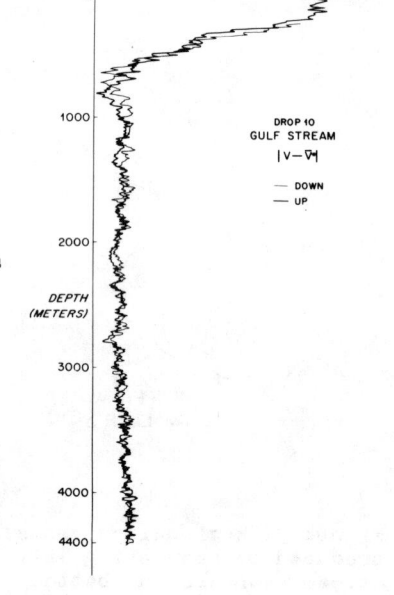

Figure 8. (right) Profile of total flow speed measured by free-fall instrument.

References

1. T.B. Sanford and W.J. Schmitz, Jr., 1971. A comparison of direct measurements and G.E.K. observations in the Florida Current off Miami. J. Marine Research 29, 347-359.

2. H.M. Stommel, 1965. The Gulf Stream, (2nd ed.), University of California Press, Berkeley, 243 pp.

3. M. Faraday, 1832. Bakerian Lecture - Experimental researches in electricity. Phil. Trans. Roy. Soc. London, 1832; Part I, 163-177.

4. Any standard college physics text.

5. H.U. Sverdrup, M.W. Johnson, and R.H. Fleming, 1942. The Oceans, Prentice-Hall, Englewood Cliffs, N.J., 1087 pp.

6. J.C. Larsen, 1966. Electric and magnetic fields induced by oceanic tidal motion. PhD. Thesis, University of California, San Diego, 99 pp.

7. M.S. Longuet-Higging, 1949. The electrical and magnetic effects of tidal streams. Monthly Notices Royal Astro. Soc., Geophys. Suppl. 5, 285-307.

8. P.C. Mangelsdorf, Jr., 1968. Gulf Stream transport measurements using the salt-bridge GEK and Loran-C navigation, (Abstract). Trans. Am. Geophys. U. 49, 198.

9. R.A. Horne, 1969. Marine Chemistry, Wiley-Interscience, New York, 568 pp.

10. G.K. Wertheim, 1954. Studies of the electric potential between Key West, Florida, and Havana, Cuba, Trans. Am. Geophys. Union, 35, 872-882.

11. J.E. Pillsbury, 1891. The Gulf Stream. Methods of the investigation and results of the research. U.S. Coast & Geod. Serv., Rept. for 1880. Appendix no. 10, 461-620.

12. T.B. Sanford, personal communication.

13. P.C. Mangelsdorf, Jr., 1962. The world's longest salt bridge, in Marine Sciences Instrumentation, Vol. 1, Plenum Press, New York, 173-185.

14. K.F. Bowden, 1956. The flow of water through the Straits of Dover, related to wind and differences in sea level, Phil. Trans. Roy. Soc., A248, 517-551.

15. K.F. Bowden and P. Hughes, 1961. The flow of water through the Irish Sea and its relation to wind. Geophys. J. Royal Astro. Soc. 5, 265-291.

16. P. Hughes, 1969. Submarine cable measurements of tidal currents in the Irish Sea. Limnol. & Oceanogr. 14, 269-278.

17. T. Teramoto, 1968. Day-to-day to monthly variations in oceanic flows estimated from cross-stream differences in electric potential. Unpublished document, Ocean Res. Inst., Univ. of Tokyo.

18. W. Düing, 1965. Strömungsverhältnisse im Golf von Neapel., Pubbl. staz. zool. Napoli 34, 256-316.

19. R.M. Morse, M. Rattray, Jr., R.G. Paquette, and C.A. Barnes, 1958. The measurement of transports and currents in small tidal stream by an electromagnetic method. Tech. Report No. 57, Dept. of Oceanography, U. of Washington, Seattle. 70 pp.

20. H. Stommel, 1954. Exploratory measurement of electrical potential differences between widely spaced points in the North Atlantic Ocean. Arch. Meteor. Geophys. Bioklim, Series A, 7, 292-304.

20. J.C. Larsen and C.S. Cox, 1966. Lunar and solar daily variation in the magneto-telluric field beneath the ocean. J. Geophys. Res., 71, 4441-4445.

21. C.S. Cox, J.H. Filloux, and J.C. Larsen, 1968. Electromagnetic studies of ocean currents and electrical conductivity below the ocean floor. in The Sea, Vol. 4, Part I, A.E. Maxwell, ed; Wiley-Interscience, New York, 637-693.

22. H. Stommel, 1948. The theory of the electric field induced in deep ocean currents. J. Mar. Res. 7 (3), 386-392.

23. W.V.R. Malkus and M.E. Stern, 1952. Determination of ocean transports and velocities by electromagnetic effects. J. Mar. Res. 11, 97-105.

24. M.S. Longuet-Higgins, M.E. Stern, and H. Stommel, 1954. The electrical field induced by ocean currents and waves with applications to the method of towed electrodes. Pap. Phys. Oceanogr. Meteor., M.I.T. and W.H.O.I., Vol. XIII, 1-37.

25. T.B. Sanford, 1971. Motionally induced electric and magnetic fields in the sea. J. Geophys. Res. 76, 3476-3492.

26. W.S. von Arx, 1950. An electromagnetic method for measuring the velocities of ocean currents from a ship under way. Pap. Phys.

Oceanogr. & Meteorol., M.I.T. & W.H.O.I., Vol. XI, (3), 1-61.
(See also ref. 25, Chapter 9).

27. W.S. von Arx, 1962. An Introduction to Physical Oceanography, Addison-Wesley, Reading, Mass., 422 pp.

28. J.A. Knauss and J.L. Reid, 1957. The effects of cable design on the accuracy of the G.E.K., Trans. Amer. Geophys. Union, 38, 320-325.

29. F. Chew, W.S. Richardson, and G.A. Berberian, 1971. A comparison of direct and electric current measurements in the Florida Current. J. Marine Res. 29, 339-346.

30. R.G. Drever and T.B. Sanford, 1970. A free-fall electromagnetic current meter - instrumentation, Proc. of I.E.R.E. Conf. on Electronic Eng'ng. in Ocean Technol., I.E.R.E., London, 353-370.

31. R.M. Pytkowicz, 1962. Effect of gravity on distribution of salts in sea water. Limnol. & Oceanogr. 7, 434-435.

32. P.C. Mangelsdorf, Jr., F.T. Manheim, and J.M.T.M. Gieskes, 1970. Role of gravity, temperature gradients and ion-exchange media in formation of fossil brines. Am. Assoc. Petr. Geol. Bull. 54, 617-626.

33. J. Martin, 1964. Experiences de G.E.K. vertical dans le Detroit de Gibraltar. Cahiers Oceanographiques XVI, 377-392.

34. V.G. Bororov, R.M. Demeniskaya, A.M. Gorodnitskiy, M.M. Kazanskiy, V.M. Kontorovich, E.M. Litvinov, N.N. Trubyatchinskiy, and V.D. Fedorov, 1967. Character and causes of the vertical variation of the natural electric field in the ocean. Oceanology 9, 622-626.

35. R. Margalef, 1967. Significado de los diferencias verticales de potencial electrico en el mar. Inv. Pesq. (Barcelona), 31, 259-263.

36. T.B. Sanford, 1967. Measurement and interpretation of motional electric fields in the sea. PhD Thesis, Mass. Inst. of Technology, Cambridge, 161 pp.

37. E.C. Bullard and R.L. Parker, 1968. Electromagnetic induction in the oceans. Chapter 18 in The Sea, Vol. 4, Part I, A.E. Maxwell ed., Wiley-Interscience, New York, pp. 695-730.

38. J.C. Larsen, 1968. Electric and magnetic fields induced by deep sea tides. Geophys. J. Royal Astr. Soc. 16, 47-70.

39. S.R.C. Malin, 1969. The effect of the sea on lunar variations of the vertical component of the geomagnetic field. Planet Space Sci. 17, 487-490.

40. S. Chapman and P.C. Kendall, 1970. Sea tidal generation of electric currents and magnetic fields: applications to five stations within the British Isles. Planet. Space Sci. 18, 1597-1605.

41. P. Lorrain and D.R. Corson, 1970. Electromagnetic Fields and Waves, 2nd edition, W.H. Freemen & Co., San Francisco, 706 pp.

42. E.C. Bullard and R.G. Mason, 1963. The magnetic field over the oceans. Chapter 10 in The Sea, Vol. III, M.N. Hill, Ed., Wiley-Interscience, New York, pp. 175-217.

43. J.R. Heirtzler, 1970. Magnetic anomalies measured at sea. Ch. 3 in The Sea, A.E. Maxwell, ed., Wiley-Interscience; New York, pp. 85-128.

44. F.J. Vine and D.H. Matthews, 1963. Magnetic anomalies over oceanic ridges. Nature 199, 947-949.

45. N.P. Opdyke, 1970. Paleomagnetism. Ch. 5 in The Sea, A.E. Maxwell, ed., Wiley-Interscience; New York, pp.

46. A. Cox, R.R. Doell, and G.B. Dalsymple, 1963. Geomagnetic polarity epochs. Science 143, 382-385.

47. B.L. Isacks, J. Oliver, and L.R. Sykes, 1968. Seismology and the new global tectonics. J. Geophys. Res. 73, 5855-5899.

THE EFFECTS

OF DEEP SEA ENVIRONMENT

ON ELECTROCHEMICAL SYSTEMS

Chairmen:
Mario Banus
Marcus Hotz

SURVEY OF TECHNIQUES TO STUDY AQUEOUS SYSTEMS
AT PRESSURES OF 1-3 KBAR

W.A. Adams
and
A.R. Davis

Water Science Subdivision
Inland Waters Directorate
Department of the Environment
562 Booth Street
Ottawa, Ontario, K1A 0E7
Canada

ABSTRACT

Techniques are described for the simulation in the laboratory of the deep water, high-pressure environment. The effects of pressure on the equilibrium and transport properties of aqueous solutions can only be understood in terms of the unique structural properties of water. Spectroscopic methods are illustrated by a high-pressure laser-Raman spectroscopic study of ionic equilibria. Conductivity measurements up to 3 kbar on the aqueous $MgSO_4$ system between 0 and 25°C are presented as an example of the utility of high-pressure conductivity techniques. The piezooptic coefficient of sea water and several electrolyte solutions has been measured by a laser-interferometric technique using automatic electronic fringe counting. These studies cover the range 0-2 kbar and 0-25°C. Static permittivity measurements on water, and other types of measurements made at high pressures of electrochemical interest are reviewed.

1. INTRODUCTION

The surface waters of the world, both inland and marine, are an environment in which pressures range from atomspheric at the air-water interface up to 1.2 kbar in the deep oceans (1 bar = 1 x 10^6 dyne cm^{-2} = 1 x $10^{-5} Nm^{-2}$ = 0.987 int. atm. = 14.5 p.s.i.). This high-pressure environment is now an area of increased activity as undersea mining, the use of military and commercial submarines, ocean waste disposal, and the scientific investigation of the ocean depths accelerates. Inland lakes give rise to pressures of 40 to 50 bars, e.g. the Great Lakes. These pressures have significance since many of the properties of a body of water are determined by what happens at the sediment - water interface. It is these interfacial phenomena (including biological activity) that exhibit the greatest pressure sensitivity. Deep well disposal of wastes is another area of interest, since high pressures in the disposal zone formation, besides causing mechanical effects, produce changes in the chemistry, e.g. unexpected precipitates can form reducing the efficiency of the well. Simulation in the laboratory of the conditions of temperature and pressure found in the environments described above, permits a controlled study of the parameters affecting solubility, transport and equilibria involving the natural aqueous system.

Since the properties of aqueous solutions which make them a unique class of solutions depend on the behaviour of pure water, much energy has been expended in the last decade to devise a theory of water. Several books have recently appeared on water, most of which contain chapters on the properties of water and aqueous solutions under high pressure (1-5). Numerous conferences are being held to discuss water and aqueous solutions (four between June and September, 1972, in North America). It is clear, however, that until accurate and extensive experimental data is available, which covers a wider range of temperature and pressure, it will be impossible to formulate an accurate structural and dynamic theory for aqueous solutions.

There are many good sources of general information on high-pressure techniques available in review articles or in books, often containing articles by specialists in various areas of high-pressure research (6-22). The collected works of Bridgman (23) are worth special mention. There is a journal devoted to high pressure called "High temperatures·High pressures" published by Pion Ltd., London and a high pressure bibliography edited by L. Merrill, of the High Pressure Data Center, Brigham Young University, Utah. Articles or reviews of a more specialized nature will be referred to in the next section. The range of pressures chosen for this survey is limited to from 1 atm. to 3 kbar. It can be seen from the shaded area on figure 1, which represents the range of pressure and temperature found in the surface waters of the earth, that this range is adequate for most applications research involving oceanography or limnology. Some indication of the complexity of phenomena to be expected of aqueous solutions is suggested by the eleven solid phases of pure water now known to exist under different conditions of temperature and pressure (25). The pressure range of this survey is by no means a limit in terms of the natural sciences, since geological and geophysical studies are concerned with pressures which extend to 3.64×10^3 kbar at the earth's centre (26) and pressures of 10^{15} kbar are predicted by astrophysicists to exist in white dwarf stars (13). However, inexpensive standardized apparatus is readily available from commercial manufacturers for the pressure range of this survey. Furthermore, it is not necessary to have extensive experience in mechanical engineering and design to assemble and operate high-pressure equipment up to 3 kbar. For pressures in excess of 3 kbar and in particular above 7 kbar, the techniques are not standardized and experimental difficulties are greater. The selection of topics and the depths of coverage to be given in this survey are arbitrary, representing more the interests of our laboratory rather than any attempt at completeness.

II. HIGH PRESSURE TECHNIQUES

A. Thermodynamic Measurements

Volume and compressibility are the most fundamental thermodynamic properties of an aqueous system required for the analysis and interpretation of many types of physical-chemical pressure measurements. The review article by Kell (27) summarizes the volume and compressibility data available for water up to 1971. Although the approximation that the solution density is equal to that of pure water is often made, for concentrated solutions or for very precise work this is not acceptable. Comparative measurements with water or mercury as reference liquids are often made, so that the accurate data available at high pressure for these liquids can be used to calibrate the apparatus. In general, techniques that do not require the disassembly of high-pressure equipment at each pressure at which a density is desired are preferable. A volumometer technique using a commercial injector will provide rapid results to about ± 0.5% (28) while for more precise volume data, a glass piezometer cell, such as that shown in figure 2, can be used (29). There are several methods for detecting the compression of a liquid in a piezometer. Platinum electrical contacts can be inserted in the tube wall so that as a mercury column rises with compression, the shorting of the contact is detected. In the cell illustrated in figure 2, the position of the mercury at different pressures is indicated by a very precise capacitance bridge. The various techniques used for compressibility measurements have been reviewed up to 1972 by Whalley (30), by Steele and Webb (31), the Russian literature by Tsiklis (9), and at low temperatures for gases and liquids by Sengers (32). The specific volume of sea water as a function of temperature and pressure has been estimated by velocity of sound techniques by Crease (33), by a differential transformer volumetric system by Wilson and Bradley (34), (similar to the system used by Duedall (35) for differential compressibility), and by Lipple and Millero (36) using a piezometric technique. The densities of NaCl solutions at 25°C relative to the densities of water have been measured up to 1.2 kbar by Millero et al (37) using a magnetic float densitometer. A precision of ~ 10 p.p.m. is obtained with a sensitivity of 0.5 p.p.m.

Isothermal pressure changes of enthalpy, entropy, internal energy and heat capacity depend on volume and can be calculated from the equation of state of the liquid and properties of the dilute vapor (27).

B. Spectroscopic Measurements

(i) Instrumentation

Spectroscopic studies of materials under high pressure are limited by the difficulties of incorporating transparent windows into the pressure cell design. The ancillary high-pressure equipment and spectrometers used in high-pressure spectroscopy are the same as those used in standard work. For pressures in excess of 10 kbar a review article by Whatley and Van Valkenburg (38) is useful, especially in the fields of X-ray diffraction and optical microscopy. Weigang and Robertson (39) present theoretical considerations useful in the design and subsequent interpretation of high-pressure U.V., visible and I.R. spectroscopic investigations of molecular liquids. Spectroscopic high-pressure techniques can be found in references in the above reviews as well as in the review by Oksengorn, Vu and Vodar (40). Several investigators have described high-pressure cell designs for spectrophotometric work (41-44) some of which are well suited to corrosive liquids such as sea water (45-47). A two window high-pressure cell manufactured by Harwood Engineering, Walpole, Mass. is shown in figure 3. One sapphire window and packing is shown removed in the foreground. The cell is designed for 10 kbar Laser-Raman measurements using the backscattering technique of Davis and Adams (48) in which a small mirror coated on the sapphire is used to reflect back the exciting laser beam allowing collection of Raman scattering at 180° to the incident beam. Walrafen has used the Cary Model 81 "image slicer" entrance lens system to collect a maximum amount of Raman scattered light with in-line, two window, high-pressure cells for pressures up to 4 kbar (49) and up to 7.2 kbar (50) with aqueous systems. In more recent work, Walrafen (51) has used the same "image slicer" technique with a two window in-line high-pressure cell and a 90° two window high-pressure cell to obtain liquid water and ice VI laser-Raman spectra to 10 kbar. Frank and Linder (52) have also studied pressure effects on the Raman spectra of water up to several kbar. A laser-Raman cell useful to about 100 bar has been described by Cavagnat *et al* (53) which makes use of an o'ring sealed quartz tube. A miniaturized diamond-anvil cell useful to very high pressures (up to 150 kbar) has been described by Weir *et al* (54). This type of cell has been used to obtain laser-Raman of liquids and solids under pressure by Melveger *et al* (55, 56), but suffers the disadvantage that the pressure on the sample may not be hydrostatic and is not easily determined. Other laser-Raman pressure studies (57-60) are concerned primarily with the solid state and higher pressures than the 1-3 kbar range of interest in this survey. There is work in progress at C.E.R.L. (61) concerned with high-temperature, high-pressure aqueous solutions. Franck (62) has described an ingenious high-pressure cell for I.R. work to 4 kbar in which one sapphire window is used in conjunction with a mirror spring-loaded against the inside of the window to avoid variation in path length with pressure for the thin samples needed in aqueous studies. The I.R. studies (63, 64) and the Raman studies (52) of high-temperature and high-pressure at the University of Karlsruhe, Germany, have been discussed by Tödheide (65). An I.R. cell for following reaction kinetics to 200 bar has been describe by Noack (66).

(ii) Laser-Raman

Irish (67) has reviewed the recent literature (up to 1970) concerning vibrational spectral studies of electrolyte solutions and has presented as well a valuable historical introduction to Raman spectroscopy of aqueous solutions. The effect of phase and pressure changes on vibrational spectra has been reviewed by Davis (68) up to 1970. One of the limitations of the Raman technique is that it is limited to concentrations greater than 0.1 molar if the intensities of Raman bands are to be used to obtain the concentration of the scattering species responsible for the observed bands. A second limitation is that it is only possible to observe the Raman spectrum of polyatomic species. However, since the Raman scattering intensity, measurable to about 2%, is directly proportional to the concentration of the scattering species it is possible to obtain information about complex equilibria in aqueous solutions often not available using any other technique. It is particularly well suited to pressure studies of aqueous ionic equilibria since the laser has become available as a Raman source. This is because of the high intensity of exciting radiation that can be concentrated in a narrow laser beam making large optical windows unnecessary.

The three-window cell shown in figure 4 was constructed by the American Instrument Company, Silver Spring, Md. from Hastelloy C to withstand corrosive solutions. It has sapphire windows and a working pressure of 2.5 kbar. The valves and fittings are also Hastelloy C. A strain-gauge pressure transducer manufactured by the Viatran Corporation, Buffalo, N.Y. (on the right side of the cell) is used to monitor directly the pressure on the solution being studied. A Heise bourdon pressure gauge (not shown) is used on the pump or hydraulic fluid side of the isolator. The isolator is a piston which travels in a cylinder to separate the hydraulic oil from the aqueous solution and is shown above the valve to the left of the cell. A drilled-out aluminum plate is bolted to each side of the cell so that thermostating can be accomplished by circulating from a constant-temperature bath. In figure 5, the cell is shown mounted on an optical track with a Jarrell-Ash Model 450 Raman spectrometer (far left) which has been set up to leave ample space for high-pressure apparatus. A Spectra-Physics 164 Argon ion laser is placed on a rack below the vibration isolation table and the exciting laser light (1.5W at 4880Å) is brought up at the end of the table, along under the optical track and up through the in-line windows by means of easily adjusted magnetically mounted mirrors. The Raman scattering is collected by a Mamiya camera lens from the 90° window and focussed on to the slit of the monochromator to the left. Figure 6 is an example of the pressure dependence of two Raman bands, the right band due to $SO_4^=$ and the left band due to HSO_4^-. It can be seen that in this 2.0 M aqueous solution of $KHSO_4$ the concentration of HSO_4^- decreases while the concentration of $SO_4^=$ increases when pressure is applied at 26.0°C. Figure 7 shows a diagram of an ionic equilibrium in which two oppositely charged ions are in equilibrium with a neutral species. By le Chatelier's principle, pressure will cause the equilibrium to shift in the direction of the least partial molal volume, i.e. toward the free ions since these have the greatest amount of electrostriction. That this principle is qualitatively correct, can be seen from the fact that the ratios of dissociation constants at 1000 atm. to their values at 1 atm. are greater than unity. The integrated intensities of the $SO_4^=$ and HSO_4^- bands for several temperatures up to 1 kbar have been plotted as a ratio in figure 8. In conjunction with the data of Irish and Chen (74), it is possible to convert these intensity ratios to concentration quotients. If the variation of the activity coefficient ratio with pressure is neglected, then the equation given in figure 9 can be used to calculate the volume change associated with equilibrium given below:

$$HSO_4^- \rightleftharpoons H^+ + SO_4^= \tag{1}$$

The effect of different cations on the ΔV values recorded in the table in figure 9 are interpreted tentatively as due to cation hydration effects on the secondary hydration of the species involved in equation (1) (75). The ΔV for H_2SO_4 is in satisfactory agreement with that obtained using a conductivity technique by Horne et al (72). Although there is definite Raman evidence in 2M aqueous $MgSO_4$ of $MgSO_4$ ion pairing from the shape of the sulphate symmetrical stretching band, it has not been possible to obtain a reliable estimate of ΔV for the ion-pair equilibrium from Raman pressure studies because of the small proportion of bound $SO_4^=$ in solution. Work is continuing on this problem.

The Raman spectrum of CO_2 in the region 1250 to 1450 cm^{-1} is shown in figure 10 for (a) 0.17 M CO_2 dissolved at 5 atm in H_2O, (b) 0.17 M CO_2 dissolved at 5 atm in D_2O, and (c) pure CO_2 gas at 5 atm pressure as found by Davis and Oliver (76); (a), (b), and (c) refer to the top, middle and bottom spectra in figure 10. The CO_2 Raman band at 1388 cm^{-1} is not detectable with current techniques in air-saturated water at 1 atm. The Raman spectra of aqueous carbonate and bicarbonate solutions of alkali metals have been studied by Oliver and Davis (77). Pressure Raman studies of the bicarbonate-carbonate equilibrium of aqueous $KHCO_3$ indicated a slight increase in the $[CO_3^=]/[HCO_3^-]$ ratio with pressure up to 1.2 kbar (78). The $CO_3^=$ symmetrical stretching band at 1064 cm^{-1} has been detected at high-pressures in H_2O and D_2O solutions saturated with liquid CO_2 at 25°C. The OH and OD vibration regions in these high-pressure CO_2-H_2O and CO_2-D_2O solutions have also shown very pronounced shifts toward the band shapes characteristic of solid H_2O or D_2O (78). A white precipitate, $CO_2 \cdot 6H_2O$ or $CO_2 \cdot 6D_2O$, first reported by Wroblewski (79) and recently investigated by Larson (80), is produced by rapid decompression of a saturated aqueous CO_2 solution. The effects of the CO_2 on the water spectra are thought to be due to the formation of a cage of strongly bonded water molecules which hydrate the CO_2 in solution.

Using a four-window pressure cell designed for 7 kbar and supplied by Pressure Products Industries, Hatboro, Pa., the laser-Raman spectrum of water has been recorded at 4880Å for five pressures up to 2.5 kbar (see figure 11). The most obvious characteristic of the pressure effect on the spectrum of H_2O, is that it is not very large at 20°C. There is a decrease in the intensity of the intermolecular restricted translational mode, ν_T, between 150 and 200 cm^{-1}, as pressure is applied. This has been interpreted by Walrafen (81) from a study of the pressure effect on the combined intermolecular hydrogen-bond bending and restricted translational intensities, to indicate support for a multistate structural model of liquid water. A decrease in the intensity of intermolecular bands with pressure, indicates that at least up to 2.5 kbar, there is a breakdown of the short range order which persists up to distances of 8Å (82) at atmospheric pressure. This would account for the decrease in viscosity observed at high pressures (83, 84) and the pressure enhanced ionic mobility (85). There is a small effect also observed by Walrafen (51) in the OH intramolecular stretching region between 3000 and 4000 cm^{-1}. The low frequency shoulder between 3200 and 3300 cm^{-1} increases with pressure while the maximum of the band close to 3400 cm^{-1} decreases. Although Walrafen (81) interprets this effect to pressure enhanced intermolecular coupling, because of the uncertainty in the computer resolution of the complex band envelopes and in the spectra themselves, very careful intensity-pressure studies will be required to establish the pressure coefficients of these band intensities.

In figure 12 is shown a laser-Raman spectrum of sea water. The only solute feature visible is the band due to $SO_4^=$, the other features represent bands due to H_2O.

C. Transport Measurements

(i) Conductivity

A comprehensive review up to 1970 of the effect of pressure on the conductance of aqueous solutions has been given by Brummer and Gancy (86). A description of experimental high-pressure conductivity techniques and the treatment of data is included in this work. Horne (87) has discussed the interpretation of the pressure and temperature coefficients of the electrical conductivity of aqueous solutions in a review of the literature up to 1968. The problems in the design and use of conductivity cells has been discussed by Gancy and Brummer (88). One of the experimental difficulties, causing leaking cells and cracking at the platinum electrode-glass joint, is the difference in the compressibility of glass and platinum. This problem has been overcome by the use of platinum electrodes sealed into slightly larger glass tubes with an elastic non-leaching sealant that bonds to both glass and platinum (89). The cell shown in figure 13 has been used with aqueous KCl to 7 kbar with no leakage or cracking at the electrode seals. A high pressure vessel containing 15 of these type of cells arranged in a manner similar to that of Adams and Laidler (90) has been used to study the effect of pressure on the conductivity of aqueous $MgSO_4$ solutions up to 3.0 kbar at 0°, 10°, 18° and 25°C (91). In figure 14 are plotted the ratios of the resistance of solutions at 1 atm to the resistance at pressure, P, for two $MgSO_4$ solutions, 1×10^{-3} M and 2×10^{-2} M. This is the unprocessed conductance data with no corrections for solvent conductivity or density. The conductance measurements require about three hours for each point in order to reach thermal equilibrium after compression. The agreement is seen to be good between rising pressure (open points) and falling pressure (closed points). The ratios of Fisher (73) appear to differ somewhat from the present work, especially at pressures above 1 kbar. It can be seen that pressure produces a greater enhancement in the conductivity of the more concentrated solution. This is due to the pressure enhanced dissociation of the $MgSO_4$ ion pairs in the more concentrated solution. The conductivity of aqueous $MgSO_4$ solutions at 25°C has also been measured by Inada, Shimizu and Osugi (92) up to 1.2 kbar. Our data at 25°C are in close agreement with the data of these workers. Pytkowicz (93), Fisher (94) and Millero (95) have commented on the effect of pressure on the sulphate ion association in sea water. This question relates closely to the establishment of the concentrations of free and associated species present under various conditions of pressure and temperature in sea water. Our conductance data for aqueous $MgSO_4$ at four temperatures up to 3.0 kbar is plotted in figure 15 for 1×10^{-3} M $MgSO_4$. There is a large temperature effect. The pressure effect on the conductance ratio at 0°C is about twice as large as at 25°C. We have also measured the pressure effect on the conductance of NaCl, Na_2SO_4, and $MgCl_2$ at the same four temperatures. Using this data the effect of pressure on the association constant of $MgSO_4$ in water can be determined over the temperature range of interest to oceanographers, 0-25°C.

(ii) Transport Numbers

Transport numbers as a function of temperature and pressure are extremely important for gaining an understanding of the mechanism of ionic mobility. To apply the kinetic theory of ionic mobility proposed by Brummer and Hills (96, 97) rigorously, requires transport numbers so that single ion mobility data can be obtained as a function of temperature and pressure. From these data are derived the ionic activation parameters, energy at constant volume, E_v and energy at constant pressure E_p, volume, ΔV^*, and entropy ΔS^* which are of great significance in the interpretation of liquid transport properties (98). The early transference number methods for pressure studies (99 - 101) have been improved by Kay *et al* (102) who have developed

an electronic bridge method of detecting a moving boundary to such a
degree that transport numbers can be now measured under pressure with
as much precision as electrical conductivity, ±0.01% (103). Considerable progress in the molecular interpretation of the temperature and
pressure coefficients of ionic conductivity can be expected as the
transport number data at high pressures becomes available. This will
be of particular significance in the case of aqueous solutions in the
region from 0-25°C and 0-3 kbar where the structural effects of water
are most pronounced.

(iii) Diffusion

The work by Richardson, Bergsteinsson et al (104) is the only
experimental study of the effect of high pressure on the rate of
diffusion of an electrolyte in "sea water" (3.5 percent by weight
solution of $NaCl-H_2O$). This investigation was not precise enough to
determine the sign of the pressure coefficient. A diaphragm cell with
built-in conductance electrodes in each compartment to obtain the
change in concentration with time from conductivity was used. Barton
and Speedy (105) have reviewed the high-pressure diffusion literature up
to 1971 and conclude that diffusion coefficient accuracy at high pressures range from ±5% to ±10%. Significantly better accuracy, ±1% is
claimed for a diaphragm high-pressure cell developed by Collings et al
(106). Results on the self-diffusion coefficient of liquid benzene
up to 1.5 kbar (107) would seen to verify these claims.

(iv) Viscosity

The effect of pressure on the viscosity of water has been
studied with a rolling ball viscometer by Horne and Johnson (108)
between 2 and 20°C up to 2 kbar. They have reviewed the earlier
high-pressure liquid viscosity literature up to 1966. The effect
of pressure on the viscosity of sea water (109) and $NaCl-H_2O$ (110) was
also studied. Accurate pressure viscosity data is required if hydrodynamic theories of ionic conductivity are to be tested by conductivity-
pressure experiments. The viscosity of water at high temperatures
and high pressures has been measured by Kerimov et al (111) and by
Rivkin et al (112) using a capillary method. A promising new technique
involving the damping effect of a viscous liquid on a torsionally
vibrating quartz crystal has been used by Collings and McLaughlin (113)
to measure the effect of pressure on the viscosity of several liquids to
7 kbar. The high-pressure viscosity data, with an estimated uncertainty
of 1 %, agree well with the literature for several organic liquids.
The high conductivity of aqueous solutions would present experimental
difficulties if this method were to be used for viscosity studies of
these solutions.

D. E.M.F. Measurements

A review of the effect of pressure on electrode processes has
been given by Hills (114) with coverage of the literature up to 1968.
The effect of pressure on the electrical double layer has been studied

by Hsieh (115). Since the effect of pressure on the E.M.F. of a cell under isothermal conditions depends on the volume of activation, ΔV^*, of the electrode processes, the interpretation of pressure effects depends on our understanding of the ΔV^* parameter. In the case of ionic reactions occurring in solution, some progress is being made. ΔV^* is comprised of two components: the volume change intrinsic to the reacting species and the volume change of solvent associated with the reacting species (116). The solvent component usually dominates (117). If gas evolution is involved, the effect of pressure on the reaction should be much greater. The interpretation of ΔV^* from E.M.F. pressure studies is not yet as advanced as the theoretical work on ΔV^* for solution kinetics (118) since electrochemical cells involve heterogeneous reactions. However, despite the lack of theoretical progress, Distèche and Distèche (119) have studied the effect of pressure on the dissociation of carbonic acid with reference to sea water equilibria using buffered glass electrode cells; Kester and Pytkowicz (120) have used a pressure-compensated sodium ion sensitive electrode to study $SO_4^=$ ion pairing with Na^+ and Mg^{++} in sea water; Ben-Yaakov and Kaplan (121) have devised a pH sensor for deep sea measurements;and van Everdingen (122) has developed a pH cell for use in bore holes for in-situ ground water studies.

E. Dielectric Measurements

(i) Static Permittivity

The pressure coefficient of the static permittivity of the solvent is required to evaluate the limiting slope of the concentration dependence of the apparent volumes of strong electrolytes (123). It is also an important term required in electrostriction theories (124). The literature up to 1964 has been discussed by Whalley (125) who points out that conventional techniques used at ambient pressure can be used in high-pressure vessels except at microwave frequencies. The use of three-terminal impedance cells with transformer ratio-arm bridges to determine dielectric properties of liquids, (126-131), has become well established since the work of Cole and Gross (132). The high conductivity of aqueous solutions presents problems connected with the balancing network capacitance of this type of bridge. These have been solved with a ratio-transformer conductance network by Kay and Pribadi (133). The pressure dependence of the static permittivity of water to 1 kbar has been measured most accurately by Owen et al (134). These results agree well with the results of Lees (135), that covered a wider range of pressure, pressures up to 12 kbar, and the results of Dunn and Stokes (136), pressures up to 2 kbar.

A high-pressure three-terminal cell for static permittivity measurements on liquids is shown in figure 16. This cell has been used to measure the permittivity of water to 3 kbar (137). The electrodes are platinum fused to glass which gives it all the superior features at high pressures of the cell design for atmospheric pressure given by Vidulich and Kay (138). Its pressure limit is the freezing of the mercury. Mercury is used to separate the liquid under study from the pressure-transmitting fluid, to equalize the pressure inside and outside the glass cell, and to make electrical contact with the electrodes.

The dielectric properties of liquid water have been reviewed recently by Hasted (139) who has critically discussed the various theoretical interpretations of the static permittivity. The Kirkwood theory (140) for polar liquids involves a correlation factor, g, given below:

$$g = 1 + \sum_i z_i <\cos \gamma > \qquad (2)$$

where z_i is the number of neighbouring molecules in the ith coordination shell from a central molecule and $<\cos \gamma >$ is the mean of the cosines of the angles, γ, between the dipoles of the central molecule and the moments of the neighbours. For a tetrahedral H-bonded network, g is about 3. The experimental variation of g for water with pressure and temperature can be seen in figure 17. The calculation of g from structural models of the liquid presents a rigorous test of the models (139).

(ii) Refractive Index

There are at least three methods that have been used to determine the change of refractive index of liquids under pressure (125): the interferometric method (141-146), the maximum deviation method (147-150), and the immersion method (151). The most recent measurements of the isothermal piezooptic coefficient, $(\partial n/\partial P)_{T,\lambda}$, are those of Stanley (145) on water, and those of Josephs and Minton (152), on several organic liquids using an ultracentrifuge and a schlieren optical system. The relation between the density and refractive index, n, of solutions can be expressed to a first approximation (153, 154) by the Lorentz and Lorenz equation (155,156):

$$\frac{n^2 - 1}{n^2 + 2} = \frac{4\pi N \rho \alpha}{3M} \qquad (3)$$

where N is Avogadro's number, M is the molecular weight, ρ is the density, and α is the scaler average molecular electronic polarizability. With this equation, the pressure dependence of the refractive index of liquids can be used to estimate their density at high pressures. The piezooptic coefficient is also of theoretical interest especially in the case of water since the variation of the polarizability with density is related to structural changes (153, 154, 157).

A Fabry-Perot interferometer used to measure the variation of refractive index of aqueous solutions with pressure is shown in place in a two-window 3 kbar pressure vessel in figure 18. A He-Ne laser at 6328Å is used to produce an interference fringe pattern. When the pressure is varied fringes passing a pin hole are counted by an automated fringe-counting system (158). The change in the refractive index, Δn, for a variation of pressure, ΔP, and a number of fringes, ΔN, is given by the equation below:

$$\Delta n = \frac{\lambda \times \Delta N}{2 \times h \times \cos \theta} \qquad (4)$$

where λ is the frequency of light, h is the separation between the etalons, and θ is the angle of incident light on the etalon (90°). The pressure gauges and transducers used in this work were calibrated against a Harwood Engineering controlled clearance piston balance and are accurate to 1 bar. The results for H_2O at 1.5, 10.0 and 20.0°C up to 2.0 kbar are shown in figure 19. The measured change in n for H_2O is about 1% per kbar with precision of ± 0.001% and an accuracy of better than 0.01%. Our results agree better with those of Waxler et al (143, 144) and Rosen (149) than with the more recent results of Stanley (145). It can be seen from figure 19 that the change of refractive index with pressure decreases with increasing temperature and with added electrolyte. The refractive index change of sea water with pressure is shown on figure 20. The scatter has been reduced in our most recent experiments. There appears to be an unusual variation of the relative refractive index pressure change with temperature. This is being investigated with further experiments. Data is now available for the pressure dependence of refractive index at 6328Å , from 1 atm to 2.0 kbar: for H_2O at 25°, 20°, 18°, 10°, and 1°C; for D_2O at 25°, 20°, 18°, 10°, and 4°C. These results are being fitted to the empirical equation proposed by Reisler and Eisenberg (159):

$$\frac{n^2 - 1}{n^2 + 2} = A \rho^B \exp^{-CT} \qquad (5)$$

where A, B and C are the characteristic constants of the liquid depending only on wave length.

Interferometric techniques have a wide applicability because of their great sensitivity, e.g. studies of the freezing of sea ice (160). O'Brien (161) has discussed many applications including diffusion of gases under pressure into water.

F. Surface and Interfacial Tension Measurements

The change in surface tension with pressure has been studied by only a few investigators (162-166). The surface tension, γ, is related to the total system volume, V, and the area of the interface, A, by the following equation (167):

$$\left(\frac{\partial \gamma}{\partial P}\right)_{A,T} = \left(\frac{\partial V}{\partial A}\right)_{P,T} \qquad (6)$$

The interfacial tension between two liquid phases, or between liquid and solid phases, is an important parameter controlling the interaction of hydrocarbon-aqueous systems in subsurface formations. Guarnaschelli and Adams (168) have studied the interfacial tension of the water-benzene-silica and water-n-hexane-silica systems as a function of temperature from 1 atm to 2.5 kbar (or the freezing pressure). A diagram of the apparatus used for this experiment is shown in figure 21. There seems to be little pressure effect on the contact angle (<4%) in the water-n-hexane-silica system, while the water-benzene-silica system shows an increase of ~ 30% in the contact angle at 25°C for a pressure of 0.8 kbar.

ACKNOWLEDGEMENTS

We thank Dr. B. G. Oliver for his constructive advice and Dr. M. C. B. Hotz whose encouragement and support made this work possible.

IV. REFERENCES

1. F. Franks, ed. "Water, A Comprehensive Treatise - Volume 1, The Physics and Physical Chemistry of Water", Plenum Press, New York, (1972).
2. R.A. Horne, ed., "Water and Aqueous Solutions, Structure, Thermodynamics, and Transport Processes", Wiley-Interscience, New York, (1972).
3. D. Eisenberg and W. Kauzmann, "The Structure and Properties of Water", Oxford at the Clarendon Press (1969).
4. R.A. Horne, "Marine Chemistry, The Structure of Water and the Chemistry of the Hydrosphere", Wiley-Interscience, New York (1969).
5. A.W. Lawson and A. J. Hughes in "High Pressure Physics & Chemistry", Vol. 1, ed. R.S. Bradley, Academic Press (1963), p. 207.
6. R.S. Bradley, ed., "Advances in High Pressure Research", Vol. 1, Academic Press, New York (1966).
7. R.S. Bradley, ed., "Advances in High Pressure Research", Vol. 2, Academic Press, New York (1969).
8. D.S. Tsiklis, "Handbook of Techniques in High-Pressure Research and Engineering", (trans. from Russian) Plenum Press, New York (1968).
9. R.S. Bradley, ed., "High Pressure Physics and Chemistry", Vol. 1 and 2, Academic Press, New York (1963).
10. R.S. Bradley and D.C. Munro, "High Pressure Chemistry", Pergamon Press, Oxford (1965).
11. M.G. Gonikberg, "Chemical Equilibrium and Velocity of Reactions at High Pressures", Izdatel. Akad. Nauk. SSSR, Moscow (1960).
12. C.C. Bradley, "High Pressure Methods in Solid State Research", Butterworths, London (1969).
13. S.D. Hamann, "Physico-Chemical Effects of Pressure", Butterworths, London (1957).

14. S.E. Babb, Jr., "Techniques of High-Pressure Experimentation", in "Technique of Inorganic Chemistry", eds. H.B. Jonassen and A. Weissberger Vol. VI, Interscience (1966), p. 83.
15. A. Van Itterbeek ed., "Physics of High Pressures and Condensed Phase", North-Holland, Amsterdam(1965).
16. "High Pressure Engineering", Proc. Inst. Mech. Eng. 182, Pt. 3C (1967-68).
17. W.R.D. Manning and S. Labrow, "High Pressure Engineering", Leonard Hill, London (1971).
18. A.M. Zimmerman, "High Pressure Effects on Cellular Processes", Academic Press, New York (1970).
19. G.C. Ulmer, "Research Techniques for High Pressure and High Temperature", Springer-Verlag, New York (1971).
20. J.W. Stewart, "The World of High Pressure", Van Nostrand, Princeton, (1967).
21. G.B. Benedek, "Magnetic Resonance at High Pressure", Interscience, New York (1963).
22. E. Whalley, Ann. Rev. Phys. Chem. 18, 205 (1967).
23. P.W. Bridgman, "Collected Experimental Papers", 1-7, Harvard Univ. Press, Cambridge (1964).
24. E. Whalley, J.B.R. Heath, and D.W. Davidson, J. Chem. Phys. 48, 2362 (1968).
25. B. Kamb in "Water and Aqueous Solutions: Structure, Thermodynamics, and Transport Processes", ed. R.A. Horne, Wiley-Interscience, (1972), p. 1.
26. P.J. Wyllie in "High Pressure Physics and Chemistry", Vol. 2, ed. R.S. Bradley, Academic Press, New York (1963), Chap. 6, p. 2.
27. G.S. Kell in "Water, A Comprehensive Treatise", Vol. 1, ed. F. Franks, Plenum Press, New York (1972), p. 363.
28. W.A. Adams and K.J. Laidler, Can. J. Chem. 45, 123 (1967).
29. W.A. Adams. Unpublished work. National Research Council of Canada, (1968-69).
30. E. Whalley in, "Experimental Thermodynamics", Vol. 2, eds. B. Vodar and B. LeNeindre, Butterworths, London (1973). In prepartion.
31. W.A. Steele and W. Webb in "High Pressure Physics and Chemistry", Vol. 1, ed. R.S. Bradley, Academic Press (1963), p. 145.
32. J.M.H. Levelt Sengers in "Physics of High Pressures and Condensed Phase", ed. A. Van Itterbeek, North-Holland, Amsterdam (1965), p. 60.
33. J. Crease, Deep-Sea Res. 9, 1209 (1962).
34. W. Wilson and D. Bradley, Deep-Sea Res. 15, 355 (1968).
35. I.W. Duedall, Third Canadian Oceanographic Symposium (Abstracts), Burlington, Ontario (1972).
36. F.K. Lepple and F.J. Millero, Deep-Sea Res. 18, 1233 (1971).
37. F.J. Millero, J.H. Knox, and R.T. Emmet, J. Sol. Chem. 1, 173 (1972).
38. L.S. Whatley and A. Van Valkenburg in "Advances in High Pressure Research", ed. R.S. Bradley, Academic Press, New York (1966), p. 327.

39. O.E. Weigang, Jr. and W.W. Robertson in "High Pressure Physics and Chemistry", Vol. 1, ed. R.S. Bradley, Academic Press, New York (1963), p. 177.
40. B. Oksengorn, H. Vu and B. Vodar in "Physics of High Pressures and the Condensed Phase", ed. A. Van Itterbeek, North-Holland, Amsterdam (1965), p. 454.
41. S. Malmrud and S. Claesson, Acta Imeko 11, 195 (1964).
42. W. Rigby , R. Whyman and K. Wilding, J. Phys. E: Sci. Instrum. 3, 572 (1970).
43. M.J. Blandamer, M.F. Fox, M.C.R. Symons and M.J. Wootten, Trans. Faraday Soc. 66, 1574 (1970).
44. R.J. Jakobsen, Y. Mikawa and J.W. Brasch, App. Spect. 24, 333 (1970).
45. S.A. Hawley and C.E. Chase, Rev. Sci. Instrum. 41, 553 (1970).
46. H.D. Lüdemann and W.A. Mahon, High Temp.-High Press. 1, 215 (1969).
47. W.F. Giggenbach, J. Phys. E: Sci. Instrum. 4, 148 (1971).
48. A.R. Davis and W.A. Adams, Spectrochim. Acta 27A, 2401 (1971).
49. G.E. Walrafen, J. Chem. Phys. 55, 768 (1971).
50. G.E. Walrafen, J. Chem. Phys. 55, 5137 (1971).
51 G.E. Walrafen, personal communication.
52. E.U. Franck and H. Lindner, doctoral thesis of H.L., University of Karlsruhe, 1970.
53. R.C. Cavagnat, J.J. Martin and G. Turrell, App. Spect. 23, 172 (1969).
54. C.E. Weir, A. Van Valkenburg and E.R. Lippincott in "Modern Very High Pressure Techniques", ed. R.A. Wentorff, Butterworth, Washington (1962), p. 51.
55. A.J. Melveger, J.W. Brasch, and E.R. Lippincott, Mat. Res. Bull. 4, 515 (1969).
56. A.J. Melveger, J.W. Brasch and E.R. Lippincott, App. Optics 9, 11 (1970).
57. J.W. Brasch, A.J. Melveger and E.R. Lippincott, Chem. Phys. Lett. 2, 99 (1968).
58. C. Postmus, V.A. Maroni, J.R. Ferraro and S.S. Mitra, Inorg. Nucl. Chem. Lett. 4, 269 (1968).
59. O. Brafman, S.S. Mitra, R.K. Crawford, W.B. Daniels, C. Postmus and J.R. Ferraro, Solid State Comm. 7, 449 (1969).
60. M. Nicol, J. Opt. Soc. Am. 55, 1176 (1965).
61. G. Hall and D.J. Turner, Central Electricity Research Laboratory Note No. RD/L/N43/70, Leatherhead, U.K.
62. E.U. Franck, Ber. Bunsenges. Phys. Chem. 73, 135 (1969).
63. E.U. Franck and K. Roth, Disc. Faraday Soc. 43, 108 (1967).
64. K. Roth, Dissertation, Universität Karlsruhe, 1969.
65. K. Todheide in "Water, A Comprehensive Treatise", Vol. 1, ed. F. Franks, Plenum Press, New York (1972), p. 463.
66. K. Noack, Spectrochim. Acta 24A, 1917 (1968).
67. D.E. Irish in "Ionic Interactions", ed. S. Petrucci, Vol. II, Academic Press, New York (1971), p. 187.
68. J.E.D. Davies, J. Mol. Struct. 10, 1 (1971).
69. S.D. Hamann, J. Phys. Chem. 67, 2233 (1963).

70. S.D. Hamann and W. Strauss, Trans. Faraday Soc. 51, 1684 (1955).
71. R.A. Horne, B.R. Myers and G.R. Frysinger, Inorg. Chem. 3, 452 (1964).
72. R.A. Horne, R.A. Courant and G.R. Frysinger, J. Chem. Soc. Part II, 1515 (1964).
73. F.H. Fisher, J. Phys. Chem. 66, 1607 (1962).
74. D.E. Irish and H. Chen, J. Phys. Chem. 74, 3796 (1970).
75. A.R. Davis and W.A. Adams, in preparation.
76. A.R. Davis and B.G. Oliver, J. Sol. Chem. 1, 329 (1972).
77. B.G. Oliver and A.R. Davis, submitted to Can. J. Chem.
78. W.A. Adams and A.R. Davis, in preparation for publication.
79. S.V. Wroblewski, Compt. rend. 94, 212 (1882).
80. S.D. Larson, Doctoral Thesis, University of Illinois (1955).
81. G.E. Walrafen in "Hydrogen-Bonded Solvent Systems", ed. A.K. Covington and P. Jones, Taylor and Francis Ltd., London (1968), p. 9.
82. A.H. Narten and H.A. Levy in "Water, A Comprehensive Treatise", ed, F. Franks, Plenum Press, New York (1972), p. 311.
83. P.W. Bridgman, Proc. Nat. Acad. Sci. U.S. 11, 603 (1925).
84. R.A. Horne and D.S. Johnson, J. Chem. Phys. 45, 21 (1966).
85. R.A. Horne and R.P. Young, J. Phys. Chem. 71, 3824 (1967).
86. S.B. Brummer and A.B. Gancy in "Water and Aqueous Solutions: Structure, Thermodynamics, and Transport Processes", ed. R.A. Horne, Wiley-Interscience, New York (1972), p. 745.
87. R.A. Horne in "Advances in High Pressure Research", Vol. 2, ed. R.S. Bradley, Academic Press, New York (1969), p. 169.
88. A.B. Gancy and S.B. Brummer, J. Electrochem. Soc. 115, 804 (1968).
89. B.G. Oliver and W.A. Adams, Rev. Sci. Inst. 43, 830 (1972).
90. W.A. Adams and K.J. Laidler, Can. J. Chem. 46, 1989 (1968).
91. B.G. Oliver and W.A. Adams, in preparation.
92. E. Inada, K. Shimizu, and J. Osugi, Nippon Kagaku Zasshi 92, 1096 (1971).
93. R.M. Pytkowicz, Geochim. Cosmochim. Acta 36, 631 (1972).
94. F.H. Fisher, Geochim. Cosmochim. Acta 36, 99 (1972).
95. F.J. Millero, Geochim. Cosmochim. Acta 35, 1089 (1971).
96. G.J. Hills in "Chemical Physics of Ionic Solutions", eds. B.F. Conway and R.G. Barradas, Wiley, New York (1966), p. 521.
97. S.B. Brummer and G.J. Hills, Trans. Faraday Soc. 57, 1816 (1971).
98. A.F.M. Barton, Rev. Pure and Appl. Chem. 21, 49 (1971).
99. F.T. Wall and S.J. Gill, J. Phys. Chem. 59, 278 (1955).
100. F.T. Wall and J. Berkowitz, J. Phys. Chem. 62, 87 (1958).
101. G.J. Hills, P.J. Ovenden, and D.R. Whitehouse, Disc. Faraday Soc. 39, 207 (1965).
102. R.L. Kay, K.S. Pribadi, and B. Watson, J. Phys. Chem. 74, 2724 (1970).
103. R.L. Kay, B. Watson and K.S. Pribadi, Abstract of paper presented to joint Chemical Institute of Canada and American Chemical Society Meeting, Toronto, 1970.
104. W.L. Richardson, P. Bergsteinsson, R.J. Getz, D.L. Peters, R.W. Sprague, "Sea Water Mass Diffusion Coefficient Studies", Philco Aeronutronic Div., Publ. No. U-3021, W.O. 2053 (Dec 1964), Office of Naval Res. Contract. No. Nonr-4061 (00) (Unclass).

105. A.F.M. Barton and R.J. Speedy, High Temp.-High Pres. 2, 587 (1970).
106. A.F. Collings, D.C. Halls, M.A. McCool, and L.A. Woolf, J. Sci, Inst. Phys. E 4, 1019 (1971).
107. M.A. McCool, A.F. Collings and L.A. Woolf, Trans. Faraday Soc. I 68, 1489 (1972).
108. R.A. Horne and D.S. Johnson, J. Phys. Chem. 70, 2182 (1966).
109. R.A. Horne and D.S. Johnson, J. Chem. Phys. 44, 2946 (1966).
110. R.A. Horne and D.S. Johnson, J. Phys. Chem. 71, 1147 (1967).
111. A.M. Kerimov, N.A. Agaev and A.A. Abaszade, Teploenergetika 16, 87 (1969).
112. S.L. Rivkin, A. Ya. Levin and L.B. Izrailevskii, Teploenergetika 17, 79 (1970).
113. A.F. Collings and E. McLaughlin, Trans. Faraday Soc. 67, 340 (1971).
114. G.J. Hills in "Advances in High Pressure Research", Vol. 2, ed. R.S. Bradley, Academic Press, New York (1969), p. 225.
115. S.A.K. Hsieh, "Thermodynamics of the Electric Double Layer and of Electrode Kinetics in Solution", Dissertation, University of Southampton, U.K. (1969).
116. E. Whalley, Adv. Phys. Org. Chem. 2, 93 (1964).
117. D.L. Gay, Can. J. Chem. 49, 3231 (1971).
118. K.J. Laidler and R. Martin, Int. J. Chem. Kinet. 1, 113 (1969).
119. A. Distèche and S. Distèche, J. Electrochem. Soc. 114, 330 (1967).
120. D.R. Kester and R.M. Pytkowicz, Geochim. Cosmochim. Acta 34, 1039 (1970).
121. S. Ben-Yaakov and I.R. Kaplan, "Design and Application of a Deep Sea pH Sensor", Abstract No. 312, 142nd National Meeting Electrochemical Society (1972).
122. R.O. van Everdingen, Ground Water Subdivision, Calgary, Environment Canada, Personal communication.
123. H.S. Harned and B.B. Owen, "The Physical Chemistry of Electrolyte Solutions", 3rd Ed., Reinhold, New York (1958).
124. J.E. Desnoyers, R.E. Verrall, and B.E. Conway, J. Chem. Phys. 43, 243 (1965).
125. E. Whalley in "Advances in High Pressure Research", Vol. 1, ed. R.S. Bradley, Academic Press, New York (1966), p. 143.
126. G.A. Vidulich, D.F. Evans, and R.L. Kay, J. Phys. Chem. 71, 656 (1967).
127. C.G. Malmberg, J. Res. Nat. Bur. Stand. 60, 609 (1958).
128. F. Mopsik, J. Res. Nat. Bur. Stand. 71A, 287 (1967).
129. F. Mopsik, J. Chem. Phys. 50, 2559 (1969).
130. W.G.S. Scaife and C.G.R. Lyons, Rev. Sci. Instrum. 41, 625 (1970).
131. W.G. Scaife, J. Phys. A: Gen. Phys. 5, 897 (1972).
132. R.H. Cole and P.M. Gross, Jr., Rev. Sci. Instrum. 20, 252 (1949).
133. R. L. Kay and K.S. Pribadi, Rev. Sci. Instrum. 40, 726 (1969).
134. B.B. Owen, R.C. Miller, C.E. Milner, and H.L. Cogan, J. Phys. Chem. 65, 2065 (1961).
135. W.L. Lees, Dissertation, "Dielectric Constants, as Functions of Pressure and Temperature, of Water and Three Alkyl Halides", Dept. of Physics, Harvard University, Massachusetts (1949).

136. L.A. Dunn and R.H. Stokes, Trans. Faraday Soc. 65, 2906 (1969).
137. W.A. Adams and E. Whalley, Abstract of Papers, 52nd Canadian Chemical Conference, Montreal (1969).
138. G.A. Vidulich and R.L. Kay, Rev. Sci. Instrum. 37, 1662 (1966).
139. J.B. Hasted in "Water, A Comprehensive Treatise", ed. F. Franks, Plenum Press, New York (1972), p. 225.
140. J.G. Kirkwood, J. Chem. Phys. 4, 592 (1936).
141. M.J. Jamin, C.R. Acad. Sci., Paris 45, 892 (1857).
142. V. Raman and K.S. Venkataraman, Proc. Roy. Soc. A171, 137 (1939).
143. R.M. Waxler and C.E. Weir, J. Res. Nat. Bur. Stand. 67A, 163 (1963).
144. R.M. Waxler, C.E. Weir, and H.W. Schamp, J. Res. Nat. Bur. Stand. 68A, 489 (1964).
145. E.M. Stanley, J. Chem. Eng. Data 16, 454 (1971).
146. D.W. Langer and R.A. Montalvo, J. Chem. Phys. 49, 2836 (1968).
147. T.C. Poulter, C. Ritchey, and C.A. Benz, Phys. Rev. 41, 366 (1932).
148. W.J. Lyons and F.E. Poindexter, J. Opt. Soc. Am. 26, 146 (1936).
149. J.S. Rosen, J. Opt. Soc. Am. 37, 932 (1947).
150. G.J. Besserer and D.B. Robinson, Can. J. Chem. Eng. 49, 651 (1971).
151. R.E. Gibson and J.K. Kinkaid, J. Am. Chem. Soc. 60, 511 (1938).
152. R. Josephs and A.P. Minton, J. Phys. Chem. 75, 716 (1971).
153. H. Eisenberg, J. Chem. Phys. 43, 3887 (1965).
154. E. Reisler, H. Eisenberg, and A.P. Minton, Trans. Faraday Soc. II 68, 1001 (1972).
155. H.A. Lorentz, Ann. Physik. Chem. Wied. 9, 641 (1880).
156. L.V. Lorenz, Ann. Physik, Chem. Wied. 11, 70 (1880).
157. A.P. Minton, J. Phys. Chem. 76, 886 (1972).
158. W.A. Adams, A.R. Davis, G. Seabrook and W. Ferguson, in preparation.
159. E. Reisler and H. Eisenberg, J. Chem. Phys. 43, 3875 (1965).
160. R. Farhadieh and R.S. Tankin, J. Geophys. Res. 77, 1647 (1972).
161. R.N. O'Brien in "Physical Methods of Chemistry", Part IIIA, eds. A. Weissberger and B.W. Rossiter, Wiley, New York (1972).
162. A.S. Michaels and E.A. Hauser, J. Phys. Chem. 55, 408 (1951).
163. R.R. Harvey, J. Phys. Chem. 62, 322 (1958).
164. O.K.Rice, J. Chem. Phys. 15, 333 (1947).
165. M.E. Hassan, R.F. Nielsen, and J.C. Calhoun, Trans. Soc. Petrol. Eng. AIME 198, 299 (1953).
166. D.S. Tsiklis, "Handbook of Techniques in High-Pressure Research and Engineering", Plenum Press, New York, N.Y. (1968). Chap. XI.
167. J.C. Eriksson, Svensk Kem. Tidskr. 78, 739 (1966).
168. C. Guarnaschelli and W.A. Adams, to be published.

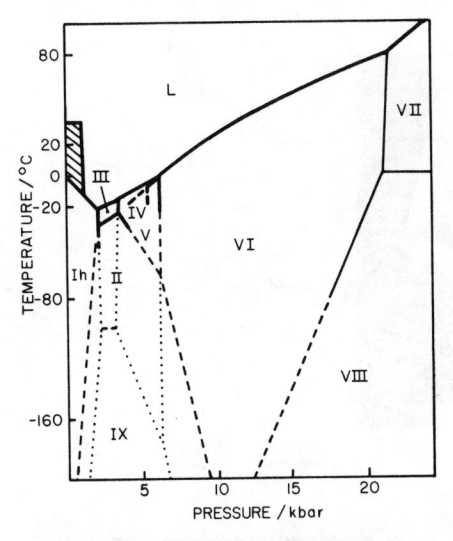

Figure 1. Solid-liquid phase diagram of water (Whalley et al, (1968) (24)

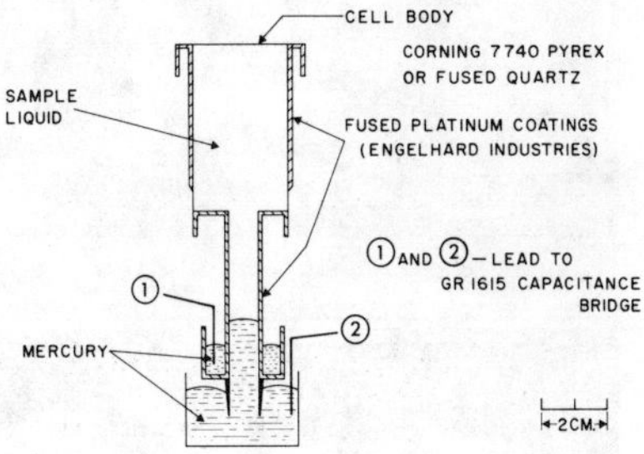

Figure 2. Mercury type capacitance piezometer

Figure 3. 2-window 10 kbar high-pressure optical cell

Figure 4. 3-window corrosion resistant high-pressure laser-Raman cell

Figure 5. High-pressure laser Raman spectrometer

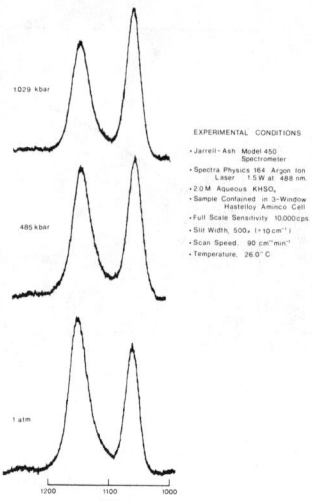

Figure 6. $SO_4^=$ and HSO_4^- Raman spectra in 2 M KHSO$_4$ at high pressures

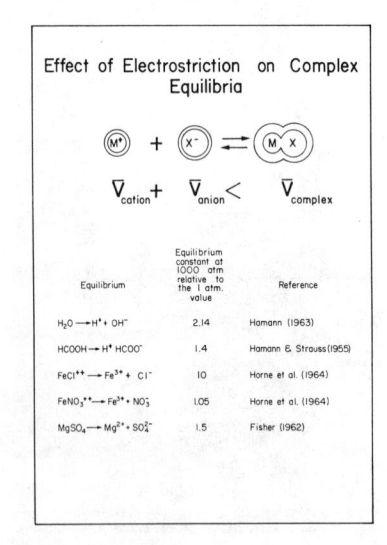

Figure 7. Effect of pressure on ionic equilibria (refs. 69-73 in order)

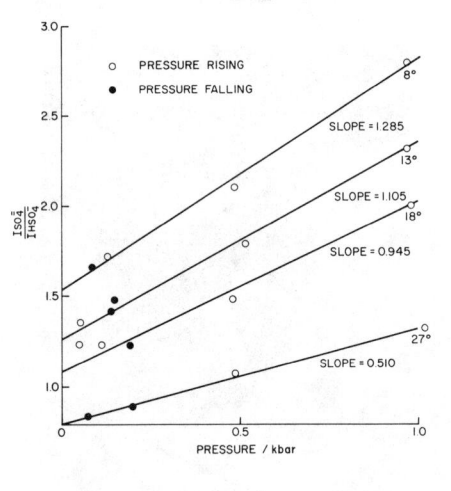

Figure 8. Effect of pressure on the sulphate: bisulphate intensity ratio

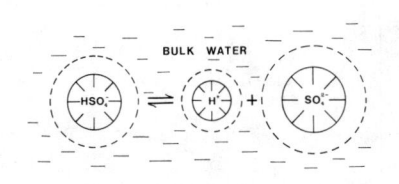

$$\Delta V = -RT\left[\frac{\partial \ln K}{\partial P}\right]_T - RT\kappa$$

SALT	ΔV /cm³ mol⁻¹ THIS WORK	HORNE(1964) (25°C)	$10^4 \bar{\kappa}$ /cm³ bar⁻¹mol⁻¹
H_2SO_4	-10 ± 3	-15 ± 3	0
NH_4HSO_4	-13 ± 3		-11
$KHSO_4$	-17 ± 1		-57

AVG. VALUES OVER TEMP. 5 - 40°C

Figure 9. Volume changes in the sulphate bisulphate equilibrium (Horne et al (72))

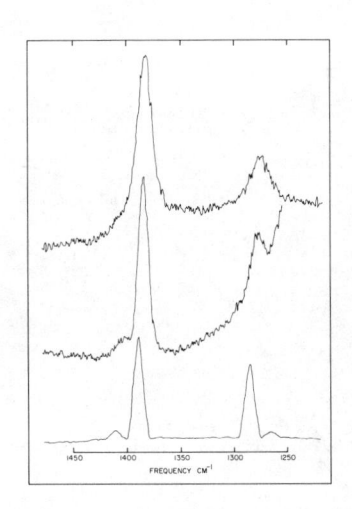

Figure 10. Raman spectrum of CO_2 top, in H_2O; middle in D_2O; bottom, in pure CO_2, all at 5 atm.

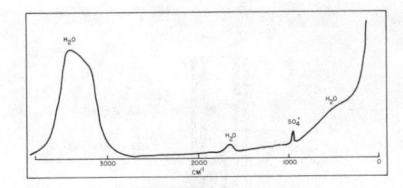

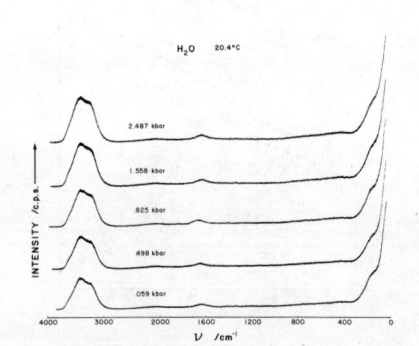

Figure 11. Laser-Raman spectra of H_2O at high pressure

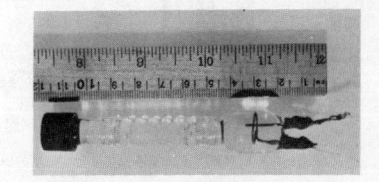

Figure 12. Laser-Raman spectrum of sea water

Figure 13. High-pressure conductivity cell

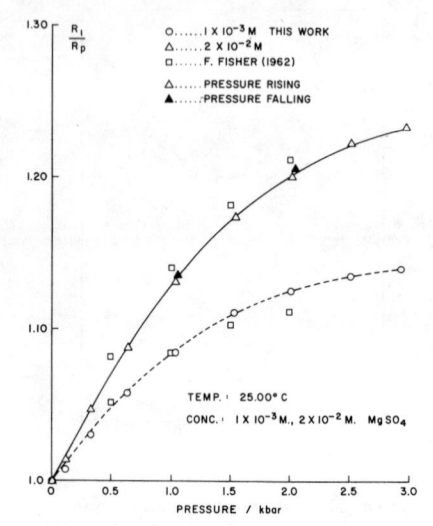

Figure 14. Effect of pressure on the conductance ratio of 2 concentrations of aq. $MgSO_4$

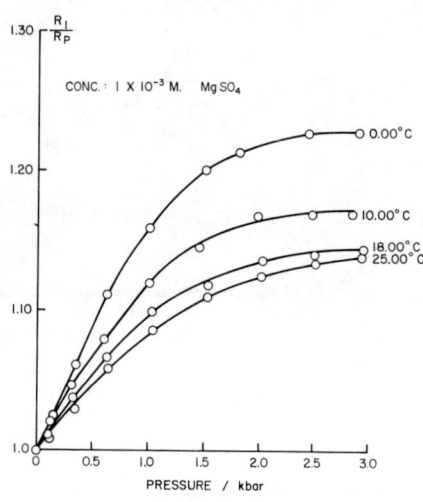

Figure 15. Effect of temperature and pressure on the conductance ratio of aq. $MgSO_4$

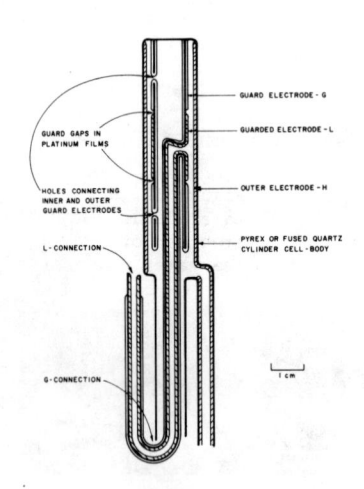

Figure 16. High-pressure permittivity cell for liquids

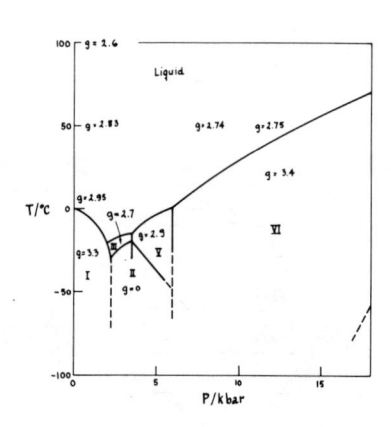

Figure 17. Solid-liquid phase diagram of water showing Kirkwood correlation factors

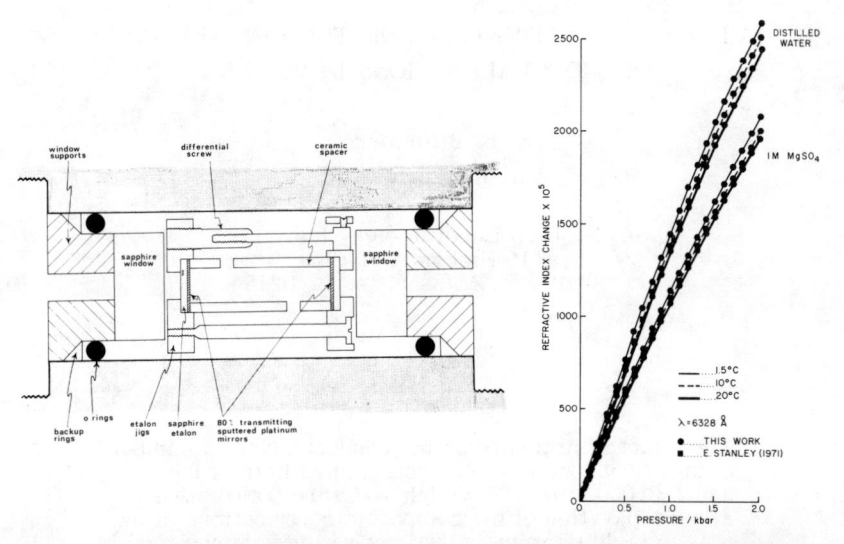

Figure 18. High-pressure interferometer for refractive index studies of liquids

Figure 19. Pressure effect on the refractive index of water and aq. $MgSO_4$

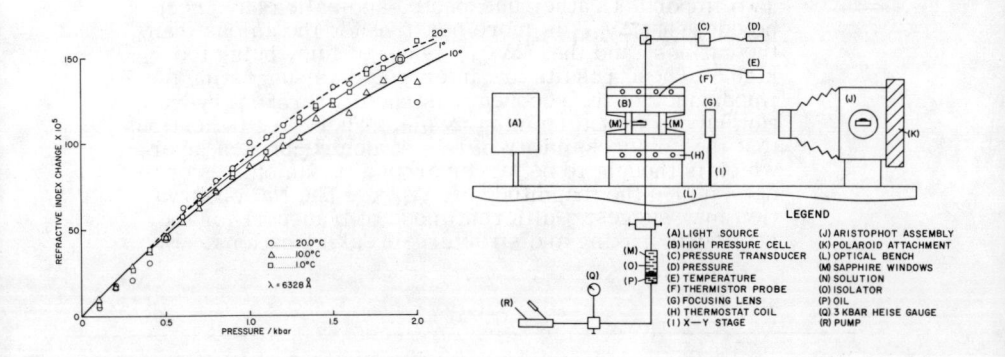

Figure 20. Pressure effect on the refractive index of sea water

Figure 21. High-pressure apparatus to measure interfacial tension

VOLUMES OF ACTIVATION FOR THE CONDUCTANCE OF MONOVALENT IONS IN WATER

S. B. Brummer
A. B. Gancy[*]

Tyco Laboratories, Inc.
16 Hickory Drive
Waltham, Massachusetts 02154

ABSTRACT

The effect of pressure on the conductance of a number of univalent ions has been determined in the range 3-55°C and 1-2,000 atm. New high pressure transport data allow separation of the component contributions of the ions to yield their individual activation volumes, ΔV_{ion}^*. ΔV_{ion}^* is negative under most conditions of interest, which is unusual behavior associated with the structural interactions between the ions and water. ΔV_{ion}^* increases with increase in ionic size and, collaterally, with increasing structure-breaking affinity.

Two important rather unexpected anomalies have been noted: That ΔV_{ion}^* is more negative for the anions than the cations, and that ΔV_{Na^+} is out of line, being too high. These results are interpreted as suggesting the importance of the secondary (structure-broken) hydration layers in determining ΔV_{ion}^*. There are indications that the compressibility of this structure-broken layer, which is thought to be larger around the anions, grossly determines the magnitude of ΔV_{ion}^*. The Na^+ observation may suggest a different mode of transport for the structure-making and structure-breaking cations.

---

[*]Present address: Industrial Chemicals Division, Allied Chemical Corporation, Solvay, New York 13209

INTRODUCTION

The effect of pressure on the conductance of ions in aqueous solutions has been the subject of a number of detailed investigations in recent times.[1-5] That work confirmed the early findings[6,7] of the anomalous, positive pressure coefficient of conductance. Horne's work, in particular,[5] has emphasized the information on the structural properties of the local water around the ions which can be obtained from such studies. However, the full range of interpretation of such data has not been possible up to now because of lacks in accuracy, detail of study, range of ions investigated, ability to extend to infinite dilution, and the availability of accurate high pressure transport number data.

Because of recent advances in experimental technique,[8,9] which have allowed an accuracy after extrapolation to infinite dilution of ~0.1% in $\Lambda°$,[9] we have carried out a detailed investigation of the effect of pressure on the conductance of a number of simple 1:1 salts in the ranges 3-55°C and 1-2,000 atm.[10] Recent accurate measurements of transport numbers for KCl up to 2,000 atm at 25°C.[11] have allowed us to make the first definitive separation of the individual ionic contributions to the observed pressure coefficients for a number of simple salts. This separation, for the alkali metal ions, the halide ions, and for NH_4^+ and NO_3^-, is presented in this paper.

TREATMENT OF DATA

It has become common in analyzing the effect of pressure on reaction kinetics to use the volume of activation formalism.[12] In all of our recent work[13-17] we have, following Stearn and Eyring,[18] also used this method of analysis for the variation of $\lambda°$ with P. Viz,

$$\Delta V_{ion}^* = -RT\left(\frac{\partial \ln \lambda°}{\partial P} - \frac{2}{3} \cdot \frac{\partial \ln V}{\partial P}\right)_T . \qquad (1)$$

Here, ΔV_{ion}^* is the volume (change) of activation, $\lambda°$ is the ionic conductance at infinite dilution, and the term $\partial \ln V_{solvent}/\partial P$ arises because of the assumption of a quasi-lattice model of migration. That is, it is assumed that the ion jumps from position to position in a lattice comprising the solvent molecules, with a spacing given by the approximate liquid intermolecular spacing. The latter is presumed to vary with P as $\partial V^{1/3}/\partial P$. This is clearly an unrealistic model of the migration process in a fluid, but it does serve to express the effect in an easily comprehended form suitable for intercomparison between the various ions of interest, and it will be used in this work.

Values of ΔV_{ion}^* were determined as follows: $\Lambda_P°/\Lambda_1°$ data for each salt were expressed as third order polynomials in P. Since in all early examinations of ΔV^*'s appropriate transport numbers were not available, quasi-salt ΔV^*'s were calculated(13-17, 19) from these polynomials according to:

$$\Delta V^*_{salt} = -RT \left(\frac{\partial \ln \Lambda^\circ}{\partial P} - \frac{2}{3} \cdot \frac{\partial \ln V}{\partial P} \right)_T . \qquad (2)$$

The transport data for KCl[11] are now available at 1, 500, 1000, 1500, and 2000 atm, and these allow determinations of ΔV^*_{ion}. Initially, for interpolations and convenience of handling, these were expressed as a third order polynomial in P, to obtain their derivatives, and values for $\Delta V^*_{K^+}$ and $\Delta V^*_{Cl^-}$ were calculated according to:

$$\Delta V^*_{ion} = \Delta V^*_{KCl} - RT \left(\frac{\partial \ln t_{ion}}{\partial P} \right)_T . \qquad (3)$$

Subsequently, vide infra, the last term in Eq. 3 was extracted graphically.

For other salts containing Cl^- or K^+, respectively, values of t_{Cl^-} or t_{K^+} were determined (at 1, 500, 1000, 1500, and 2000 atm) using the previous $\Lambda^\circ_P / \Lambda^\circ_1$ polynomials to determine Λ°_P, and the KCl data to fix $\lambda^\circ_{Cl^-}$ and $\lambda^\circ_{K^+}$ at the required pressures. As before, the transport data were expressed as third order polynomials for convenience of interpolation. ΔV^*'s were then calculated using the analogues of Eq. 3. The final ΔV^*_{ion} values were determined, as above, using graphical derivations of $(\partial \ln t_{ion} / \partial P)_T$.

It is hard to estimate the accuracy of this overall procedure. The Λ°'s are thought to be accurate to 0.1%, even after interpolation. The transport data are even better than this (~0.05%). The differentiations required to obtain the single ion ΔV^*'s significantly degrade the accuracy, however, and, based on the scatter, we believe that their accuracy is no better than ± 0.1 ml/mole in ΔV^*_{ion}.

RESULTS AND DISCUSSION

Some of the properties of ΔV^*_{salt} were analyzed in our recent paper.[19] We reported that ΔV^*_{salt} is negative, i.e., anomalous, for all salts in the region of room temperature. It was more negative at lower temperatures, lower pressures, and with those salts having smaller ions. These results suggest, in common with the conclusions of other investigators, that negative values of ΔV^*_{salt} are contributed by the unusual structural properties of the water adjacent to the ions. From the experimental point of view, excellent agreement was found for KCl between our data and those of Ovenden.[3] This agreement suggested a precision of about ± 0.05 ml/mole in ΔV^*_{salt}.

In analyzing the possible relationships between ΔV_{salt}^* and various ion properties such as the Jones and Dole B-coefficient[20] and the ionic radii, it appeared that there is a step-wise gradation of ionic properties.[19] This, it was suggested, might imply that only discrete hydrated ion sizes are stable for transport. Based on our present analysis, where we have separated ΔV_{salt}^* into the individual ion contributions, we now believe that that conclusion was not warranted. In particular, we believe that little or no transport-mechanistic information may be derived from values of ΔV_{salt}^*, because this quantity bears no simple or even direct relationship to the individual ionic volumes of activation, ΔV_{ion}^*.⨆ Of course, many of the conclusions about ionic properties which can be drawn from study of the qualitative variations of ΔV_{salt}^*[5], as mentioned above, are substantially correct.

Figure 1 shows some approximate values of ΔV_{ion}^* as a function of solvent volume (pressure) at 25°C. These values are to be considered approximate because they were obtained from the polynomial differentiation of the transport data, as indicated in Eq. 3. We believe that the transport data are inadequate in quantity to make this a reliable procedure to obtain absolute values for ΔV_{ion}^*, especially near the ends of the experimental pressure range. These values are typically 0.1 to 0.2 ml/mole less negative than the values we report on below, which were obtained by graphical differentiation of the transport data. They are useful to allow us to examine the trend of ΔV_{ion}^* with pressure and ion type.

Thus we see that ΔV_{ion}^* becomes less negative at higher pressures. This is in contrast with observations in more normal solvents(14-17), where ΔV^* decreases at higher pressure. This behavior in relatively unstructured solvents such as DMF[15] is expected only if the transition state is more compressible than the initial state. This could imply some breakdown of electrostriction in the transition state but this assumption is not necessary, as can be seen in Fig. 2 which is a representation of the transition state. We see that the transition state would be expected to have more "free volume" than the initial state, even though the ion is in contact with solvent molecules S_5 and S_7 at the saddle point. Because of this excess volume, the transition state would be expected to be more compressible than the initial state.

In a highly structured solvent such as water, ΔV^* is negative because some of the "ice-like" water structure near the ion is destroyed as the ion jumps from place to place. At higher pressures, there is

---

⨆The relationship between ΔV_{salt}^* and the individual ionic ΔV^*'s is

$$\Delta V_{salt}^* = \frac{1}{2}\left(\frac{\Delta V_+^*}{\lambda_+^\circ} + \frac{\Delta V_-^*}{\lambda_-^\circ}\right)$$

less of this structure to be lost from the initial state during the transition, and hence ΔV^* becomes less negative. Eventually, at sufficiently high pressures or temperatures, ΔV^* will become "normal," i.e., positive.

A rather more surprising effect in Fig. 1 is the substantial difference between the effects of the anions and the cations. It is generally thought that cations are more strongly hydrated,[21] i.e., they influence the local water structure more than anions. Yet our data show much more anomalous, i.e., negative, values of ΔV_{ion}^* for the anions. This greater anomaly for the anions gives us information about the relative contributions to the transition state of the various hydration co-spheres of Frank and Wen.[22]

Before attempting an analysis of the effect, it is helpful to summarize current views on the differences in hydration between the cations and anions. It is well known--see, for example, (21) and (38) --that small cations in particular attract relatively large numbers of water molecules into their immediate hydration layers. Larger cations, and many anions, exhibit what Samoilov[39] has described as "negative hydration," or what Frank and his co-workers call "structure-breaking." Such ions exhibit Stokes radii which are smaller than their crystal radii[40,41], and their viscosity B-coefficients tend to be negative.[42] An explanation for this kind of behavior has been proposed by Frank and Evans[43] and Gurney[44]: The field near small ions is strong enough to orient the nearby water molecules radially to the center of the ion (Bockris' (45) "primary hydration"). Once this orienting field is of the same order as the normal structural effects of water molecules, the solvent layer becomes highly disordered. In this structure-broken region the local viscosity is low, and kinetic probes of hydration such as conductance will tend to express a large influence of the structure-broken layer. An ion will appear as an over-all "structure maker" or "structure breaker" depending on which type of hydration is the larger.

According to Samoilov,[39,46] the most adequate description of hydration is in terms of the rate of exchange of hydrated H_2O's with the bulk. He argues that except for Li^+, Na^+, and F^-, H_2O is more rapidly exchanged in the vicinity of the ion than in pure water. This latter effect is more pronounced for the anions than for the cations [see, for example, (47)].

Indeed, there are a number of properties which are basically more responsive to changes of the anions' properties than of the cations'. For example, infrared, Raman,[52,53] and nmr studies of solutions seem to show very much greater sensitivity to anion properties than to cation properties.[48] Kavanau[21] describes the matter as follows: Anions tend to attract the protons in the adjacent water molecules. In this situation, the possible configurations of the nearest-neighbor water molecules apparently are more sharply dependent on the size of the ion than they are for cations.[23] It seems that the water molecules tend to orient with one of the O−H bonds normal to the ion surface, the interaction between the anion and water being similar to a hydrogen bond. The most important point is that this leaves the anion-bound water molecule freer to rotate than a cation-bound water molecule, and permits it to form hydrogen bonds to three other water molecules. Indeed, Walrafen's Raman results suggest considerable structure formation around the anions.[52b, 52c] Such a view is also supported by dielectric constant measurements.[24-26]

Basically, though, it seems to be agreed that there is net structure-breaking[27] as, for example, shown by early investigations of infrared spectra,[28] nmr studies,[29] and thermodynamic calculations.[30] In Hindman's view,[29] based on nmr, only F^- among the halide ions forms a hydrate in the chemical sense. The larger halide ions break down the water structure in their immediate vicinity.

Another example of a strongly anion-dependent (rather than cation-dependent) property is the glass transition temperature for aqueous solutions. Angell and Sare[31] report a strong sensitivity of T_g on the type of anion, and very little sensitivity with respect to the cation. They suggest that the cations are shielded behind a sheath of tightly held water molecules and thus present essentially similar exteriors to the next nearest neighbor (water) particles.

We see then that there is a range of solution properties which is more dependent on the nature of the anion than on the nature of the cation and, to summarize the general view of this, we would say that those properties which are most sensitive to the labile character of the "secondary hydration"[45] show this dependence on anion rather than on cation properties.

For all this, the model of the ion-water environment which describes our observations is not easy to come by. Since ΔV_{anion}^* is so very negative, the transition state would appear to be much less structured than the initial state, i.e., the transition state possesses many fewer normally oriented H-bonds. The more structure present in the initial state, the more there is to lose when the ion enters the transition state; hence ΔV^* is less negative for the structure-breaking I^- than for the structure-making F^-. The problem is that the environment of the anions is generally thought to be so much less organized than that of cations, while ΔV_{anion}^* is so much more anomalous than ΔV_{cation}^*. The latter implies clearly that the transition process for both kinds of ions involves predominantly changes in the secondary, structure-broken water layer. Then, the above dilemma can be resolved by assuming that ΔV_{anion}^* is so negative not only because of structural effects, but also because of <u>compressibility effects</u> in this structure-broken water layer.

As Hertz[49] has pointed out, there are good physical reasons for supposing that the arrangement of water molecules near a structure-breaking ion is relatively isotropic. This isotropic layer should then be relatively readily compressed as the translating ion "squeezes" through the saddle point. Hence, ΔV^* would be unexpectedly negative. We do not, perhaps, obtain the space creation indicated for transitions in more normal solvents (Fig. 2), because of the more open initial state than in those solvents. ΔV_{anion}^* should then change more with pressure for I^- than for F^-. In fact, we observe no relative effect of pressure for the different anions (ΔV^* changes by 2.0_5, 2.1_5, 2.1_7, 2.0_7 ml/mole as we increase P from 70 to 1950 atm for the F^-, Cl^-, Br^- and I^- ions). At least this result argues against the likelihood that the main reason ΔV_{anion}^* is so negative is because of structure loss. Under that circumstance, we would expect a much larger effect of P for F^-. It may be then that with increase of P we see the net of structure-breaking and compression, particularly of the initial state; the former decreases as we proceed for F^- to I^-, but the latter increases.

With cations we must, as with anions, fix our main attention on the structure-broken water layer. This layer is different in character than for anions, principally in that it does not in general extend right to the ion's surface. Also, the layer has the opposite orientation, viz., the water protons are generally towards the ion. This layer must be less compressible than the structure-broken layer around the anions, presumably because of the details of its substructure. This must be the predominant reason that ΔV_{cation}^* is greater than ΔV_{anion}^*. As with the anions, there is also an effect of structure loss (from the primary hydration layer?) in proceeding from the initial to the transition states. Hence $\Delta V_{Li^+}^*$ is less than $\Delta V_{Cs^+}^*$.

As before,[19] we have attempted to correlate ΔV_{ion}^* with other well-known ionic properties such as size and structural parameters. Most of the data presented were determined at 70 atm, corresponding to a solvent specific volume of 1.00 ml/g.[32]

Figure 3 shows ΔV_{ion}^* for the alkali halides as a function of (Pauling) crystallographic radii. As noted above, ΔV^* is very much more negative (anomalous) for the anions than the cations, but less so as we proceed from the smaller, more structure-making ions to the structure breakers such as I^-. It is interesting that even for I^-, a large structure-breaking ion, ΔV^* is still more negative than for Li^+, a small structure-making ion. This effect disappears at higher pressure (see Fig. 4). In both of these cases ΔV_{ion}^*, as we have mentioned, derives basically from changes in the structure-broken layer. Evidently, these changes are much larger for I^- than for even the most anomalous cation. Also, the extent of this structure-broken layer with I^- is such that it is relatively compressible compared with other forms of bound and free H_2O. This is why I^- is relatively "normal" at the highest pressure and its ΔV_{ion}^* is more dependent on pressure than say for Li^+.

The situation with the cations is even less satisfactory. Thus, in proceeding from Li^+ to Cs^+, there is little doubt that ΔV^* becomes substantially less negative, as with the anions. Indeed, were it not for the Na^+ data, one might see a similar flattening in the cation curve as one passes from Rb^+ to K^+ to Li^+ as is seen in the anion curve in passing from Br^- to Cl^- to F^-. The Na^+ data are out of line, however, and this has been acknowledged by connecting the Li^+ and Na^+ points. This would indicate that there are three families of ions: the anions, the structure-breaking cations, and the structure-making cations. At high pressure, where more of the local water structure around the ion is destroyed, the separation of the data into these three families of behavior is most convincing (Fig. 4).

This separation between the different kinds of cations is by no means so apparent when the ΔV_{ion}^* data are correlated with the partial molal ionic volume.[33] We see that there might be a shallow increase of ΔV^* with increasing $\bar{V}^°$ as we proceed from Li^+ to Cs^+, particularly if we note the relatively shallow slope of the anion curve as we advance from F^- to I^-. However, both Na^+ and NH_4^+ are sufficiently out of line to give us serious pause, and at this stage we do not believe we can discern the trend for the cations. It is difficult to isolate structural effects on $\bar{V}^°$ from packing and intrinsic size effects, however.[33]

Correlations with ionic structural properties such as ionic entropy[34] and the Jones and Dole B-coefficient[35] seem once again to affirm that there are two families of curves for the cations (Figs. 6 and 7). It is by no means clear, however, that correlations with these structural properties of the ions lead to any greater insight than does simple correlation with a size parameter such as the crystallographic radius.

It turns out that there are some solution properties, both kinetic and thermodynamic, which peak at Na^+ in ascending the alkali metal ion series. For example, Fig. 8 shows the correlation of ΔV_{ion}^* with the conventional heat of transport of the ion.[36] Figure 9 shows a similar correlation against the absolute entropy transported by a moving ion.[37] Those transport data show a strong anomaly at Na^+, and hence the correlation effectively removes our present discrepancy. The implication is then that the ΔV_{ion}^* probe of the ion-water co-sphere is similar to that determined in the heat of transport and entropy of transport experiments. Desnoyers and Jolicoeur[50] have pointed out a number of thermodynamic properties of the alkali chlorides which peak at NaCl.

It seems then that there is precedent for the anomalous position of Na^+. Our view in the present context would be that the

Na$^+$ ion, which is distinctly structure-making, has a relatively small amount of disordered H$_2$O in its secondary hydration layer, i.e., its primary hydration layer can mesh very effectively with the surrounding co-sphere of normal water. There are other data to support this view. Thus we may note that Samoilov finds that the water exchange rate near Na$^+$ is very similar to that in pure water.(47) Similarly, dispersion of supersonic sound is almost unaffected by Na$^+$ ions.(51)

SUMMARY AND CONCLUSIONS

The present data argue convincingly against our previous conclusion that there are discrete ion-solvent co-sphere sizes involved in the ions' movement.(19) They do raise some new, and interesting points, however.

In particular, it comes as something of a surprise to find that ΔV_{ion}^* is more anomalous (negative) for the anions than for the cations. There are a number of other "kinetic hydration probes" which agree with this kind of result.

Also disturbing is the position of Na$^+$, which may suggest a different mode of transport for the structure-making and structure-breaking cations. If verified, this would be an important result. A number of other hydration studies support the position of Na$^+$ found here.

Despite the observation that ΔV_{ion}^* correlates equally well with ion-size parameters as it does with ion-structural parameters, the general behavior of ΔV_{ion}^* with T and P continues to suggest that the prime determining factors on ΔV_{ion}^* are structural rather than purely geometric. The gross magnitude of ΔV_{ion}^* appears to be determined by changes in the structure-broken secondary hydration layer. The compression of this layer as the ion translates probably accounts for the major difference between ΔV_{anion}^* and ΔV_{cation}^*.

The most immediate requirements for further work are:

(1) To verify the data for Na$^+$.

(2) To extend the transport number data to more pressures at 25°C, and to other temperatures.

(3) To extend all the studies to higher pressures, to pursue Horne's thesis that complete ion dehydration occurs.

ACKNOWLEDGMENTS

We are pleased to acknowledge support of our high pressure conductance work by the Office of Saline Water under contracts OSW-14-01-0001-425 and OSW-14-01-001-966. The support and encouragement of Dr. W. H. McCoy of that office is appreciatively noted. We are pleased also to thank Helmut Lingertat for his painstaking care in performing most of the experimental work. A helpful discussion with Professor W. Y. Wen is gratefully acknowledged.

REFERENCES

1. W. A. Zisman, Phys. Rev., 39, 151 (1932).
2. L. H. Adams and R. E. Hall, J. Phys. Chem., 35, 2145 (1931).
3. P. J. Ovenden, Ph.D. Thesis, University of Southampton, England (1965).
4. See the review by S. D. Hamann in his book, Physico-Chemical Effects of Pressure, Butterworths, London, 1957.
5. (a) R. A. Horne, B. R. Myers, and G. R. Frysinger, J. Chem. Phys., 39, 2666 (1963); (b) R. A. Horne, Nature, 200, 418 (1963); (c) R. A. Horne, R. A. Courant, and G. R. Frysinger, J. Chem. Soc., 1964, 1515; (d) R. A. Horne, R. A. Courant, and D. S. Johnson, Electrochim. Acta. 11, 987 (1966); (e) R. A. Horne and J. D. Birkett, Electrochim. Acta. 12, 1153 (1967); (f) R. A. Horne and R. P. Young, J. Phys. Chem., 71, 3824 (1967); (g) R. A. Horne and R. P. Young, J. Phys. Chem., 72, 1763 (1968); (h) R. A. Horne and R. P. Young, J. Phys. Chem., 72, 376 (1968); (i) R. A. Horne, D. S. Johnson, and R. P. Young, J. Phys. Chem., 72, 866 (1968); (j) Advances in High Pressure Research, Vol. 2, Ed., R. S. Bradley, Academic Press, London, 1969, Chapter 3.
6. (a) E. Cohen, Piezochemie Kondensierter Systeme, Akademische Verlagsgesellschaft, Leipzig, 1919; (b) E. Cohen, Physico-Chemical Metamorphosis and Problems in Piezo-Chemistry, McGraw-Hill, New York, 1926.
7. J. Fink, Ann. Phys., 26, 481 (1885).
8. A. B. Gancy and S. B. Brummer, J. Electrochem. Soc., 115, 804 (1968).
9. A. B. Gancy and S. B. Brummer, J. Phys. Chem., 73, 2429 (1969).
10. A. B. Gancy and S. B. Brummer, J. Chem. Eng. Data, 16, 385 (1971).

11. R. L. Kay, K. S. Pribadi and B. Watson, J. Phys. Chem., 74, 2724 (1970).
12. S. Glasstone, K. E. Laidler and H. Eyring, The Theory of Rate Processes, McGraw-Hill, New York, N.Y., 1940.
13. S. B. Brummer and G. J. Hills, Trans. Faraday Soc., 57, 1816 (1961).
14. S. B. Brummer and G. J. Hills, Trans. Faraday Soc., 57, 1823 (1961).
15. S. B. Brummer, J. Chem. Phys., 42, 1636 (1965).
16. F. C. Barreira and G. J. Hills, Trans. Faraday Soc., 64, 1359 (1968).
17. G. J. Hills, P. J. Ovenden and D. R. Whitehouse, Disc. Faraday Soc., 39, 207 (1965).
18. A. E. Stearn and H. Eyring, J. Phys. Chem., 44, 976 (1940); A. E. Stearn and H. Eyring, J. Chem. Phys., 5, 113 (1937).
19. S. B. Brummer and A. B. Gancy, Water and Aqueous Solutions, Ed., R. A. Horne, Wiley-Interscience, New York, N.Y., 1972, p. 745.
20. G. Jones and M. Dole, J. Amer. Chem. Soc., 51, 2950 (1929).
21. J. L. Kavanau, Water and Solute-Water Interactions, Holden-Day, San Francisco, 1964, p. 64.
22. H. S. Frank and W.-Y. Wen, Disc. Faraday Soc., 24, 133 (1957).
23. R. M. Noyes, J. Amer. Chem. Soc., 84, 513 (1962).
24. J. B. Hasted, D. M. Ritson and C. H. Collie, J. Chem. Phys. 16, 1 (1948).
25. G. H. Haggis, J. B. Hasted and T. J. Buchanan, J. Chem. Phys., 10, 1452 (1952).
26. F. E. Harris and C. T. O'Konski, J. Phys. Chem., 61, 310 (1957).
27. J. L. Kavanau, Water and Solute-Water Interactions, Holden-Day, San Francisco, 1964, p. 59.
28. R. Suhrmann and F. Breyer, Z. Physik. Chem., B20, 17 (1933); R. Suhrmann and F. Breyer, Z. Physik. Chem., B23, 193 (1933).
29. J. C. Hindman, J. Chem. Phys., 36, 1000 (1962).
30. J. Padova, J. Chem. Phys., 39, 1552 (1963).
31. C. A. Angell and E. J. Sare, J. Chem. Phys., 52, 1058 (1970).
32. G. S. Kell and E. Whalley, Phil. Trans. Roy. Soc. London, 258 565 (1965).

33. F. J. Millero, Water and Aqueous Solutions, Ed., R. A. Horne, Wiley-Interscience, New York, N.Y., 1972, p. 519.
34. Data after R. W. Gurney, Ionic Processes in Solution, McGraw-Hill, New York, N.Y., 1953, p. 175.
35. Ibid., p. 168.
36. J. N. Agar, Advances in Electrochemistry and Electrochemical Engineering, Vol. 3, Ed., P. A. Delahay, J. Wiley & Sons, Inc., New York, N.Y., 1963, p. 96.
37. Ibid., p. 109.
38. J. E. Desnoyers and C. Jolicoeur, Modern Aspects of Electrochemistry, Vol. 5, Ed., B. E. Conway and J. O'M. Bockris, Plenum Press, New York, N.Y., 1969, p. 1.
39. O. Ya. Samoilov, Structure of Electrolyte Solutions and the Hydration of Ions, Consultants Bureau, New York, N.Y., 1965.
40. R. A. Robinson and R. H. Stokes, Electrolyte Solutions, Buttersworths, London, 1959, Ch. 6.
41. E. R. Nightingale, J. Phys. Chem., $\underline{63}$, 1381 (1959).
42. W. M. Cox and J. M. Wolfenden, Proc. Roy. Soc., London, $\underline{145\text{-}A}$, 475 (1934).
43. H. S. Frank and M. W. Evans, J. Chem. Phys., $\underline{13}$, 507 (1945).
44. Data after R. W. Gurney, Ionic Processes in Solution, McGraw-Hill, New York, N.Y., 1953, Chs. 9 and 16.
45. J. O'M. Bockris, Quart. Rev., $\underline{3}$, 173 (1949).
46. O. Ya. Samoilov and T. A. Nosova, J. Struct. Chem. (USSR), $\underline{6}$, 767 (1965).
47. O. Ya. Samoilov, Water and Aqueous Solutions, Ed., R. A. Horne, Wiley-Interscience, New York, N.Y., 1972, p. 597.
48. J. E. Desnoyers and C. Jolicoeur, Modern Aspects of Electrochemistry, Vol. 5, Ed., B. E. Conway and J. O'M. Bockris, Plenum Press, New York, N.Y., 1969, p. 44 et seq.
49. H. G. Hertz, Angew. Chem. Int., Ed., $\underline{9}$, 124 (1970).
50. J. E. Desnoyers and C. Jolicoeur, Modern Aspects of Electrochemistry, Vol. 5, Ed., B. E. Conway and J. O'M. Bockris, Plenum Press, New York, N.Y., 1969, p. 60.
51. A. V. Karyakin, A. V. Petrov, Yu. B. Gerlit and M. Ye. Zubrilina, Teor. i. Eksp. Khim., $\underline{2}$, 494 (1966).
52. (a) G. E. Walrafen, J. Chem. Phys., $\underline{36}$, 1035 (1962); (b) ___, ibid., $\underline{40}$, 3249 (1964); (c) ___, ibid., $\underline{44}$, 1546 (1966).
53. T. T. Wall and D. F. Hornig, J. Chem. Phys., $\underline{47}$, 784 (1967).

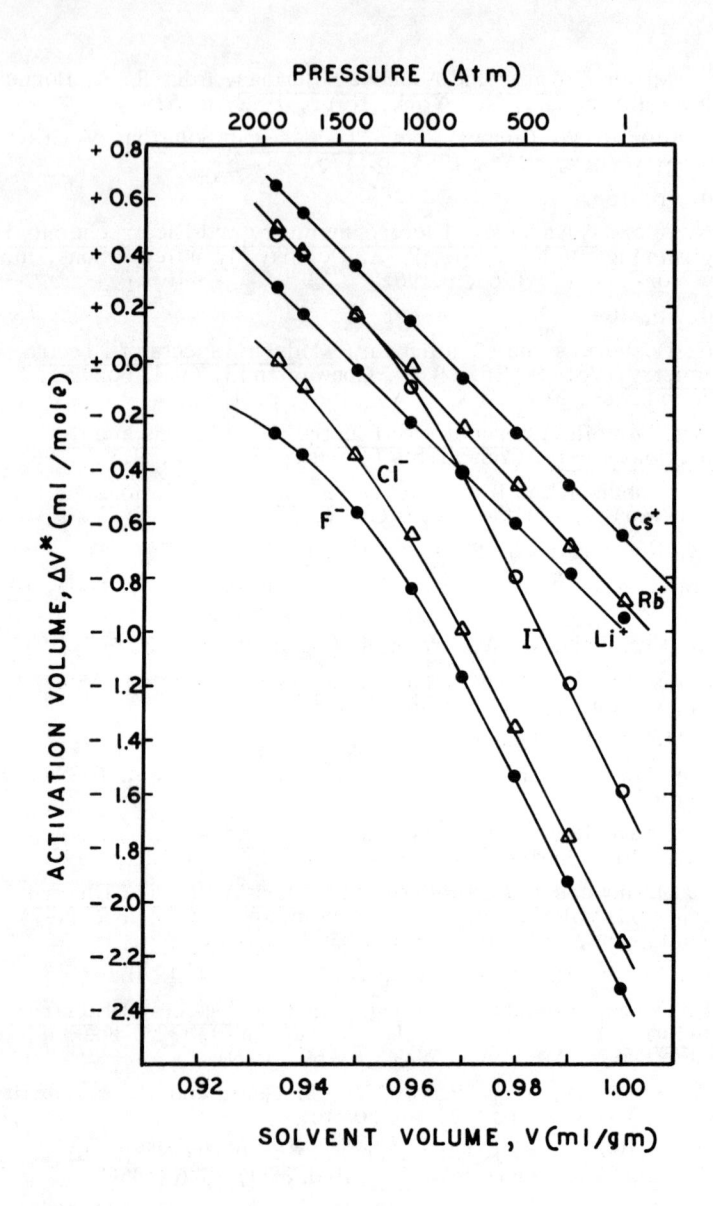

Fig. 1. Activation volume as a function of pressure.

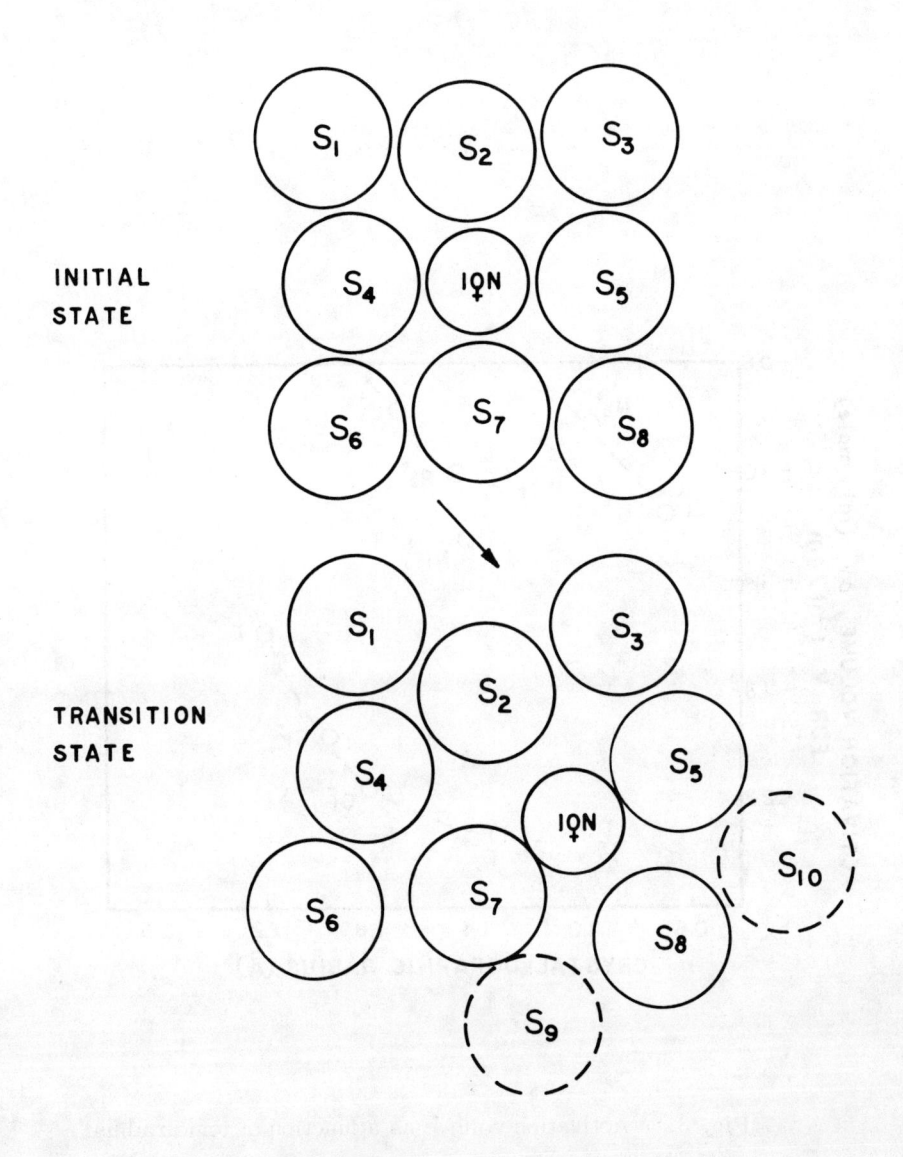

Fig. 2. Representation of the transition process in a normal, unstructured solvent.

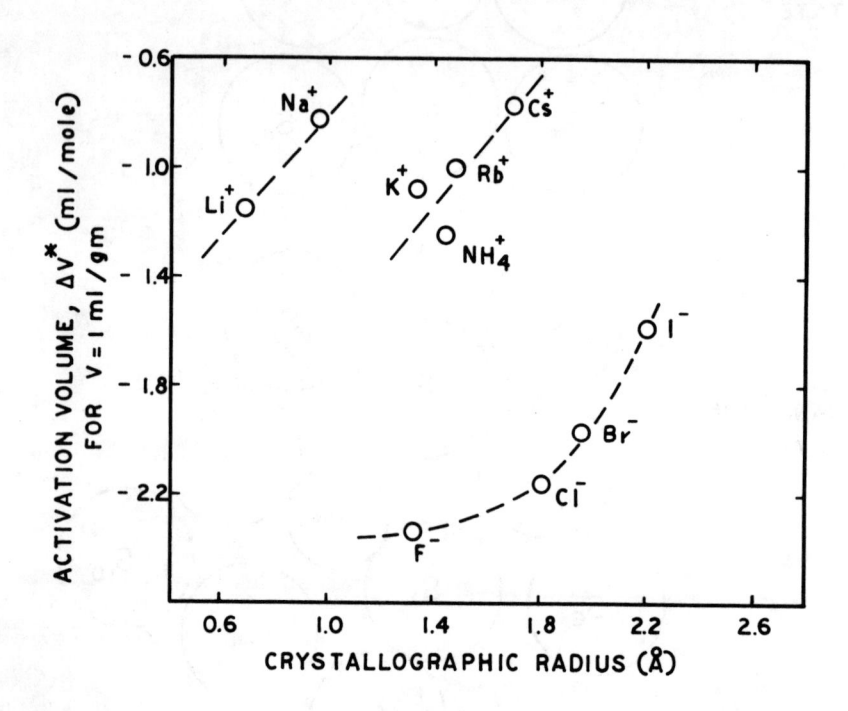

Fig. 3. Activation volume as a function of ionic radius.

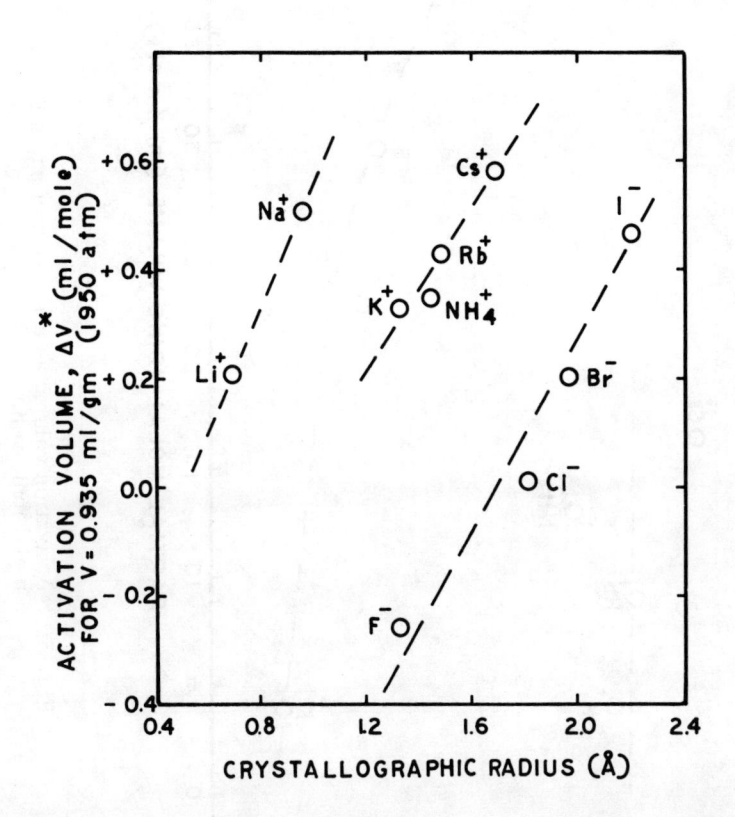

Fig. 4. High-pressure activation volume as a function of ionic radius.

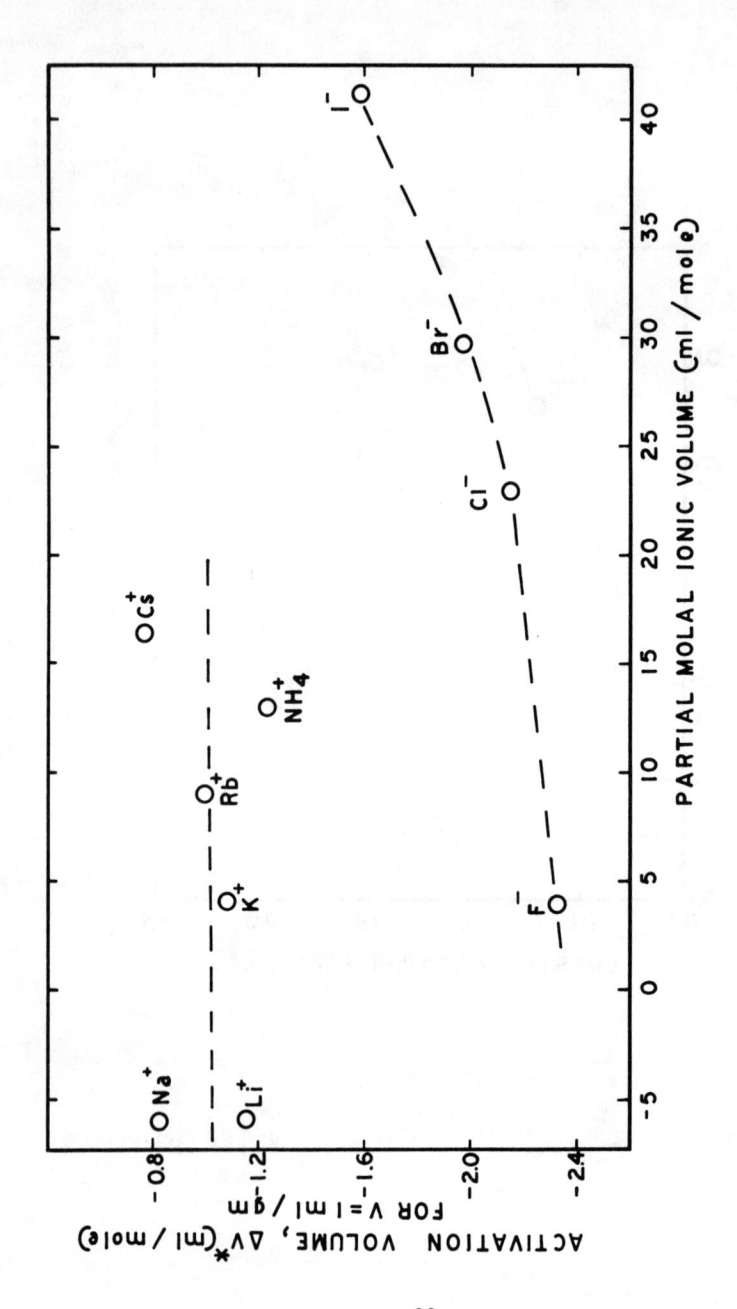

Fig. 5. Activation volume as a function of partial molal ionic volume.

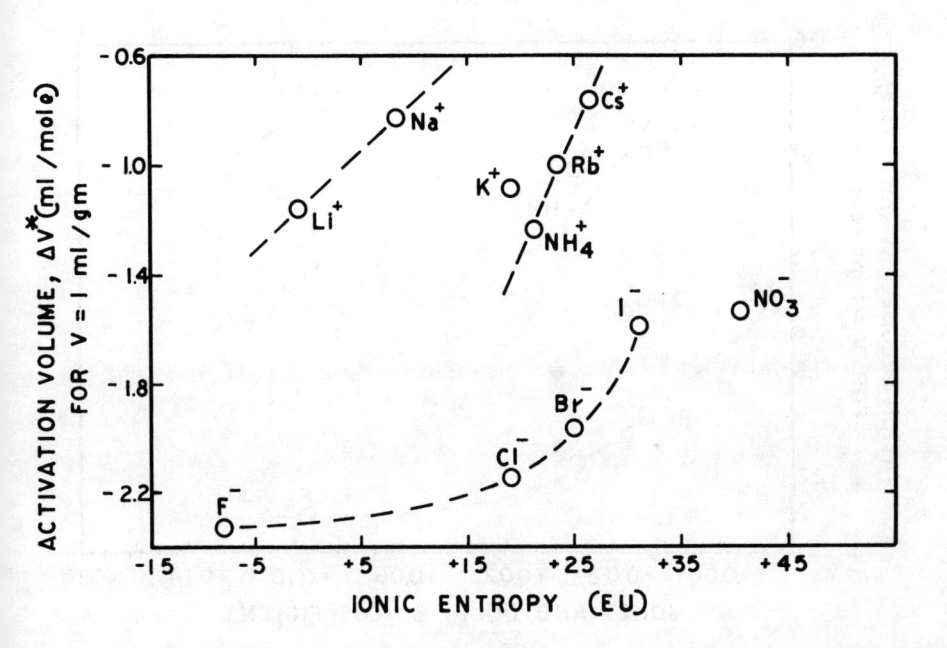

Fig. 6. Activation volume as a function of ionic entropy.

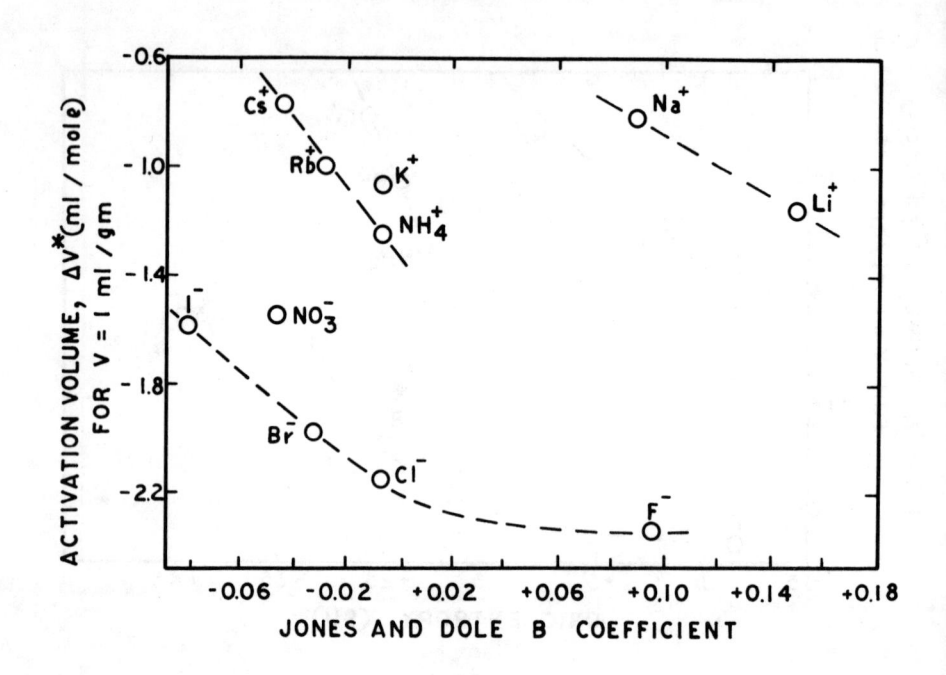

Fig. 7. Activation volume as a function of the Jones and Dole B-coefficient.

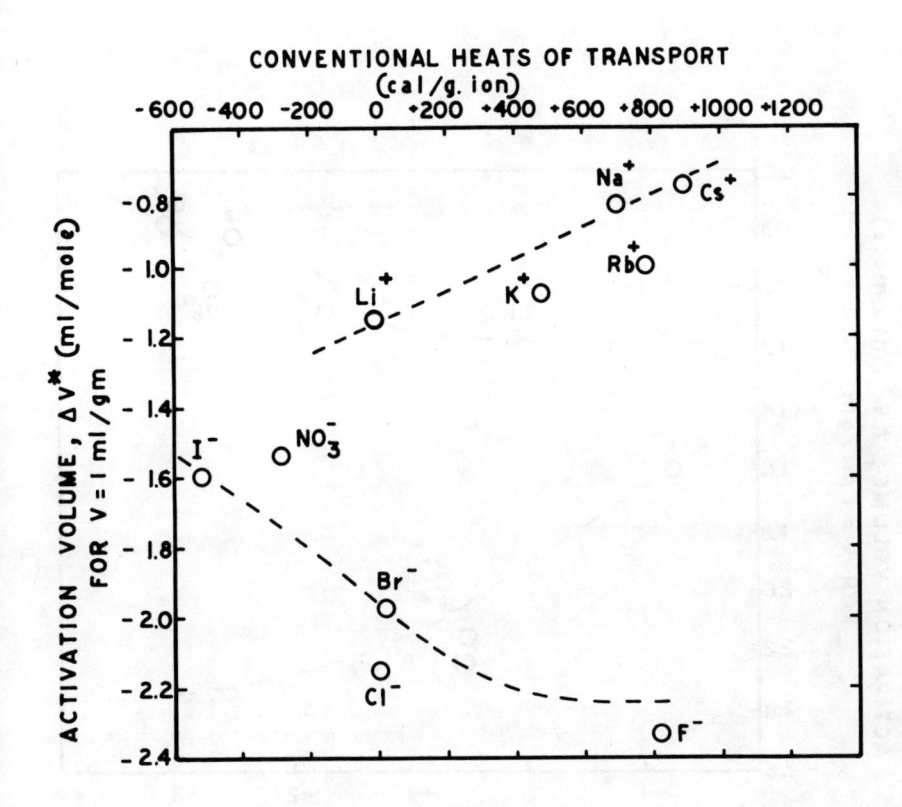

Fig. 8. Activation volume as a function of the conventional heat of transport.

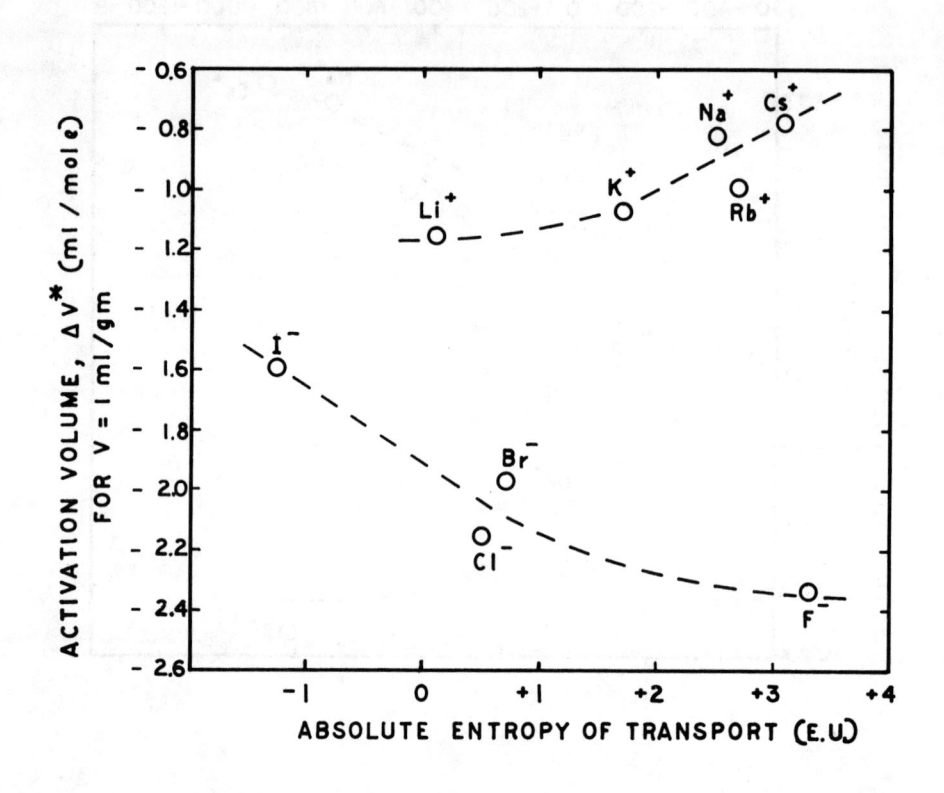

Fig. 9. Activation volume as a function of the absolute entropy of transport.

DISCUSSION

R.N. O'Brien, Chemistry Department, University of Victoria, Victoria, B.C., Canada: Making no assumptions about what you mean by structure making and breaking can I ask did you consider the evidence from freezing potentials? That is the potentials development when solutions typically below 10^{-3} μ are frozen quickly, where the advancing ice interface is moving about an order of magnitude faster than the mobility of the ion and the ice formed has a potential in relation and its solution. Typically NH_4^+ gives ∿ 250 volts, the ice being positive and F^- gives about 40 volts. It seems to me that really all that can be said about these ions under these conditions is that they are not structure breaking.

W.A. Adams and A.R. Davis, Water Science Subdivision, 562 Booth Street, Ottawa, Ontario, Canada: Low frequency intermolecular Raman bands of liquid water have been found in our work to show large intensity variations with pressure at temperatures close to 0°C. This contrasts with the insensitivity of the intramolecular region of the water spectrum that seems to vary little with pressure. Perhaps this observation will lead to further understanding of the lower viscosity and enhanced mobility of ions produced by pressure especially at temperatures close to 0°C.

DESIGN AND APPLICATION OF A DEEP SEA pH SENSOR

S. Ben-Yaakov and I.R. Kaplan

Department of Geology, UCLA
Los Angeles, California 90024

ABSTRACT

A rugged, high pressure pH sensor was constructed and successfully applied to oceanographic studies at depth to 4000 meters. The sensor comprises an equilibrated glass electrode, an incrementally pressurized reference electrode and is incorporated in an in situ instrumentation system. The sensor was applied to studies related to the carbonate system in the ocean. A close relation between pH and dissolved O_2 as measured and the North Eastern Pacific appears to be near saturation with respect to calcium carbonate down to 3500 meters.

INTRODUCTION

Marine pH measurements have traditionally been made by bringing samples to the surface in collecting bottles and equilibrating them to a fixed temperature. However, since the PH of sea water is a non-conservative parameter, pH values determined on the surface are not equal to the in situ pH. The correct pH can be calculated by taking into account the temperature and pressure dependence of the apparent dissociation constants of carbonic acid (Ben-Yaakov, 1970a). In situ pH measurements, therefore, offer an attractive method to overcome some of the inherent inaccuracies of conventional pH measurements, due to uncertainty in temperature and pressure coefficients of the constants used to correct shipboard values, and possibly due to errors caused by CO_2 exchange while the samples are handled on board ship.

Direct measurements of pH were made by Manheim (1961) to a depth of 16 m in the Baltic Sea using a robust glass electrode and a reference electrode floating at the water's surface. Disteche (1959) built pressure-compensated glass and reference electrodes which were used to a depth of 2350 m from the French Bathyscaphe Archimede (Disteche, 1964). However, in the latter case, the electrodes were not used as a remote sensor and no attempt was made to calibrate them and to translate the measured potentials into pH units.

This paper describes the design and application of a deep sea pH sensor developed at UCLA by the authors. As details of the design and results of specific studies made with the in situ system were given elsewhere (see reference list), an attempt will be made here to give an overview and to summarize our experience with the in situ pH sensor, which could be considered as a model for deep sea specific ion electrode measurements.

SENSOR CONSTRUCTION AND CALIBRATION

The glass membrane electrode can be used for in situ oceanic pH measurements because it maintains its sensitivity to pH at high hydrostatic pressure when both sides of the membrane are exposed to the same pressure (Disteche, 1959). Besides the basic requirement of pressure equilibration to avoid breakage of the delicate glass membrane, a practical design of an in situ pH electrode must, also ensure good electrical insulation of the electrodes output lead. The internal resistance of glass electrodes may reach 10^8 ohm (and even higher values at low temperatures) so that an insulation resistance of at least 10^{12} ohm must be achieved in order to prevent leakage attenuation. These requirements were met in the present design (Figure 1), by constructing the in situ glass electrode from two compartments; an outer silicone oil compartment and the internal solution compartment. The compartments are separated from each other, and from the ambient sea water, by flexible walls (Tygon^R tubing) which transmit the hydrostatic pressure to the inside of the pH sensitive glass bulb (Figure 1). Electrical insulation of the output signal is obtained by the silicone fluid compartment, which surrounds the output lead, and by a specially constructed high pressure feed-thru. This inlet is built around a stainless steel pipe fitting and consists of a platinum wire sealed into glass and cemented to the fitting with epoxy resin. The insulation resistance of this inlet is between 10^{12} and 10^{13} ohms.

The potential of the glass electrode is measured against a silver-silver chloride (Ag/AgCl) reference electrode with a leaky junction. This electrode (Figure 2) is also composed of two compartments; the upper silicone fluid compartment and the lower KCl filling solution compartment. The electrolyte flow is controlled by a rubber bulb which is slightly inflated, thereby maintaining a hydrostatic pressure always higher than the ambient pressure. This ensures an outflow of the KCl solution through the nylon wick which serves as the liquid junction. The junction end of the reference electrode also serves as a filling port. Filling is accomplished by unscrewing the teflon cap and replacing it with a special cap connected to a syringe. The solution is forced in, the tubing which leads to the rubber bulb is temporarily clamped to prevent loss of solution, and the original cap is replaced.

The internal reversible electrodes (Ag/AgCl in 2.7 M KCl) of the glass and reference electrodes are symmetrical, so as to improve the thermal tracking of the two cells. The internal solution of the glass electrode is also made 0.1 N in HCl to obtain a reference pH which is practically independent of hydrostatic pressure (Harned and Owen, 1958). The asymmetry potential shift of electrode pairs was checked in a high pressure chamber (Ben-Yaakov, 1970b) and was typically found to be approximately 1 mv for a 10,000 psi pressure range. Since the error due to this shift is small, it is generally neglected when converting the output potential of the pH sensor to pH units.

Conversion of the output signals to pH units is accomplished by preparing a calibration curve for each electrode pair. The curve is obtained by recording the output potential of the electrode when placed in a temperature controlled phosphate buffer (6.86 at 21°C; Bates, 1964). The buffer is first cooled to 0°C, and then slowly warmed to room temperature while the output potential of the electrode pair is registered for a number of temperature points. This information, along with the buffer test data obtained on board ship prior to each ocean measurement, is later used in the computation. The calculation is made under the assumption that the slope of the pH electrode approximately follows the Nernst slope. This is periodically verified by a conventional two-buffer test.

IN SITU INSTRUMENTATION

The high internal resistance and the low DC output of the pH sensor demand in situ signal conditioning. Since analog transmission over long electrical cables may result in appreciable error due to attenuation along the line, we have used a frequency modulation technique to transmit the signals from the in situ assembly (Ben-

Yaakov and Kaplan, 1968b). A simplified block diagram of the probe is given in Figure 3. The output signal of the pH sensor is buffered and converted to a proportional frequency signal by a current to frequency converter (Ben-Yaakov, 1968a). The output signal is then either sent to the surface via an electrical cable (Ben-Yaakov and Kaplan, 1968b) or recorded by an in situ recorder (Ben-Yaakov, 1968b). A more detailed schematic diagram of the amplifier used to buffer the output signal of the pH sensor is given in Figure 4. The high input resistance is achieved by a commercially available electrometer type operational amplifier (Analog Devices Type 311J) which is connected as a non-inverting amplifier. The reference electrode is connected to the "ground" of the electronic circuit which is floating with respect to the case and connected to it by a by-pass capacitor (C_b). The DC drift of the operational amplifier is monitored by shorting relay B (Figure 4) and monitoring the output voltage of the amplifier. Relays A and B are special dry reed relays with an insulation resistance of 10^{15} ohms (Compac series 10-1A-HIR).

The overall resolution of the instrumentation system used in our in situ oceanic studies is $\pm$ 0.1 mv (Ben-Yaakov, 1970b). The overall accuracy of the measurements is estimated to be 0.1% after correcting for drifts. Corrections are facilitated by routinely monitoring in situ internal standards built into the circuit (Figure 3). Accuracies stated above apply to the measurements of the raw signals and not to the overall accuracy of in situ measurements. The latter depend not only on the accuracy of the electronic system but also on errors introduced by the pH electrodes, the precision of the calibration curve and the accuracy of the on board buffer standardization. It will be demonstrated below, that direct deep sea pH measurements with an overall reproducibility to within $\pm$ 0.02 pH units are feasible.

FINE DETAILS OF OCEANIC pH PROFILES

Typical shallow pH and temperature profiles for the ocean off southern California at a station approximately 32 km west of San Diego are shown in Figure 5. In general, the value of pH at the surface of the sea was near 8.3 and a maximum value was reached between 20 and 40 m below the surface. In general, the temperature and pH curves follow the same trend below the pH maximum. Interestingly, pH and temperature were found to vary linearly, approximately 0.1 pH unit per 1°C. This dependence is about 10 times larger than the pH temperature dependence of sea water with fixed composition. It was found (Ben-Yaakov and Kaplan, 1968c) that the pH drop is controlled by respiration and oxidation of organic carbon. These processes consume oxygen and produce CO_2 thereby lowering the pH of the waters.

The insert of Figure 3 demonstrates the ability of the in situ measuring assembly to follow the fine details of the pH profile. The magnitude of this subsurface maximum was found to vary with time of day, seasons, and location. Since the maximum is located within the euphotic zone, it is probably associated with consumption of CO_2 by photosynthesis (or primary production) on the one hand and to the rate of atmosphere water exchange and mixing on the other. This pH layering should be maintained whenever the latter rate is smaller than the former. It is suggested, therefore, that detailed pH profiles, especially if recorded diurnally, may be used to estimate primary production, if the photosynthetic compensation depth is below the well-mixed layer.

IN SITU CARBONATE SATUROMETRY

Apart from its application as a tool for mapping the pH in the ocean, the oceanographic pH sensor has also been used to perform in situ carbonate saturometer experiments (Ben-Yaakov and Kaplan, 1971a). These experiments are useful in testing the degree of saturation of a calcium carbonate in the ocean, and the information is important for understanding related phenomena such as the mechanism controlling the amount of $CaCO_3$ in marine sediments and the CO_2 and calcium cycles in nature.

The basic principle of the carbonate saturometer experiment is the measurement of the pH before and after a sample of sea water has been equilibrated with a solid phase of $CaCO_3$ (Weyl, 1961). A pH change will take place according to the reaction:

$$2CaCO_3 + 2H^+ \rightleftharpoons 2HCO_3^- + 2Ca^{++}$$

Hence, a pH increase indicates that the original sea water was undersaturated with respect to the carbonate mineral, whereas a pH decrease signifies supersaturated. The magnitude of the pH shift can be used to accurately calculate the degree of saturation (the ratio of ionic product) of the original sample (Ben-Yaakov and Kaplan, 1969).

The in situ carbonate saturometer (Figure 6) is composed of a high pressure glass electrode which is inserted in a plexiglass cup filled with coarse $CaCO_3$ (calcite or aragonite) and connected through a hose to a pump and a solenoid-operated valve. When the pump and valve are on, sea water is rapidly pumped through the cell to flush out any water present in the cell, and the pH electrode then registers the pH of the sea water. During the off period, the trapped sea water reacts with the $CaCO_3$, moving toward equilibrium with the solid phase. The chemical reaction between the sea water and the carbonate mineral is monitored by the pH electrode, and this information is later used to calculate the degree of saturation of the sea water at the point of measurement. A typical output of the in situ carbonate saturometer is shown in Figure 7. The measurements were taken at a depth of 1200 m in the northeastern Pacific, 100 nautical miles west of San Diego. The horizontal axis of the figure is given in pH units (0.0125 pH units per dot), whereas the vertical axis represents running time (min.). The two plots correspond to the output of the saturometer electrode (O) and to the output of an auxillary pH electrode (*). Also marked on the figure are flushing periods (pump on) during which time the output of the two electrodes should be equal. The agreement between the two independent in situ pH determinations is $\pm$ 0.02 pH units.

The saturometer electrode output (*) shifted during equilibration toward higher pH values, which implies that the carbonate material (aragonite) was dissolving, and hence the Pacific is undersaturated with respect to aragonite at that station. The degree of saturation (IP/K'sp) for this measurement was calculated to be 0.7.

CONCLUSIONS

The results presented here demonstrate the feasibility of applying ion selective electrodes for in situ measurement in the oceanic environment. Judging from the results of our field experiments it would appear that the expected inaccuracy of in situ specific ion determination should be within $\pm$ 0.02 pIon units and perhaps can be improved to $\pm$ 0.01 pIon units. The corresponding concentration uncertainty for the error is about $\pm$ 3% for univalent ions and 6% for divalent ions. As the range of variation of the major ion in sea water is much smaller than that band (Culkin, 1965), the accuracy of direct in situ determination of major ions in open sea water is probably unacceptable. However, in situ specific ion electrode measurement could be useful in estuarine or lake studies, for determining activities of minor ions, and for performing specialized experiments such as carbonate saturometry.

ACKNOWLEDGMENTS

We wish to thank Mr. Ed Ruth for his assistance in calibrating the electrodes, and carrying out field measurements. The support of the National Science Foundation, Contract GA-34104, is gratefully acknowledged.

REFERENCES

Ben-Yaakov, S. (1970a). A method for calculating the in situ pH of sea water. Limnol. Oceanogr., 15: 326-328.

Manheim, F. (1961). In situ measurements of pH and Eh in natural waters and sediments. Stockholm Contrib. Geol., 8: 27-36.

Disteche, A. (1959). pH measurements with a glass electrode withstanding 1500 kg/cm^2 hydrostatic pressure. Rev. Sci. Instru., 30: 474-478.

Disteche, A. (1964). Nouvelle cellule a electrode de verre pour la measure directe du pH aux grands profondeur sous marine. Bull. Inst. Oceanogr. Monaco, 64, 1320: 1-10.

Harned, H.S. and B.B Owen (1958). The physical chemistry of electrolytic solutions. Third edition. Reinhold, New York, 803 pp.

Ben-Yaakov, S. (1970b). An oceanographic instrumentation system for in situ measurements. Ph.D. Dissertation, University of California, Los Angeles, 343 pp.

Bates, R.G. (1964). Determination of pH, theory and practice. John Wiley and Sons, New York, 435 pp.

Ben-Yaakov, S. and I.R. Kaplan (1968a). A high pressure pH sensor for oceanographic application. Rev. Sci. Instr., 39: 1133-1138.

Ben-Yaakov, S. and I.R. Kaplan (1968b). A versatile probe for in situ oceanographic measurements. J. Ocean **Tech.** 3 (2): 25-29.

Ben-Yaakov, S. (1968a). Analog to frequency converter is simple and accurate. Electronic Design, 16: 96-98 (July 4).

Ben-Yaakov, S. (1968b). A data recording system for deep sea logging. Proc. ITC: 149-160.

Ben-Yaakov, S. and I.R. Kaplan (1968c). pH temperature profiles in ocean and lakes using in situ probe. Limnol. and Oceanogr., 13: 688-693.

Ben-Yaakov, S. and I.R. Kaplan (1969). Determination of carbonate saturation of sea water with a carbonate saturometer. Limnol. and Oceanogr., 14: 874-882.

Ben-Yaakov, S. and I.R. Kaplan (1971a). Deep sea in situ carbonate saturometry. J. Geophy. Res., 76: 722-731.

Ben-Yaakov, S. and I.R. Kaplan (1971b). An oceanographic instrumentation system for in situ applications. Marine Tech. Soc. J., 5 (1): 41-46.

Weyl, P.K. (1961). The carbonate saturometer. J. Geol., 69: 32-43.

Culkin, F. (1965). The major constituents of sea water. Chemical Oceanography, J.P. Riley and G. Skirrow, eds., Academic Press, New York 1, pp. 121-162.

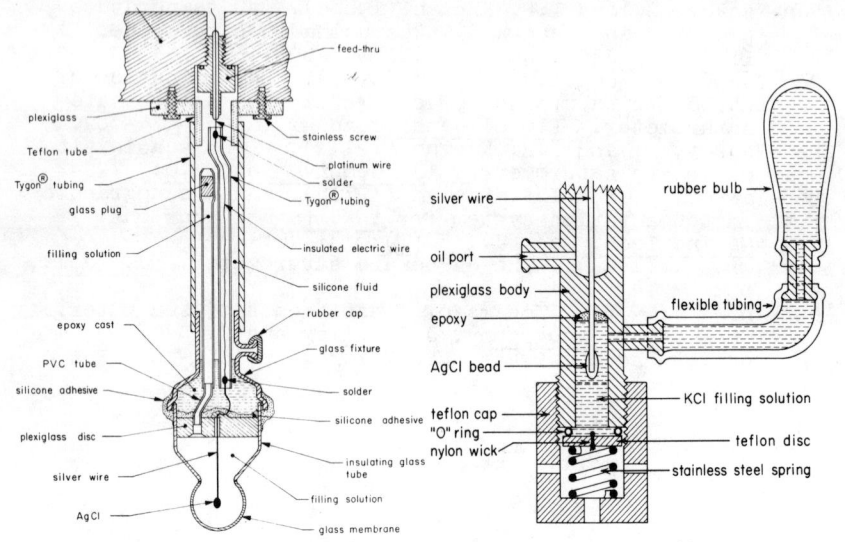

Figure 1: <u>In situ</u> pH electrode.

Figure 2: <u>In situ</u> reference electrode.

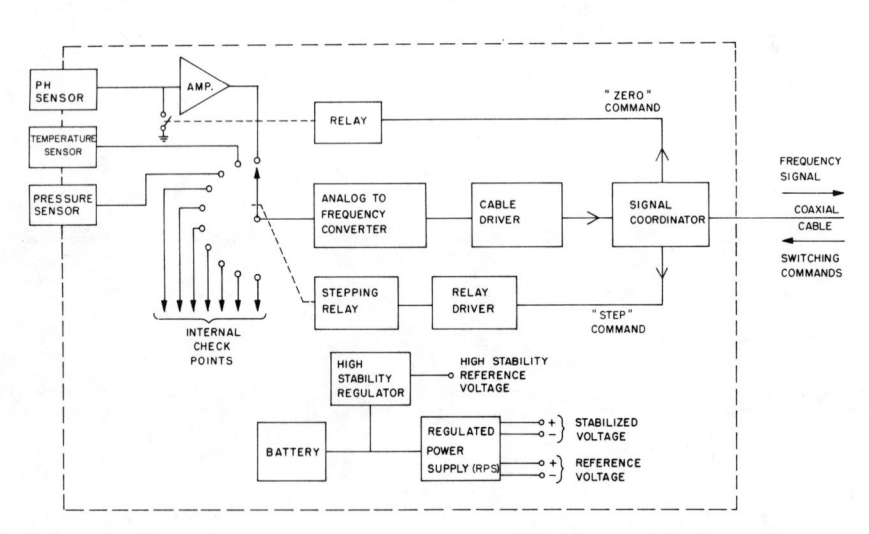

Figure 3: Block diagram of an oceanographic probe designed for <u>in situ</u> pH measurement.

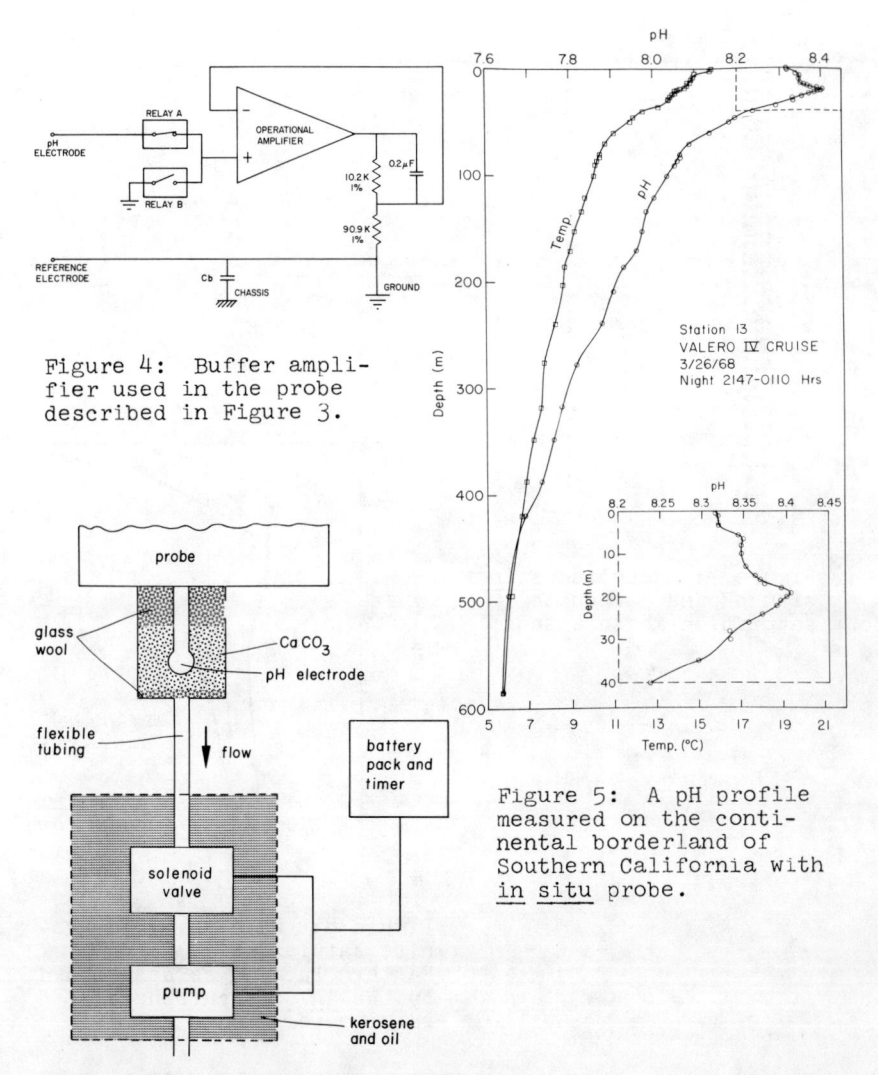

Figure 4: Buffer amplifier used in the probe described in Figure 3.

Figure 5: A pH profile measured on the continental borderland of Southern California with in situ probe.

Figure 6: Block diagram of an in situ calcium carbonate saturometer.

Figure 7: Results of in situ pH and carbonate saturation measurements.

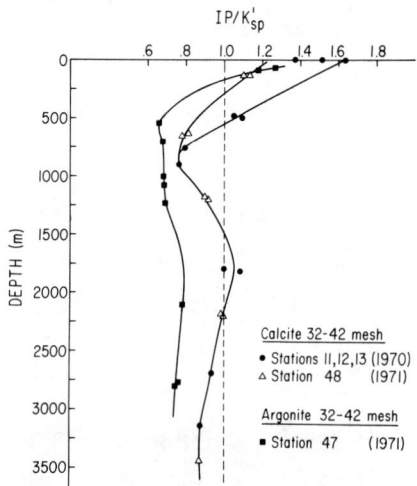

Figure 8: Calcite and aragonite saturation in North Eastern Pacific as determined by the in situ carbonate saturometer.

DISCUSSION

Paul Mangelsdorf: How does the calcium carbonate supersaturation arise? (Graph of IP/K'sp$^+$ vs depth for calcide)

S. Ben-Yaakov: The degree of calcite saturation (IP/K'sp = [Ca^{2+}][CO$_3^{2-}$]/K'sp) is by definition dependent on three parameters: total concentration of the calcium ion (Ca^{2+}), total concentration of the carbonate ion (CO$_3^{2-}$) and the apparent solubility constant of the carbonate mineral (K'sp). Since Ca^{2+} variations along oceanic profiles are relatively small, the ratio is mainly a function of the other two parameters which are pressure and temperature dependent. The apparent solubility product of calcite increases with depth due to the pressure buildup and temperature decrease. The total concentration of the carbonate ion is a function of pH which passes through a minimum at the oxygen-minimum zone (about 700 m in northeastern Pacific). At the minimum zone (CO$_3^{2-}$) concentration is therefore small, enough to cause supersaturation with respect to calcite (Figure 8). Below 3000 m depth supersaturation is again pronounced because the solubility of calcite (K'sp) increases considerably due to the high pressure and low temperature.

John W. VanLandingham, P.O. Box 7219, Ludlum Branch, Miami, Florida 33155: The comment in your paper relative to a pH maximum (subsurface) is of much interest to me. I have observed these non-conservative features in many areas of the world ocean. Phosphate, nitrate, copper, etc., usually show subsurface minimo along with the maximo of oxygen and pH.

S. Ben-Yaakov: Little attention has been paid until now, to the fine details of oceanic pH profiles which are evidently strongly associated with biological processes in the upper layer. By studying these profiles and correlating them with nutrients and DO profiles and the physical properties of the upper layer, we may improve out understanding of the processes that control the primary conductivity in the world eceans.

G. Atkinson, Dept. of Chemistry, University of Oklahoma, Norman, Oklahoma 73069: We have very good standards for defining the pH scale under laboratory conditions (1 atm). Could you tell me how you define your pH scale under pressure conditions?

S. Ben-Yaakov: The operational definition of the practical pH scale at high pressure is an extension of its definition at atmospheric pressure. That is, the pH of a sample at high pressure is defined through the potential shift of an "ideal" pH-sensitive electrode when moved from a standard

Discussion (Continued)

buffer to the sample (Bates 1956). Practical high pressure pH measurements are made with glass membrane electrodes which were shown to maintain their pH sensitivity at high pressures (Distèche 1959). Of course, one has to take into account erroneous potential shift which may be generated when glass electrodes are pressurized. These include: asymmetry potential shifts (Distèche 1959), possible variations of the inside solution of the glass electrode (Distèche 1962) and potential shifts due to a temperature difference between the standardizing buffer on the sample (Bates 1964). Once these potential shifts are corrected for (Ben-Yaakov and Kaplan 1968), the high pressure pH values (pH_x) are derived from the relationship:

$$pH_x = pH_s + S \, \Delta V$$

Where pH_s is the pH value of the standard buffer, S is the slope of the electrode (pH units/volts) and ΔV is the potential shift between standard and sample.

REFERENCES

Bates, R. G. (1956). The meaning and standardization of pH measurements. In: Symposium on pH measurements. Am. Soc. Testing Material Pub. No. 190; 1-11.

Bates, R. G. (1964). Determination of pH, theory and practice. John Wiley and Sons, New York; 435 pp.

Ben-Yaakov, S., and I. R. Kaplan (1968). A high pressure pH sensor for oceanographic application. Rev. Sci. Instr. 39; 1133-1138.

Distèche, A. (1959). pH measurements with a glass electrode withstanding 1500 Kg/cm^2 hydrostatic pressure. Rev. Sci. Instru. 30; 474-478.

Distèche, A. (1962). Electrochemical measurements at high pressures. J. Electrochem. Soc., 109; 1084-1092.

THE INCREMENTAL CONCENTRATION CELL AND ITS APPLICATION FOR STUDYING IONIC DIFFUSION IN SEAWATER

Sam Ben-Yaakov

Department of Geology, UCLA
Los Angeles, California 90024

ABSTRACT

A novel configuration of a concentration cell was used to measure liquid junction potentials generated between seawater and modified seawater. The potentials were found to be compatible with a seawater diffusion model which assumes that ionic mobilities in seawater are close to the limiting mobilities, and that speciation follows the model of Garrels and Thompson (1962). The study suggests that the mobility of sulfate complexes is low and that electrical gradients and ionic association may modify diffusion fluxes in interstitial waters of marine sediments.

INTRODUCTION

Studies on interstitial waters of marine sediments revealed that the ionic composition of the pore waters may differ from the composition of the overlying waters (Presley,1969; Initial Reports of the Deep Sea Drilling Project, 1969-1972). It is therefore probable, that the concentration gradients at the sediment-seawater interface, would result in a diffusional mass transport between oceans and sediments. Geochemical calculations which consider mass-balances of elements in the oceans must take into consideration the possible contribution of these fluxes to the overall inflow and outflow of seawater ions. Unfortunately, accurate estimates of diffusion fluxes on the ocean floor cannot be presently made, as many problems related to the diffusion process in marine sediments are poorly understood. In particular, there is insufficient experimental data on diffusion in a complex electrolyte system such as seawater, and the role of ion-sediment exchange in the diffusion of seawater ions has not yet been investigated.

In a recent study, Ben-Yaakov (1972) investigated the process of ionic diffusion from seawater into a dilute solution. It was found that the electrical coupling between the ions extensively modifies the diffusion fluxes. It was concluded, therefore, that this electrical coupling must be taken into consideration when modelling ionic diffusion in marine sediments. The present study represents an attempt to investigate another aspect of ionic diffusion; the role of ion complexing. It is generally accepted that most of the ions in seawater exist as ion-pairs (or complexes), which must be considered in explaining processes in seawater. Garrels and Thompson (1962) have proposed a thermodynamic model for the distribution of these species, and subsequent investigations have shown (Berner, 1971) that this model agrees with a number of independent tests. If the model of Garrels and Thompson (1962) is accepted, then one must conclude that the driving forces of ionic diffusion in seawater are not the concentration gradients of seawater ions, but rather, the individual gradients of the free and complexed species.

Experimental investigation of ionic diffusion in seawater is cumbersome for at least two reasons. First, diffusion processes are slow (a typical value for a diffusion coefficient is 10^{-5} cm^2/sec. or about 0.9 cm^2/day) and if the experiments are to be repeated for all seawater ions, many weeks and probably months of investigation would be involved. Second, each experimental run would require high precision analyses of all seawater ions in order to detect small concentration changes--which is a difficult and a highly time-consuming task. Application of radioactive tracer techniques for studying ionic diffusion in seawater can greatly simplify the investigation. However, diffusion of tracer ions in electrolyte solutions cannot be equated with the diffusion of the bulk ions (Robinson and Stokes, 1970). It is highly questionable, therefore, whether tracer diffusion studies of seawater ions would shed light on the process of bulk ionic diffusion in seawater.

It is generally accepted (MacInnes, 1961) that the liquid junction potential (or the potential of a concentration cell with transference) is a function--among other things--of the mobility of the ions in the junction. This dependency has been used in the past to estimate the magnitude of the liquid junction potential (Guggenheim, 1930) and to minimize the junction potential of salt bridges (MacInnes, 1961). The present study attempted to apply the dependency between ionic mobility and the liquid junction potential in the reverse direction. That is, liquid junctions were generated and their potential measured for the purpose of estimating ionic mobilities in seawater.

THEORETICAL CONSIDERATIONS

The liquid junction potential (E_j) generated between two electrolytes is related to the activity of the ions in solution by the differential equation (MacInnes, 1961):

$$dE_j = \frac{RT}{F} \sum \frac{t_i}{z_i} d\mu_i \qquad (1)$$

where R is the gas constant, T absolute temperature, F Faraday's constant, t_i, μ_i and z_i are the transference number, chemical potential and valence of a charged species (i), respectively, and the summation is over all charged species. The transference number, defined as the proportion of total charge carried by a given species (i), is related to the mobilities (U) and concentration (C, expressed in equivalents per liter) of the charged species by:

$$t_i = \frac{U_i C_i}{\sum U_i C_i} \qquad (2)$$

By combining equations (1) and (2) one obtains the relationship:

$$dE_j = \frac{RT}{F} \sum \frac{U_i C_i / z_i}{\sum U_i C_i} d\mu_i \qquad (3)$$

The magnitude of the potential of a given liquid junction can be calculated by integrating equation (3) along the junction under consideration. An analytical solution of this integration was derived by Henderson (MacInnes, 1961) for the mixture boundary junction which is of interest here. This junction is defined as a boundary in which each layer is a linear mixture of the two end solutions. Under these conditions and for the special cases where the activity coefficients and mobilites are constant along the integration path, Henderson's equation is applicable. The equation relates the total liquid potential to the concentration of the charged species of the two end solutions (designated by prime and double prime):

$$E_j = \frac{RT}{F} \frac{\sum (\frac{U_i}{z_i})(C_i'' - C_i')}{\sum U_i (C_i'' - C_i')} \ln \frac{\sum U_i C_i''}{\sum U_i C_i'} \qquad (4)$$

It should be noted that the integrated value of the junction potential depends only on the concentrations and mobilities of the charged species in the two boundary solutions.

The first basic requirements (mixture boundary) of Henderson's derivation can be easily met experimentally by allowing one electrolyte to leak into the other. If the leaking rate is made larger than the diffusion rate then migration due to diffusion can be neglected and the junction can be considered to be of the mixture boundary type --for all practical purposes. The requirement of constant mobility along the junction can also be approximated because the mobilities change only slightly with concentration (Robinson and Stokes, 1970). However, the requirement of constant activity coefficients cannot be met in the general case, because the two boundary electrolytes would have different compositions. This study attempted to overcome this difficulty by applying a novel experimental technique, the incremental concentration cell, in which the measured potentials are due to small perturbation in the test solution and the activity coefficients are kept approximately constant.

The incremental concentration cell is constructed as follows:

| Ag/AgCl | sat. KCl | test solution (1) | test solution (2) | sat. KCl | Ag/AgCl |

Initially, test solution (1) and (2) are identical (artificial seawater in the present experiments), but during the course of an experiment, a relatively small amount of salt is added to test solution (2). This perturbation will result in a shift in the potentials of the liquid junction solution (1)/solution (2) and solution (2)/saturated KCl and the sum of these shifts is measured by the two reference electrodes.

The incremental addition of salt to solution (2) will not affect the potential of the reference electrodes or that of the junction solution--(1)/KCl. Therefore, these potentials need not be considered when calculating the incremental change in the cell's potential. The initial potential between the two test solutions is of course zero, because they are identical. The potential, generated after the addition of salt to solution (2), can be calculated by Henderson's equation since the mobilities and activity coefficient would not vary markedly along the junction if only a small amount of salt is added. It is assumed of course, that the junction is of the boundary mixture type.

The incremental change in the potential of the junction seawater (2)/KCl can be expressed as:

$$E = \int_2^{KCl} \frac{RT}{F} \Sigma \frac{\bar{t}_i}{Z_i} d(\ln \gamma_i \cdot m_i) - \int_2^{KCl} \frac{RT}{F} \Sigma \frac{t_i}{Z_i} d(\ln \gamma_i \cdot m_i) \qquad (5)$$

where γ_i and m_i are activity coefficient and concentration of charged species (i), respectively, and the bar designates conditions after the salt is added. The last equation can be rewritten to the form:

$$E = \int_2^{KCl} \frac{RT}{F} \Sigma \frac{\bar{t}_i}{Z_i} d(\ln m_i) - \int_2^{KCl} \frac{RT}{F} \Sigma \frac{t_i}{Z_i} d(\ln m_i) \qquad (6)$$

$$+ \int_2^{KCl} \frac{RT}{F} \Sigma \frac{\bar{t}_i - t_i}{Z_i} d(\ln \gamma_i)$$

Henderson integration is applicable to the first two integrals if the mobilities are assumed to stay constant along the integration path. The contribution of the third integral is small because $(\bar{t}_i - t_i)$ is small and the dependence of $\ln \gamma_i$ on concentration is also small. Hence, the shift in the potential of the junction solution--(2)/KCl can be estimated by the difference between the two first integrals in the last equation. It can be concluded, therefore, that the total potential shift, which is the sum of potential shifts of this junction and of the junction between the two test solutions previously, could be closely calculated by Henderson's integration of the liquid junction potential.

The calculation can be made in three steps: (1) calculation of the liquid junction potential between the unmodified solution and saturated KCl, (2) calculation of the liquid junction potential between the modified test solution and saturated KCl and (3) calculation of the liquid junction potential between the unmodified and modified solution. The total potential shift (E_c) is given by:

$$E_c = E_3 + E_2 - E_1 \qquad (7)$$

where 1,2,3 correspond to the three computation steps.

EXPERIMENTAL

The configuration of the incremental concentration cell used in this study is shown schematically in Figure 1. The two working junctions were constructed by pressing nylon wicks between rubber stoppers and glass tubes. The leak rate through the junctions was adjusted to about 1 ml/week. The assembly was put in a temperature regulated bath and the experiments were run at a temperature of $25 \pm 0.1°C$. All the experiments were conducted with artificial seawater (Kester et al., 1967) of $35 \pm 0.2\%_o$ salinity.

The output potential of the cell was signal-conditioned by an electrometer type operational amplifier (Philbrick SP2A), measured digitally and also recorded by a strip chart recorder (1 mv full scale). Each experimental run was commenced by replacing the seawater in the beaker (250 ml) and establishing a base line for the measured potential. The absolute value of this potential was not measured as it is of no interest here. After the potential had stabilized (drift rate of less than 20 μv per hour), a small amount of a salt (0.04 or 0.08 equiv/l) was added to the beaker and the magnetic stirrer was turned on, until the salt completely dissolved. The cell was then left to thermally equilibrate until the potential stabilized again. The potential values reported here are the total shift between this stable value and the initial base line.

RESULTS AND DISCUSSION

The measured values of the potential shifts (Em) due to the incremental addition of various salts are summarized in Table 1. Each reported value is an average of 2-3 repeated runs which were found to reproduce to within $\pm 10\mu v$. The overall uncertainty band of the potential measurement is estimated to be $\pm 25 \mu v$. Experiments were conducted at two incremental additions 0.04 eq/l and 0.08 eq/l. However, the higher level could not be used with bicarbonate salts due to $CaCO_3$ precipitation whereas two of the salts (KCl and $MgSO_4$) produced too low a reading for the 0.04 eq/l additions.

The preceding theoretical discussion suggests that the shifts in the liquid junction potentials can be calculated by Henderson's integration (equation 4) of the liquid junction potential provided that the concentration of all charged species and their mobilities are known. Since this information was not known a priori, the liquid junction potential was calculated for different models of ionic speciation and mobilities. In the first calculation (Model A) it was assumed that all seawater ions are free and that their mobilities in seawater are identical to their mobilities in infinite dilute solution (Ben-Yaakov, 1972). The calculated values for NaCl, KCl and $CaCl_2$ agree with the measured ones (Table 1) to within the experimental accuracy. However, the calculated values of the liquid junction potentials shifts due to the addition of the bicarbonate and sulfate salts are not consistent with the measured values. These results probably indicate that the basic assumption concerning the value of ionic mobilities is correct, but that ion pairing (Garrels and Thompson, 1962) must be taken into account which explains why the assumption of free ions has not produced the desired results.

Ion pairing was taken into account in the second set of calculations by including in Henderson's integration (equation 4) all charged species according to the model of Garrels and Thompson (1962). The concentration of the free and associated species, of both original and spiked seawater, was calculated by an iteration method (Ben-Yaakov and Goldhaber, 1972), assuming activity coefficients and dissociation constants according to Berner (1971). Apart from a knowledge of the concentration of all charged species in solution, application of Henderson's equation also requires a knowledge of the mobilities of the charged species. The study of Ben-Yaakov (1972) as well as the calculation of Model A, described above, suggest that the mobilities of the free ions in seawater are close to mobilities at infinite dilution. However, as no data are available for the mobilities of complexed species (such as $NaSO_4^-$, $CaHCO_3^+$), an assumption had to be made regarding the value of these mobilities. The simplest assumption that can be made, is that the mobility of charged ion pairs is close to the mobility of the free anion. This is deemed to be a reasonable assumption because the degree of hydration of an ion pair is probably smaller than that of each ion in its free state due to partial cancellation of the electrical fields. It is likely, therefore, that the mobility of the ion pair is limited by its effective volume which should be close to that of the anions ($SO_4^=$, HCO_3^-).

The assumption of speciation according to the model of Garrels and Thompson (1962) and mobilities of ion pairs equal to the mobilities of the respective anions (Model B) improved the agreements between the calculated and measured potentials for the bicarbonate salts (Table 1). However, the calculated potential shifts due to the addition of the sulfate salts were still in poor agreement with the measured ones. These results probably indicate that the mobility of the charged sulfate complexes are not equal to the mobility of the sulfate ion. It was found that the agreement between the calculated and measured potentials improves if the mobilities of the charged sulfate species are assumed to be much smaller than the mobility of the free sulfate ion. In fact, the best agreement was obtained when the mobilities of the charged sulfate species was assumed to be nil (Model C). The overall agreement between the calculated potential shifts according to Model C (Table 1 and Fig. 2) is very good, considering the approximate nature of the Garrels and Thompson model (Thompson and Ross, 1966).

The agreement between the calculated potential shift according to Model C, and the measured one, suggests that the assumption regarding the low mobility of the charged sulfate complexes may be valid. The low mobility may be attributed to a large effective cross-section of the sulfate complexes which increases the drag. This would imply that the two ions forming the ion pair are not in close proximity to one another and that a number of hydration molecules are involved in this ionic interaction. This implication is in accord with the study of Eigen (1957) on general and specific ionic interaction in solution. He concluded, from sound absorption spectra of electrolytes in aqueous solution, that because the interacting ions are at a relatively larger distance (6-8 Å), sulfate ion pairs involve a number of hydration molecules.

According to the model of Garrels and Thompson (1962) approximately 60% of the sulfate ions in seawater are complexed. As the results of the present study suggest, the mobility of the sulfate complexes is low, the effective diffusion coefficient of total sulfate in seawater is probably lower by about 60% than the diffusion coefficient of the free sulfate ion. However, the actual diffusion flux of a given ion in seawater is not only a function of the concentration gradient of that ion, but also a function of the concentrations, concentration gradient and mobilities of all seawater ions (Ben-Yaakov, 1972). Therefore, it is impossible to assign a single diffusion coefficient to a given ion and the actual diffusion flux of each ion can be evaluated for each case only after taking into account electrical cross-coupling between the ions (Ben-Yaakov, 1972).

Ionic diffusion in marine sediments is far more complicated than the case of diffusion in seawater due to possible solution-sediment reaction. Ion exchange between the pore water and the solid phase of the sediment would modify the ionic fluxes (Van Schaik et al., 1966). However, since the nature of this ion-exchange process and its role in modifying diffusional fluxes are still poorly understood, it is as yet impossible to accurately calculate diffusion fluxes in marine sediments. The present study suggests, however, that the driving forces of the diffusional fluxes in marine sediments are the concentration gradients of both free ions and the complexed ionic species.

ACKNOWLEDGEMENTS

I wish to thank Professor I.R. Kaplan for critically reading the manuscript, to Ed Ruth for assistance in preparing the experimental set-up and to Christy Sandberg for assistance in the experimental and computational phases of the research. This study was supported by an AEC grant, AT(04-3)-34 P.A. 134.

REFERENCES

Garrels, R.M. and M.E. Thompson (1962). A chemical model for sea water at 25°C and one atmosphere total pressure. Amer. J. Sci., 260; 57-66.
Presley, B.J. (1969). Chemistry of interstitial water from marine sediments. Ph.D. Thesis. University of California, Los Angeles, 225 pp.
Initial Reports of the Deep Sea Drilling Project (1969-1972). Washington (U.S. Government Printing Office).
Ben-Yaakov, S. (1972). Diffusion of sea water ions--I: Diffusion of sea water into a dilute solution. Geochim. et Cosmochim. Acta, 36, in press.
Berner, R. A. (1971). Principles of chemical sedimentology. McGraw-Hill, New York, 240 pp.
Robinson, R.A. and R.H. Stokes (1970). Electrolyte Solutions. Second edition, fifth impression (revised). Butterworth and Company, London; 571 pp.
MacInnes, D.A. (1961). The principles of electrochemistry. Revised edition. Dover Publication, New York; 478 pp.

Guggenheim, E.A. (1930). A study of cells with liquid-liquid junctions. Jour. American Soc., 52; 1315-1337.

Kester, D.R., I.W. Duedall and D.N. Conners (1967). Preparation of artificial sea water. Limn. Oceanog. 12; 176-179.

Thompson, M.E. and J.W. Ross, Jr. (1966). Calcium in Sea Water by Electrode Measurement. Science 154; 1643-1644.

Eigen, M. (1957). Determination of general and specific ionic interaction in solution. Faraday Soc. Discussion No. 24; 25-36.

Van Schaik, J.C., W.D. Kemper and S.R. Olsen (1966). Contribution of adsorbed cations to diffusion in clay-water systems. Soil Sci. Soc. America Proc., 30; 17-22.

Table 1 - Experimental and theoretical values of the liquid junction potential shifts. See text for details of model calculations.

Salt	Δsalt = 0.04 eq/l Ec (mv) Model A	Ec (mv) Model B	Model C	Em (mv)	Δsalt = 0.08 eq/l Ec (mv) Model A	Ec (mv) Model B	Model C	Em (mv)
NaCl	-0.246	-0.263	-0.259	-0.229	-0.475	-0.508	-0.500	-0.479
KCl	-0.033	-0.040	-0.036	--	-0.064	-0.077	-0.069	-0.058
CaCl$_2$	-0.431	-0.449	-0.458	-0.446	-0.832	-0.867	-0.882	-0.847
NaHCO$_3$	0.047	0.101	0.109	0.129	-0.091	0.188	0.204	--
KHCO$_3$	0.260	0.325	0.334	0.324	0.503	0.625	0.642	--
K$_2$SO$_4$	0.269	0.318	0.405	0.438	0.569	0.611	0.787	0.861
MgSO$_4$	-0.129	-0.116	-0.046	--	-0.249	-0.207	0.087	0.025
Na$_2$SO$_4$	0.086	0.096	0.184	0.220	0.166	0.184	0.364	0.474

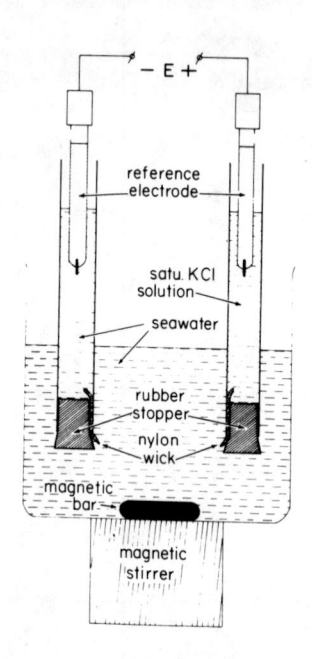

Figure 1: Practical realization of the incremental concentration cell.

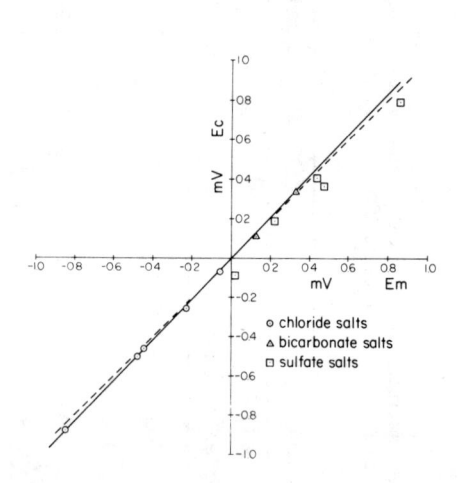

Figure 2: Measured potential shifts (Em) compared to theoretical values (Ec) calculated by Henderson's equation and assuming ionic speciation according to model C.

DISCUSSION

Professor P. Meubus, Chem. Eng. Dept., Université du Quebec, A Chicoutimi, Chicoutimi, P.Q., Canada: How far do organic colloidal material, present in sea water, interefere with the measurement of mobility and other parameters related to the measurement of ions diffusion?

S. Ben-Yaakov: Due to the relative low concentrations of colloidal suspension, their contribution to the diffusion of major ions of seawater (which were considered here) is probably negligible. However, transport of trace elements by diffusion and other processes may be affected by organic colloids due to the formation of organo-metal complexes.

COMPUTER MODELING OF INORGANIC EQUILIBRIA IN SEAWATER

Gordon Atkinson, M. O. Dayhoff, and David W. Ebdon

Department of Chemistry
University of Oklahoma
Norman, Oklahoma   73069

National Biomedical Research Foundation
3900 Reservoir Road
Washington, D. C.   20007

Eastern Illinois University
Charleston, Illinois   61920

Beginning with well-known values of the elemental composition of seawater, the standard free energies of the major species and their activity coefficients, we have calculated the equilibrium concentrations of these species for 35°/oo salinity seawater at 25°C. By use of a computer program which solves a system of linear equations and iteratively minimizes the total free energy of the system, the concentrations of the most important dissolved species have been calculated. In addition, the concentrations of several minor constituents have been calculated using estimated free energy values. Our results show some differences from those of earlier workers. These may be attributed to new values for the free energies and activity coefficients of several major species. Stoichiometric activity coefficients calculated from the computer-generated concentrations show good agreement with those determined experimentally.

Introduction

The composition of the sea has apparently remained relatively constant through the long period of animal evolution. Since life probably began in this medium, it is hardly surprising that many biological fluids bear a marked resemblance to seawater in composition. Therefore, a more detailed knowledge of the chemical species present in the sea is of interest not just to oceanographers but to biochemists and physiologists.

The determination of the speciation in seawater has drawn the attention of ocean scientists for several years. One of the first comprehensive treatments based on the ion pairing model was given by Garrels and Thompson[1], who used thermodynamic equilibrium constants and approximate single ion activity coefficients to obtain a set of equations which, upon solution, yielded the distribution of dissolved species in 19°/oo chlorinity seawater. More recently Kester and Pytkowicz[2] have used experimentally determined stoichiometric association constants for several sulfate ion pairs and arrived at a distribution of species which differs from that of Garrels and Thompson with respect to the percentages of sulfate bound in the calcium and sodium sulfate ion pairs, thus leading to a significant change in the sulfate ion concentration.

Kester and Pytkowicz' use of the ionic medium standard state eliminates the need for a single ion activity coefficient for each species. One needs only to determine concentrations of species and thereby the stoichiometric association constant. The means for doing this usually involve measurements with ion-selective electrodes, which must be calibrated by immersion in a standard solution. Since these electrodes measure <u>activities</u>, one must use some approximate means to relate the measured activity to the concentration of the ionic species studied. Often it is assumed that activity coefficients are dependent only on ionic strength; hence one takes them to be equal in different solutions of the same ionic strength. There is thus little fundamental difference between this method and that used by Garrels and Thompson, who used thermodynamic equilibrium constants, which are defined relative to the infinite dilution standard state, and approximate single ion activity coefficients, for which the assumption is made that the activity coefficient in the single salt solution is the same as that in seawater of the same ionic strength.

We have chosen to use thermodynamic association constants and mean salt values for the activity coefficients in our calculations since these association constants can usually be determined with greater accuracy than can stoichiometric constants, and the activity coefficients can be trusted to be correct to a very few percent due to the general behavior predicted by the Debye-Hückel theory and its extensions. If methods could be used to determine stoichiometric constants which did not have the shortcomings mentioned earlier, then use of these constants would indeed be preferable. But until that time we believe our method to be the more reliable.

Our approach is based on recent work[3] on the calculation of planetary atmospheres by computerized equilibrium calculations. In the present work we consider the sea and its immediate atmosphere as an equilibrium chemical system. This allows us to perform calculations using the well-proven methods of equilibrium thermodynamics. There are, however, some obvious problems we must consider. In the first stage we totally ignore any equilibria involving solids suspended in the sea or at its bottom. We do this not only because of the added complexity of such an approach but also because of the lack of precise thermodynamic data on many of the most important solids. Sillén[4] has given a provocative and imaginative discussion of the problem. Further, there is much evidence that there are processes occurring in the sea that are not governed by simple thermodynamic considerations. The nitrogen and phosphorus problems[5] are some of the more obvious examples. Finally, the data available make it possible to do complete calculations only at 25°C and 1 atm pressure. The effect of changing temperature concerns us as do the great effects caused by changing pressure in the ambient oceanic environment. Unfortunately our knowledge of pressure and temperature effects on important seawater equilibria is very limited. Kester and Pytkowicz[6] have examined the pressure dependence of the $NaSO_4^-$ equilibrium at the ionic strength of seawater, but so far a comprehensive treatment of seawater equilibria at high pressures is lacking.

In spite of the above problems, we believe the calculations described below are both meaningful and useful. There is very much evidence that the sea is a giant equilibrium system. The very constancy of many important parameters over epochs of time is most explicable in these terms. Furthermore, until we examine in some detail the predictions of an equilibrium model and check them against our experimental knowledge, it is difficult to decide on what basis to discuss nonequilibrium processes.

Calculations

Our calculations require as data the elemental composition of seawater and the free energies of formation of all species that we wish to consider. We have computed the molalities of the major constituents of seawater from data given by Culkin[7]. The free energies of neutral species and simple ions were taken from standard sources[8,9]. To calculate similar quantities for ion pairs the thermodynamic association constants are needed in addition to the free energies of the constituent ions. The majority of these constants were taken from the most complete compendium of such data[10]. The particular values chosen were the ones we judged most precise for the infinitely dilute solution at 25°C. Table I lists the logarithms of the ion pair association constants used in this work. It is worth noting that no data exist for interaction of potassium with any anion but sulfate. In the authors' opinion, this does not appear to be reasonable since sodium, a similar ion, forms ion pairs with all but fluoride. It would seem worthwhile examining anew the interaction of potassium with the carbonate species particularly.

As a typical example of the calculation of the standard free energy of formation of an ion pair, consider the following equilibrium.

$$Na^+ (aq) + SO_4^{2-} (aq) \rightleftarrows NaSO_4^- (aq)$$

$$K_1 = \frac{a_{NaSO_4^-}}{a_{Na^+} \, a_{SO_4^{2-}}}$$

This equilibrium is characterized by $\log K_1 = 0.72$ at 25°C. We calculate the standard free energy change by the equation

$$\Delta G^o = -2.303 \, RT \log K_1.$$

We then convert this to the standard free energy of formation of the $NaSO_4^-$ ion pair by

$$\Delta G^o = \Delta G^o_f (NaSO_4^-) - \Delta G^o_f (Na^+) - \Delta G^o_f (SO_4^{2-})$$

and the tabulated values[9] of the free energies of formation for the aqueous ions.

The next problem that intrudes between us and our actual calculations is the nonideality of the system. In the previous work on planetary atmospheres, it was a very good assumption to say that the system was ideal so that mole fractions could be used for fugacities. However, in seawater with an ionic strength near 0.7 molal, ideality would be a very bad assumption. The few measured activity coefficients of the ionic species in seawater do not go very far in solving the problem, for those of greatest importance are those which contain contributions from association equilibria. The normal Debye-Hückel equation would be of little value at this ionic strength, but the modification introduced by Davies[11] has been suggested[12] as a possible guide to activity coefficient values at ionic strengths up to 0.5. The equation has the form

$$\log \gamma_i = -.50 \, z_i^2 \, \frac{\mu^{1/2}}{1 + \mu^{1/2}} + b\mu$$

where z_i = charge on i^{th} ion, $\mu = 1/2 \sum_i m_i z_i^2$ and b has the value 0.2 or 0.3.

This equation says that all ions of the same charge have the same activity coefficient. At ionic strengths below about 0.2 there is good evidence to support this. At ionic strengths of the order of seawater, however, activity coefficients of electrolytes of a given charge type show quite distinct behavior[13,14], even when association is taken into account.

We therefore decided to use the mean salt method[1] for obtaining individual ionic activity coefficients. Taking KCl as our standard, for which we define

$$\gamma_{\pm \, KCl} \equiv \gamma_{K^+} \equiv \gamma_{Cl^-},$$

we calculated most of the single ion activity coefficients shown in Table II from data given by Robinson and Stokes[14]. The sulfate value was calculated from the mean ionic activity coefficients of K_2SO_4 and Na_2SO_4 after these data were corrected for ion association. The carbonate and bicarbonate data are those of Walker, Bray and Johnston[15], and the values for H^+ and OH^- are calculated from the data of Buch[16]. We have taken activity coefficients for charged ion pairs as equal to the corresponding carbonate species. These are also very close to the values generated by the Davies equation with b = 0.2. The activity coefficients of uncharged species are considered to be unity, in contrast to the value 1.13 used by Garrels and Thompson[7]. More recently Garrels and Christ[17] have concluded that the value may be unity or slightly less.

Once activity coefficients γ are obtained for each species, the standard free energy values, ΔG°_f, are modified for input into the computer program by

$$\Delta G_f^* = \Delta G_f^\circ + RT \ln 55.51 \gamma .$$

where the term $RT \ln 55.51$ corrects the free energy to the mole fraction standard state, on which basis the computer program performs the calculations, and where the term $RT \ln \gamma$ converts the free energy to the ionic medium value. The computer then takes the total elemental composition of seawater, the composition of a model atmosphere and the modified ΔG_f^* values of all the species and iteratively minimizes the total free energy of the system.

Results

Major Components

The results are given in Table III. It is immediately apparent that from the standpoint of cationic distribution the sulfate ion pairs are the only ones of major importance, others accounting for less than one percent of a given cation concentration. Thus examinations of pressure and temperature effects on seawater equilibria need center only on sulfate equilibria if the percent of free cation is the prime concern. In the anionic distribution nearly all ion pairs significantly decrease the free ion concentration. The magnesium pairs account for nearly half of the available carbonate and fluoride, all but sixteen percent of the hydroxy species and significant fractions of the sulfate and bicarbonate concentrations.

In Table IV we have compared our results to calculations and measurements of other workers. Our magnesium ion concentration shows good agreement with the measurements of Thompson[18] and of Kester and Pytkowicz[2], and our percentage of free calcium is very close to that measured by these latter authors. Our percentage of free sulfate and bicarbonate is markedly different from the two previous calculations. We have used recent ultrasonic data[18] for $MgSO_4$ equilibrium and a more realistic activity coefficient for the sodium ion so as to produce a lower percentage of free sulfate than that reported by Garrels and Thompson. Kester and Pytkowicz' stoichiometric association constants seem to be close to the correct ones for $MgSO_4$ and $CaSO_4$ since correction to infinite dilution yields values very close to accepted thermodynamic constants. Their result for $NaSO_4^-$, however, seems too large thus "overcorrecting" the free sulfate to a low value. Their constant for KSO_4^- seems incorrect in that it is approximately half that of $NaSO_4^-$. Other measurements[20,21] would put it at a value about 60% greater than that of the sodium pair. Our lower free bicarbonate concentration is due to the use of new values[22] for the sodium carbonate and bicarbonate association constants.

Also in Table IV are tabulated the measured stoichiometric or total activity coefficients for several ions in seawater and those calculated from our distribution. These calculated values were obtained by multiplying our mean salt values by the fraction of free ion for each species. Agreement is very good throughout. The bicarbonate value is within the measured range of experimental values, but more recent determinations[23] place the bicarbonate activity coefficient close to 0.55. One possible explanation[24] for discrepancy may lie in the existence of some of the bicarbonate ion pairs. The most suspect is the $NaHCO_3$ species. If one were to eliminate this ion pair and redistribute the bicarbonate among the remaining dissolved species, the resultant total activity coefficient would be 0.52. It is evident that a reexamination of the carbonate system equilibria in seawater is necessary to resolve the problem.

We have ignored borate in our calculations due to lack of sufficient data. A value of the dissociation constant of $NaB(OH)_4$ appears in the literature[25], but well-known values for the magnesium and calcium species are lacking. To gain some insight into the borate speciation, we used the known sodium value and assumed the other association constants varied approximately as those of the bicarbonate series. In the resultant distribution $NaB(OH)_4$ accounted for 41%, $MgB(OH)_4^+$ for 28%, $CaB(OH)_4^+$ for 5%, $B(OH)_4^-$ for 2.4% and $B(OH)_3$ for the remainder of the available boron. Although these numbers are at best educated guesses, they do seem to indicate that there is much less free boric acid in seawater than has previously been assumed. An experimental examination of ion pairing in the borate system is necessary before we can be more definitive concerning these equilibria.

Trace Components

The extention of this approach to minor seawater species is quite easy using the computer technique. We have calculated ΔG_f^* values for the several copper species[10] to observe how this element distributes itself in seawater. Since the Cu^{2+} concentration is quite small, to a very good approximation the concentrations of the major constituents in the present scheme will not be affected. In the calculation Cu^{2+} simply distributes itself among the different copper forms so as to minimize the total free energy of this small subset of free Cu^{2+} and CuX species.

The results are somewhat surprising in that the $Cu(OH)_2$ ion pair accounts for more than 90% of the available copper. The $CuCO_3$ species takes up 7% and other ion aggregates total slightly more than 2% leaving about 1% free copper ion! Plainly this result weighs heavily upon the free energy value for the $Cu(OH)_2$ ion pair. To take the extreme case, let us assume no such species exists. This still gives us only 9% free copper ion, again a value which would be considered "small". It is evident that our knowledge of even the minor seawater components must be extended.

Solid-Solution Interactions

A traditional problem still intrudes into the picture. Calcium carbonate is supersaturated in the above calculation. This is not surprising since many workers have demonstrated that surface oceanic waters are supersaturated with respect to calcite and aragonite[26]. It has been shown, moreover, that calcium carbonate solubility increases markedly with pressure so that deep waters are in fact unsaturated[27]. Garrells, Thompson and Siever[28] have reported that supersaturated calcium carbonate solutions produced in the laboratory are stable for periods of time sufficient to perform "equilibrium" measurements. Chave[29] has stated his belief that the apparent stability of supersaturated $CaCO_3$ in seawater is due to the coating action of polar organic molecules. This might also indicate the presence of as yet unknown calcium species in solution.

To test our model further we introduced the free energy of solid calcium carbonate into the calculation, trying first the NBS value for aragonite, - 169.53 kcal/mole. This produced a pH drop of slightly more than 0.2 units from our equilibrium value of 8.1. The measured pH drop with precipitation has been reported as 0.15 units[30]. We can reproduce this value by using a free energy of -169.42 kcal/mole only about 0.1 kcal different than the literature value. Such a small difference is within the experimental uncertainty of the free energy value for aragonite, since its determination depends on differences of the order of one-hundredths in the pH of an equilibrated aragonite-water system. This highlights the need for precise thermodynamic measurements on the carbonate system.

Pressure Effects

Preliminary estimates of the effect of increasing pressure on seawater equilibria show small increases of the order of one percent in the concentrations of sodium and potassium ions. Divalent cation concentrations remain essentially constant. The anion distribution appears to change somewhat more dramatically with unpaired sulfate increasing by about nine percent and free carbonate by slightly less than two percent. Both these shifts are principally the result of decreases in the concentrations of the sodium ion pairs. The magnesium sulfate ion pair actually appears to increase slightly in concentration, as has been noted by other workers[6], but this result could be altered by very small changes in our input data.

Conclusion

The approach described for the calculation of complex equilibrium systems is both useful and versatile. The effects of differing assumptions or new data are easily examined. Trace element speciation is readily handled. In fact, the only hindrance to much more detailed calculations is our lack of accurate thermodynamic data. In particular, we have a great lack of data on the effects of temperature and pressure on ionic equilibria.

The calculational ease of this approach makes it a very useful tool in pinpointing areas where additional data is most needed.

Acknowledgement

The initial stages of this work were supported by the National Aeronautics and Space Administration through NASA Contract 21-003-002 to the National Biomedical Research Foundation. The work is presently supported by the Office of Naval Research through contract N00014-72-A-0285-0001.

The authors gratefully acknowledge the help of the Computer Science Center of the University of Maryland, the Computer Center of Eastern Illinois University and the Merrick Computing Center of the University of Oklahoma.

References

1. R. M. Garrels and M. E. Thompson, Amer. J. Sci., 260, 57-66 (1962)

2. D. R. Kester and R. M. Pytkowicz, Limnol. Oceanogr., 14, 686-92 (1969)

3. M. O. Dayhoff, E. R. Lippincott, R. V. Eck and G. Nagarajan, NASA Report SP-3040, Washington, D. C., 1967.

4. L. G. Sillén, "The Physical Chemistry of Sea Water", in M. Sears (ed.), Oceanography, AAAS Publication No. 67, Washington, D. C., 1961.

5. H. V. Sverdrup, M. W. Johnson and R. H. Fleming, the Oceans - their Physics, Chemistry and General Biology, Prentice-Hall, Englewood Cliffs, N. J., 1942.

6. D. R. Kester and R. M. Pytkowicz, Geochim. Cosmochim. Acta., 34, 1039 (1970).

7. F. Culkin, "The Major Constituents of Sea Water", in J. P. Riley

and G. Skirrow (eds.), Chemical Oceanography, Vol. 1, Academic Press, London and New York, 1965.

8. F. Rossini, et. al., Selected Values of Chemical Thermodynamic Properties, Circular 500, National Bureau of Standards, Washington, D. C., 1952.

9. W. M. Latimer, The Oxidation States of the Elements and their Potentials in Aqueous Solution, Prentice-Hall, Englewood Cliffs, N. J., 1952.

10. L. G. Sillén and A. E. Martell, Stability Constants of Metal Ion Complexes, The Chemical Society, London, 1964.

11. C. W. Davies, Ion Association, Butterworths, Washington, D. C., (1962)

12. W. Stumm and J. J. Morgan, Aquatic Chemistry, Wiley-Interscience, New York, 1970.

13. H. S. Harned and B. B. Owen, the Physical Chemistry of Electrolytic Solutions, Third Edition, Reinhold Publishing Co., New York, 1958.

14. R. A. Robinson and R. H. Stokes, Electrolyte Solutions, Second Edition, Revised, Butterworths, London, 1965.

15. A. C. Walker, V. B. Bray and J. Johnston, J. Amer. Chem. Soc., 49, 1235 (1927).

16. K. Buch, Acta Acad. Aboensis, Math. et Physica, 11, No. 5 (1938).

17. R. M. Garrels and D. L. Christ, Solutions, Minerals and Equilibria Harper and Row, New York, 1965.

18. M. E. Thompson, Science, 153, 166 (1966). [As corrected in reference 2]

19. G. Atkinson and S. Petrucci, J. Phys. Chem., 70, 3122 (1966).

20. I. L. Jenkins and C. B. Monk, J. Amer. Chem. Soc., 72, 2695 (1950).

21. E. C. Righellato and C. W. Davies, Trans. Faraday Soc., 26, 592 (1930).

22. F. S. Nakayama, J. Phys. Chem., 74, 2726 (1970).

23. R. A. Berner, Geochim. Cosmochim. Acta, 29, 947 (1965).

24. D. Langmuir, Geochim. Cosmochim. Acta, 32, 835 (1968).

25. V. Frai and A. Ustyanovichova, Zh. Fiz. Khim., 37, 1153 (1963).

26. G. Dietrich, General Oceanography, Interscience, New York, 1963.

27. R. A. Horne, Marine Chemistry, Wiley-Interscience, New York, 1969.

28. R. M. Garrels, M. E. Thompson and R. Siever, Amer. J. Sci., 258, 402 (1960).

29. K. E. Chave, Science, 148, 1723 (1965).

30. K. E. Chave and E. Suess, Trans N. Y. Acad. Aci., Ser II, 29, 991 (1967).

TABLE I

Logarithms of Thermodynamic Association Constants at
25°C and 1 atm

	SO_4^{2-}	HCO_3^-	CO_3^{2-}	OH^-	F^-
K^+	0.96[a]	-	-	-	-
Na^+	0.72[a]	0.16[b]	0.55[b]	-0.57[c]	-
Mg^{2+}	2.21[d,e]	1.21[f]	3.28[f]	2.58[g]	1.82[h]
Ca^{2+}	2.31[i]	1.25[f]	3.92[f]	1.27[j]	1.04[h]
Sr^{2+}	2.31*	1.25*	3.92*	0.82[k]	1.04*

a. I. L. Jenkins and C. B. Monk, J. Amer. Chem. Soc., 72, 2695 (1950).

b. F. S. Nakayama, J. Phys. Chem., 74, 2726 (1970).

c. H. S. Harned and W. J. Hamer, J. Amer. Chem. Soc., 55, 2194, 4496 (1933).

d. G. Atkinson and S. Petrucci, J. Phys. Chem., 70, 3122 (1966).

e. H. S. Dunsmore and J. C. James, J. Chem. Soc., 1951, 2925.

f. I. Greenwald, J. Biol. Chem., 141, 789 (1941). (Values corrected to μ=0.)

g. D. I. Stock and C. W. Davies, Trans Faraday Soc., 44, 856 (1948).

h. R. E. Connick and M. S. Tsao, J. Amer. Chem. Soc., 76, 5311 (1954).

i. R. P. Bell and J. H. B. George, Trans Faraday Soc., 49, 619 (1953).

J. F. G. R. Gimblett and C. B. Monk, Trans Faraday Soc., 50, 965 (1954).

k. H. S. Harned and T. R. Paxton, J. Phys. Chem., 57, 531 (1953).

* Value assumed equal to that for calcium species.

TABLE II

Single Ion Activity Coefficients at 25°C and 1 atm for $\mu = 0.67$

$K^+ \equiv Cl^-$	0.63^a	H^+, OH^-	0.75^b
Na^+	0.71^a	F^-	0.68^a
Mg^{2+}	0.29^a	HCO_3^-	0.68^c
Ca^{2+}	0.26^a	CO_3^{2-}	0.21^c
Sr^{2+}	0.25^a	SO_4^{2-}	0.22^{a*}

Singly charged ion pairs 0.68^{**}

Doubly charged ion pairs 0.21^{**}

Davies equation-monovalent 0.69 (b = .2) .75 (b = .3)

Davies equation-divalent   0.23 (b = .2) .31 (b = .3)

a. Robinson and Stokes, Electrolyte Solutions, Second Edition Revised, Butterworths, London, 1965.

b. K. Buch, Acta Acad. Aboensis, Math. et. Physica, 11, No. 5 (1938).

c. A. C. Walker, U. B. Bray and J. Johnston, J. Amer. Chem. Soc., 49, 1235 (1927).

* Mean salt value when corrected for ion association.

** By assuming behavior similar to bicarbonate and carbonate.

TABLE III

Distribution of Principal Ionic Species in 35°/oo
Seawater at 25°C and 1 atm

A. Cationic Distribution

	Total Molality	% Free Ion	% MSO$_4$	% MHCO$_3$	% MCO$_3$	% MF	% MOH
K$^+$	.01026	97.6	2.4	-	-	-	-
Na$^+$	.48558	98.3	1.6	.08	.002	-	.002
Mg^{2+}	.05516	87.3	11.9	.5	.2	.06	.02
Ca^{2+}	.01068	85.4	13.2	.5	.9	.01	.001
Sr^{2+}	.00009	86.4	12.3	.4	.9	.01	.0003

B. Anionic Distribution

	Total Molality	% Free Ion	% KA	% NaA	% MgA	% CaA	% SrA
SO$_4^{2-}$	.02926	45.9	.84	25.9	22.5	4.8	.04
HCO$_3^-$	.00187	62.9	-	20.5	14.0	2.6	.02
CO$_3^{2-}$	.00026	9.2	-	3.4	49.6	37.5	.3
F$^-$	.00007	51.8	-	-	46.8	1.3	.01
OH$^-$	1.23x10^{-5}	14.6	-	.6	83.9	0.9	.002
Cl$^-$+Br$^-$	.56666	100					

TABLE IV

Comparison of Current Results with Those from Other Sources

A. Percentage Unassociated Ions

	[Na$^+$]	[Mg^{2+}]	[Ca^{2+}]	[SO$_4^{2-}$]	[HCO$_3^-$]	[CO$_3^{2-}$]
GT (calc)	99	87	91	54	69	9
KP (calc)	97.7	89.0	88.5	39.0	70.0	9.1
TR (meas)	-	88	82±2	-	-	-
KP (meas)	97.7	88.1	86.3	-	-	-
This work (calc)	98.3	87.4	85.6	45.8	63.1	9.3

GT - R. M. Garrels and M. E. Thompson, Amer. J. Sci., <u>260</u>, 57 (1962).

KP - D. R. Kester and R. M. Pytkowicz, Limnol. Oceanogr., <u>14</u>, 686 (1969)

TR - M. E. Thompson, Science, <u>153</u>, 966 (1966); M. E. Thompson and J. W. Ross, ibid., <u>154</u>, 1643 (1966). [As corrected in KP.]

B. Total Activity Coefficients

Ion	γ_T* (meas)	γ_T (calc)
Na$^+$	0.68- 0.73	.70
Mg^{2+}	-	.25
Ca^{2+}	0.22, 0.24	.22
SO$_4^{2-}$	0.11	.10
HCO$_3^-$	0.36- 0.55	.42
CO$_3^{2-}$	0.019-0.020	.019

* F. J. Millero, "The Physical Chemistry of Multi-Component Salt Solutions", in H. R. Elden (ed.), A Treatise on Skin, Vol. 1, Wiley, Wiley/Interscience, New York, 1969.

BATTERIES FOR DEEP

SUBMERGENCE USE

Chairmen:
A. Himy
Paul Howard

HIGH CAPACITY SALT WATER BATTERIES UTILIZING FUSION-CAST LEAD AND CUPROUS CHLORIDE CATHODES

Dr. T.J. Gray, Director and Dr. J. Wojtowicz, Research Associate,
Atlantic Industrial Research Institute
Halifax, Nova Scotia, Canada.

Abstract

Medium and high capacity salt water batteries have been constructed utilizing either magnesium or aluminum alloy anodes against $PbCl_2$: Cu_2Cl_2 cathodes. Electrodes of excellent mechanical strength with cast-in current collection screens are fabricated by fusion casting. Sustained performance at 20-50 A/ft^2 at 1.0 - 0.85 volts has been obtained with better than 90% cathodic efficiency.

Fundamental properties are analyzed against composition, porosity, build-up of metal at the cathode and nature of the anode over the temperature range 0 - 25°C.

Introduction

In an attempt to design a sea-water battery utilizing an inexpensive cathodic material which could successfully compete on the overall economy - performance basis with the conventional magnesium-silver chloride batteries, a new type of lead chloride and cuprous chloride cathode has been devised and tested. The usual procedure for forming the plates has been by compressing the powdered active material together with binding and conducting agents. Performance characteristics for lead chloride cathodes have been published by Vyselkov[1] who also obtained a patent for such electrodes[2]. Coleman[3] introduced several major improvements but many problems still remained. Such electrodes are fragile both initially and during discharge and the amount of active material is reduced by presence of the conductive phase and the binder. The electrodes to be described were prepared from molten chlorides by casting into the suitable forms. By this simple method, plates of high capacity can be more easily obtained, while the performance of electrodes made solely from the active component is superior to that of electrodes fabricated by the low temperature method, where electrochemically inert ingredients must be incorporated in the mass and where certain initial porosity is unavoidable.

A large number of electrodes has been fabricated by this new technique and submitted to detailed evaluation. Separate experiments on cathodic reduction of solid chlorides, as well as the development of a simple theoretical model for the system, has helped in a better understanding of the mode of operation and of the characteristic features of this type of electrode.

The conditions for a reaction like $PbCl_2 + 2e^- = Pb + 2Cl^-$ to proceed in a continuous manner are:

i) unhindered contact between the electrolyte and the active solid phase,
ii) a path of low resistance for the electrons arriving from the external branch of the circuit to the solid phase undergoing conversion.

As the lead (or cuprous) chloride is a very poor conductor (specific conductivity at room temperature is probably as low as 10^{-8} ohm^{-1} cm^{-1}), the charge transfer reaction can occur only in the immediate vicinity of a current collector; in other words, at the ternary interface between electrolyte, solid chloride and the electronic conductor. Previously a conductor such as graphite powder has been introduced into the active mass. The current can then flow from the reaction sites through a chain of the graphite grains to the current collector proper. However, when the current collector such as copper gauze is placed at the surface of the cathode, no special means of "increasing the electrode conductivity" are necessary. Successful operation of the electrode in this configuration depends entirely on the appreciable difference between the molar volumes of the chlorides used as active materials and those of the corresponding metals. Owing to this difference, reduction of $PbCl_2$ to the metal results in a residual porosity of about 61.5%. In the case of cuprous chloride, the porosity of the copper matrix formed is even higher (74.6%). The access of the electrolyte to the still unreduced chloride is thus assured, and the process can proceed continuously from the surface of the plate inward. The current collector must be perforated (in practice a wire gauze is used) with openings large enough to ensure that the blocking effect does not become significant.

Reduction starts on the surface of the electrode at the line of contact between the chloride, the bare wires of the current collector screen and the electrolyte. The reaction spreads gradually over the whole surface while simultaneously, owing to the porous structure of the metal formed, the reaction moves into the body of the electrode. The profile of the reaction front gradually approaches a plane parallel to the electrode's surface. Since the reaction occurs at the ternary interface, the length of this boundary determines the charge transfer and lattice transformation overvoltages. Ideally, the overvoltage (except the part connected with concentration changes) becomes practically constant under galvanostatic conditions as soon as the reaction front ceases to differ significantly from the plane.

The changing configuration of the reaction profile is reflected in the working potential as measured with the reference probe placed at the surface of the electrode. At the moment of closing the circuit, a momentary jump from the initial open-circuit value to some highly negative value occurs. However, owing to the increasing length of the reaction interface, the potential immediately starts to increase. This trend lasts approximately as long as the decrease of the overvoltage predominates over the effects connected with the growth of the pores. After a period of time which depends on the current density and the structure of the current collector, the potential attains a maximum and begins to decrease at a very low rate, thus reflecting the increasing voltage drop in the electrolyte in the pores and the increasing concentration polarization. The elctrode has now entered the main phase of the discharge, which comprises the predominant part of the whole process, typically over 90%.

Finally, the rate of fall of the potential increases sharply and the discharge is soon terminated. This may reflect the complete exhaustion of the active material, or the concentration of salts (mainly NaCl) might have risen up to the saturation level at the blind ends of the pores, causing them to be effectively blocked by the precipitation of crystals.

A model for cathodic reduction of lead chloride cast electrode

The ultimate thickness and capacity of a $PbCl_2$ cathode can be estimated theoretically. The reduction of a solid block of $PbCl_2$ can proceed only as long as the pores do not become blocked by the crystallizing salt. The moment when, under the galvanostatic conditions, solution in the pores attains saturation level can be calculated on the basis of the following simplified model of the electrode.

Let us assume that reduction of a solid chloride plate results in the formation of a matrix of porous metal in which pores are uniformly distributed and equally accessible to the electrolyte, and the porosity equals the theoretical value

$$\sigma = (V_{PbCl_2} - V_{Pb})/V_{PbCl_2}$$

where V-s are molar volumes of indicated substances.

Under steady state conditions a linear distribution of concentration of chloride ions along the pores may be assumed although rigorous treatment would necessitate integration of the diffusion equation for moving boundary conditions. The simplification should not introduce any significant error in view of the comparative slow growth of the pores.

The concentration $\bar{c}$ of Cl^- at the dead end of a pore (reaction site) can now be obtained from the conservation equation:

$$A = B + C$$

where A = quantity of Cl^- (gramions) produced in the electrode process during the time dt

B = quantity of Cl^- present in the volume of pores created during the time dt

C = quantity of Cl^- leaving the pores and diffusing into the inter-electrode space during the time dt.

We thus have:

$$(I/F)dt = \bar{c}\,dv + \bar{D}\frac{\bar{c} - c_0}{\beta L}s_0 dt \quad \text{or} \quad I/F = \bar{c}(dv/dt) + \bar{D}\frac{\bar{c} - c_0}{\beta L}s_0$$

where c_0 is the concentration in the bulk, v the pore volume, $\bar{D}$ the effective diffusion constant, L the thickness of the porous layer, s_0 the cross-sectional area of pores, β a tortuosity factor, and other terms have their usual significance.

Introducing the porous layer thickness $L = VJt/zF$, with J denoting the apparent current density and $z = 2$ (for $PbCl_2$) and expressing the rate of pore volume increase by $dv/dt = s_0 \beta dL/dt = \beta s_0 VJ/zF$, noting that $s_0 = \sigma s/\beta$ and $J = I/s$, with s as apparent area of the electrode surface, the following relation is obtained:

$$\frac{\bar{c}}{c_0} = \frac{\frac{\beta^2 VJ^2}{z\bar{D}F^2 \sigma c_0}t + 1}{\frac{\beta^2 V^2 J^2}{z^2 \bar{D} F^2}t + 1}$$

Taking for the bulk concentration (sea water) $c_0 = 0.5 \times 10^{-3}$ gion.cm^{-3} and introducing numerical values of constants (F = 26.8 Ah.$equiv^{-1}$, V = 47.54 cm^3.mol^{-1}, σ = 0.616, β = 1[*], $\bar{D}$ = 0.2cm^2.h^{-1}[**]), the final formula obtains

---

[*] Tortuosity factor is taken here at its limiting value $\beta = 1$. Although its true value is not known it may be noted that on the photographs of the discharged electrodes cross-sections (Fig. 1) the metal occupies approximately the same area as the voids. With the porosity of 0.6 this would indicate that β does not differ appreciably from 1.

[**] Assuming that the electrolyte consists only of NaCl solution, the effective diffusion constant can be expressed as $\bar{D} = D/(i - t_-)$, where t_- denotes the transference number of Cl^- ions, which for concentrated solutions equals about 2/3. Under these conditions there is $D = 1.8.10^{-5} cm^2.s^{-1}$, which gives $\bar{D}$ = 0.2 $cm^2.h^{-1}$.

$$\bar{c} = \frac{537\, J^2 t + 1}{4\, J^2 t + 1} \cdot 0.5 \cdot 10^{-3}$$

From it the time at which the solution becomes saturated ($\bar{c} = 5.5 \cdot 10^{-3}$) can be readily calculated:

$$t^* = \frac{0.02}{J^2} \quad \text{hours}$$

The corresponding specific capacity is

$$Q^* = J t^* = \frac{0.02}{J} \quad \text{Ah.cm}^{-2}$$

and the limiting electrode's thickness

$$L^* = V J t^* / zF = \frac{0.018}{J} \quad \text{cm}.$$

The working potential of a porous cathode differs from the initial open circuit potential[*] by the voltage drop along the pores, the concentration polarization and the overvoltage. The first two terms may be calculated as follows:

Concentration polarization E_c is given by

$$E_c = \frac{RT}{F} \ln \frac{\bar{c}}{c_o} \quad \text{and substituting for } \bar{c}$$

$$E_c = 0.06 \log \frac{537 J^2 t + 1}{4 J^2 t + 1} \quad \text{(at room temperature)}.$$

To calculate the voltage drop in the pores a relation between conductivity and concentration must be known. An analytical solution of the problem is only feasible when a linear dependence is assumed: $\varkappa = \lambda \cdot c$. The value of the average molar conductivity is taken as $\lambda = 72 \text{ ohm}^{-1} \text{mol}^{-1} \text{cm}^2$. This results from the linearization of the experimental data, as seen in Fig. 2. Because of the assumed linear distribution, concentration at a distance x from the outlet of the pore is given by

$$c = c_o + (\bar{c} - c_o)\, x / \beta L.$$

---

[*] Electrodes which have copper as the only metal phase in contact with $PbCl_2$ and the electrolyte, exhibit an open circuit potential different (and not too well defined) from that which is characteristic for $Pb-PbCl_2/Cl^-$ system. However if even a very small amount of $PbCl_2$ has been reduced to lead, and the circuit has again been opened, the potential soon attains the value which is close to the appropriate equilibrium potential. This value has been used as "the open circuit potential".

The required voltage drop E_R can now be obtained as follows:

$$dE_R = \frac{1}{\varkappa} \cdot \frac{I}{s_0} \, dx = \frac{\beta J}{\sigma \varkappa} \, dx = \frac{\beta^2 L J \, dx}{\sigma \lambda (\bar{c} - c_0) x + \beta L c_0}$$

$$E_R = \int_0^{\beta L} dE_R = \frac{\beta^2 L J}{\sigma \lambda (\bar{c} - c_0)} \ln (\bar{c}/c_0)$$

Substituting for L and $\bar{c}$ the final relation results:

$$E_R = 0.172 \, (4J^2 t + 1) \log \frac{537 J^2 t + 1}{4 J^2 t + 1}$$

Reduction of lead chloride: the main period

Galvanostatic reduction of $PbCl_2$ has been investigated independently from other processes in the cell. It was desired to estimate the ultimate practical specific capacity of $PbCl_2$ cathodes at different current densities to establish the trend of potential change and to determine to what extent the model described can predict their actual performance.

Electrodes of high specific capacity in the shape of cylindrical rods were made by casting into glass tubes (dia. 4 mm). A copper wire (dia. 0.25 mm) passing through the tube served as the current collector. The frontal face of a rod, which constituted its active surface, was ground on fine emery (400), while the rest of the rod's surface was insulated with transparent metacrylate lacquer.

The reference electrode (calomel, saturated) Luggin capillary probe was attached to be flush with the active surface. Cathodes were placed in 1 liter beakers filled with artificial sea water, at a distance of ca. 6 cm from AZ31 anodes. Constant current was supplied either by a battery of 90V with a suitable resistor in series, or by a transistorised power source. The potential was monitored by means of a Keithley Electrometer, whose output was fed to a recorder.

Tests were conducted at room temperature and usually lasted until the evolution of hydrogen and precipitation of $Mg(OH)_2$ at the cathode's surface was observed.

The results for $PbCl_2$ cathodes at different current densities are presented in Fig. 3 in the form of curves $\Delta E = f(t)$, where ΔE is the difference between the actual working potential and its open circuit value.

Because of the shortness of the initial reaction interface (ca. 0.625 cm. per cm^2 apparent area) the first phase of the discharge is prolonged and not fully reproducible. For the predominant part the curves may be compared with the results obtained on the basis of a simplified model and for this purpose the experimental curves for 20 $mA.cm^{-2}$ and 40 $mA.cm^{-2}$ have been redrawn in Fig. 4, together with the corresponding "theoretical" curves which represent the dependence of $(E_c + E_R)$ on time. It is seen that while during the first half of the process the two curves lay close to each other (in spite of the fact that the calculated values do not include overvoltage), during the later stages of the discharge the potential decreases much more rapidly than predicted by the model. On the other hand there exists good agreement between the observed and calculated values for electrode life-times; the time when an abrupt decrease of potential is observed agrees well with the time at which, according to the model, the precipitation of salt should occur.

	Current density, $mA.cm^{-2}$		
	20	40	80
Calculated	50.6h	12.5h	3.2h
Observed	50.0h	15.2h	3.8h

The apparent inconsistency of the model may be attributed primarily to the non-uniform distribution of pores in respect to their diameter. Some of them will become blocked much earlier than others, causing the voltage drop in the electrolyte as well as overvoltage to rise. The final breakdown may still occur at the predicted time. It is also reasonable to expect that Cl^- ion concentration influences the kinetics of transformation $PbCl_2$ (solid)→Pb (solid) in a direct manner and that the potential depends more strongly on Cl^- concentration than given by Nernst's equation. Finally, the over-simplified conductivity-concentration relation (see Fig. 2) adopted in the model must lead to too large voltage drops in the early stages of the discharge while being partly responsible for the deviation in the opposite direction which is observed when the concentration becomes high.

Incubation period

The cathodic reduction of solid $PbCl_2$ can initially occur only at the interface between the current collector, the chloride and the electrolyte. This is clearly illustrated by the photographs (Figs. 5 a,b,c,d) of an electrode's surface at different times from the start of the discharge. The cylindrical electrode was of 8 mm diameter with a conductor of 1.7 mm diameter and operated at a constant current of 2.5 mA, i.e. a current density of 5.2 mA/cm^2, current to initial interface ratio 4.7 mA/cm. The potential time-curve during the first 10 minutes is shown in Fig. 6. The rate of surface reduction decreases rapidly with time, while the interface is being developed in the body of the electrode.

As large as possible a ratio of the interface length to the current is a decisive factor for the fast rise of the potential to its least negative value, as illustrated in Fig. 7, where the potential-time curves have been recorded for PbCl$_2$ electrodes with current collector screens of different mesh. Cathodes were dipped into electrolyte, thus closing the circuit and initiating current flow. While for special applications it may be beneficial to use screens of 120 or even higher mesh, the 15 x 18 mesh regularly used for fabricating most of the electrodes is quite satisfactory.

Cathodes containing cuprous chloride

Since the equilibrium potentials of Pb-PbCl$_2$/Cl$^-$ and Cu-CuCl/Cl$^-$ differ by about 400 mV while the voltage drop in the pores and concentration polarization should be similar in both cases, cells with cuprous chloride cathodes develop higher voltage and in spite of the higher cost of CuCl may appear preferable in certain applications. The main drawback is susceptibility to oxidation in a humid atmosphere, although this is much less pronounced in the case of cast plates than in the case of powdered chloride. Electrodes with only part of PbCl$_2$ substituted by CuCl may in practice prove the most economical and consequently they deserve study.

The system PbCl$_2$-CuCl forms a eutectic (0.5 mol fraction CuCl : Mpt 285°C). Reduction of a mixed electrode may be considered to proceed as follows: On the surface of a galvanostatically polarised electrode only one chloride, CuCl, will undergo the reduction. Only under the conditions of extreme irreversibility at very high current densities could PbCl$_2$ be simultaneously reduced right from the start. While the selective reduction of CuCl proceeds, the voltage drop in the developing pores increases, as does the concentration polarization. As soon as the irreversible part of the actual working potential exceeds the difference between the equilibrium potentials of the two processes, the reduction of PbCl$_2$ starts. From this moment on, three separate zones will appear in the electrode. Starting from the electrode's surface, zone 1 constitutes a porous matrix of copper and lead, without any chlorides left. Zone 2, of lower porosity, contains copper crystals embedded in the still unconverted lead chloride. Finally, zone 3, nonporous, is simply a mixture of crystals of the two chlorides.

The discharge curve of a PbCl$_2$-CuCl electrode of eutectic composition is shown in Fig. 8. A rather sharp slope during the first 4 hours corresponds to the reduction of CuCl, when current density in the pores is high. The photograph (Fig. 9) shows a view of the electrode after 27 hours of reduction at a current of 24mA.cm^{-2}. The three zones are clearly visible.

The effect of varying proportions of CuCl in the cathodes on the overall performance of cells with AZ31 magnesium anodes can be assessed from the graph in Fig. 10 which shows specific energy yields (energy delivered per unit mass of the cathode, K, or of the cell, K'). Cells of the construction described below were discharged at 25°C with 20mA.cm^{-2} until the voltage dropped down to 700mV.

Construction and performance of cells

Cathodic plates 2.5" x 3", designed for use in cells of nominal capacity 15Ah, were fabricated by melting the appropriate chloride composition in a porcelain or graphite crucible in a muffle furnace and casting at the temperature ca. 150°C above melting point into demountable steel moulds. Owing to the considerable contraction (4%) of the chloride during solidification, no difficulties were encountered in removing the plates from the forms.

The current collecting screens of copper gauze were initially placed in the steel form on either side of the cavity with a corrugated gauze between and the molten chloride poured into the mould. While successful on a long-term basis, these electrodes developed maximum potential slowly. The preferred technique subsequently adopted was to retain the corrugated gauze which imparts considerable mechanical strength to the electrode and then to apply the surface gauzes by direct fusion into the surface in a separate step. The screens are then bonded together and secured to the main lead of heavy wire.

Cells were constructed with a view to securing the best possible conditions for convection currents, so that removal of products of anodic corrosion and equilibration of electrolyte concentration would be facilitated. The cathode was sandwiched between two anodes with the separation (1/8") being maintained by small pieces of plexiglass. No supporting frames were used and the inter-electrode space was open on all sides. The assembled plates were secured by means of rubber bands. Cells were submerged in the artificial sea water (ASTM D114152) in one-liter beakers in thermostats at 25° and 0°C. Discharge was performed using a transistor constant current loading circuit with continuous record of the cell voltage and individual electrode potentials when appropriate.

A summary of results of constant current discharge at 25°C with cathodes of different composition is presented in Table 1 and Fig. 11. In all cases AZ31 anodes were used. Similar time performance data of different anodic materials have been determined for cells having the same cathodes (80% PbCl$_2$ + 20% CuCl) but different anodes, including magnesium - AP65 and aluminum - A6, at room temperature and also at $0°$ (Fig. 12). The cathodes were fabricated by the original technique, hence the comparatively long pick-up times. In each case, discharge time was limited by the capacity of the cathode.

Conclusion

It has been clearly established that cast cathodes of lead chloride or mixtures thereof constitute a definite improvement over the conventional electrodes produced by the pressed powder technique. The method is technologically simple and no serious problems should be encountered in fabrication on a commercial scale.

It has also been established, theoretically and experimentally, that a definite limit to the specific capacity of a cathode exists, which can be calculated approximately by means of the formula $Q^* = 0.02/J$ where the apparent current density J is in $A.cm^{-2}$ and specific capacity Q^* in $Ah.cm^{-2}$. Because of the excessive voltage losses in the case of very thick electrodes, a practical limit in most applications will be probably nearer to half of this value. However, in some cases the capacity of a cell will be limited by other factors (such as extensive accumulation of corrosion products) rather than by the cathode itself.

Cast cathodes containing cuprous chloride perform considerably better than lead chloride cathodes (up to about twice as high specific energy yield, K'). However, they withstand oxidation during exposed storage only if the humidity is sufficiently low. During 4 months open storage in the laboratory during winter no signs of deterioration were observed, while cathodes fabricated when humidity was over 80% rapidly suffered visible oxidation.

ACKNOWLEDGMENT

This research program is being performed under the sponsorship of the Defence Research Board, Ottawa, whose continuing interest and support are most gratefully acknowledged.

REFERENCES

1. Vyselkov A.A., I.T. Kopyev & V.A. Naumenko, Elektro Tekh 12 50, 1969
2. Vyselkov A.A. & F.I. Rogova, USSR Patent #109,345 of Feb. 1958.
3. Coleman J.R., J. App. Electrochem 1 65, 1971.

Composition [% wt.]		End Voltage [mV]								
		1000		900		800		700		600	
PbCl$_2$	CuCl	η [%]	K¹ Wh/Kg	η [%]	K¹ Wh/Kg	η [%]	K¹ Wh/Kg	η [%]	K¹ Wh/Kg	η [%]	K¹ Wh/Kg
100	0	-	-	-	-	22.7	57.5	84	99	88	100
80	20	-	-	7.6	26.7	76.3	116	89.6	118	90	118
60	40	36.3	105	43.4	108.5	84.3	134.5	95	139	-	-
40	60	60	145	64.3	160	86	168	91	167	96.6	167
20	80	78.4	176	82.4	177	92	180	94	174	-	-
0	100	91	179	95.7	181	96.5	182	97.5	182	-	-

Table I. Degree of utilization of cathodic material (η) and specific energy yield (K¹)

Fig. 1
Cross-section of a discharged lead chloride cathode (polarized light, x 750).

Fig. 2

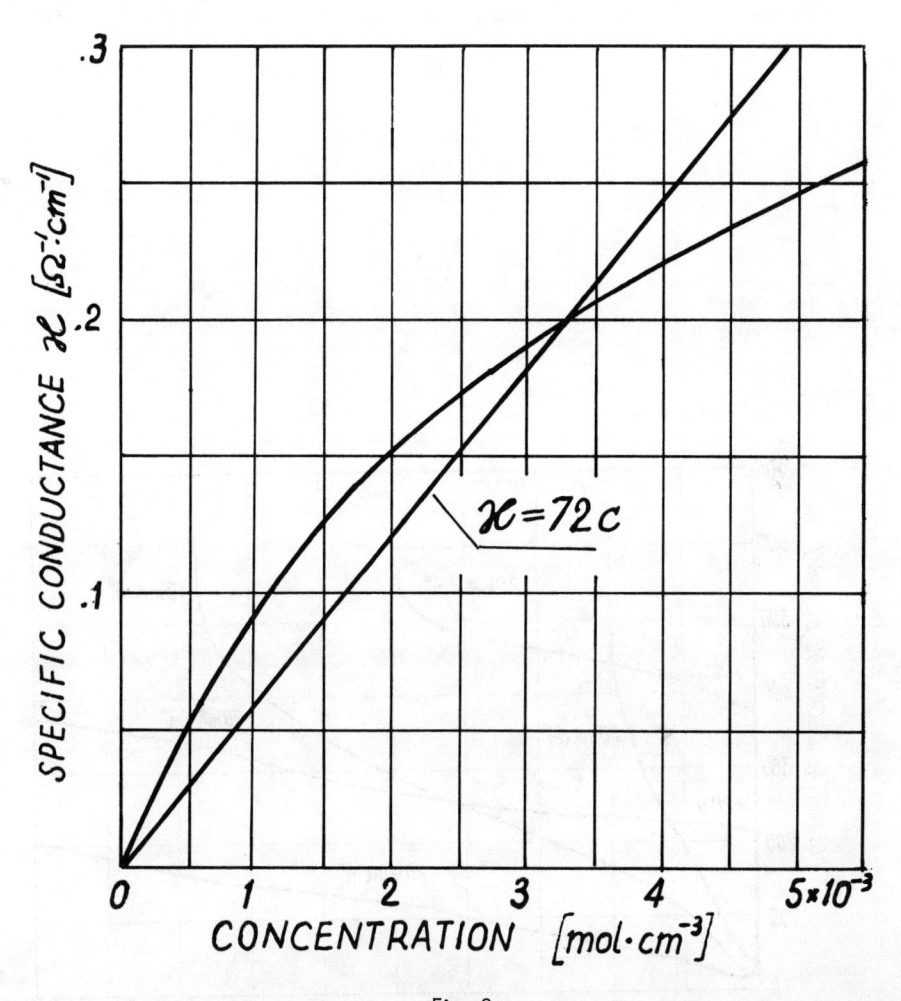

Specific conductivity of sodium chloride solutions at 25°C.

Fig. 3

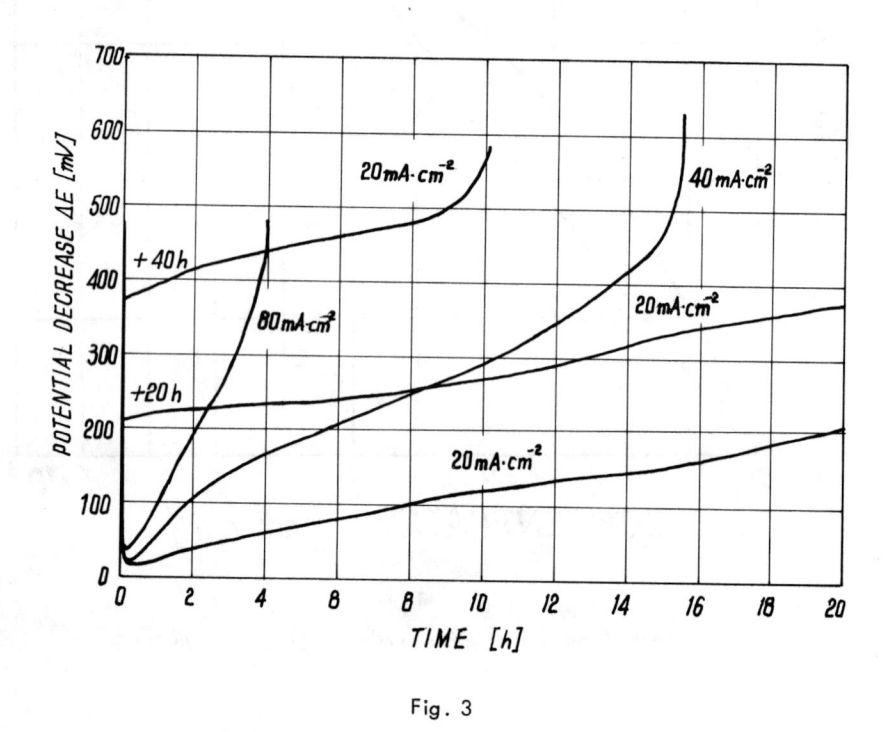

Galvanostatic reduction of lead chloride at different current densities (25°).

Fig. 4

Galvanostatic reduction of lead chloride:

———— ΔE experimental
------ $(E_R + E_c)$ theoretical

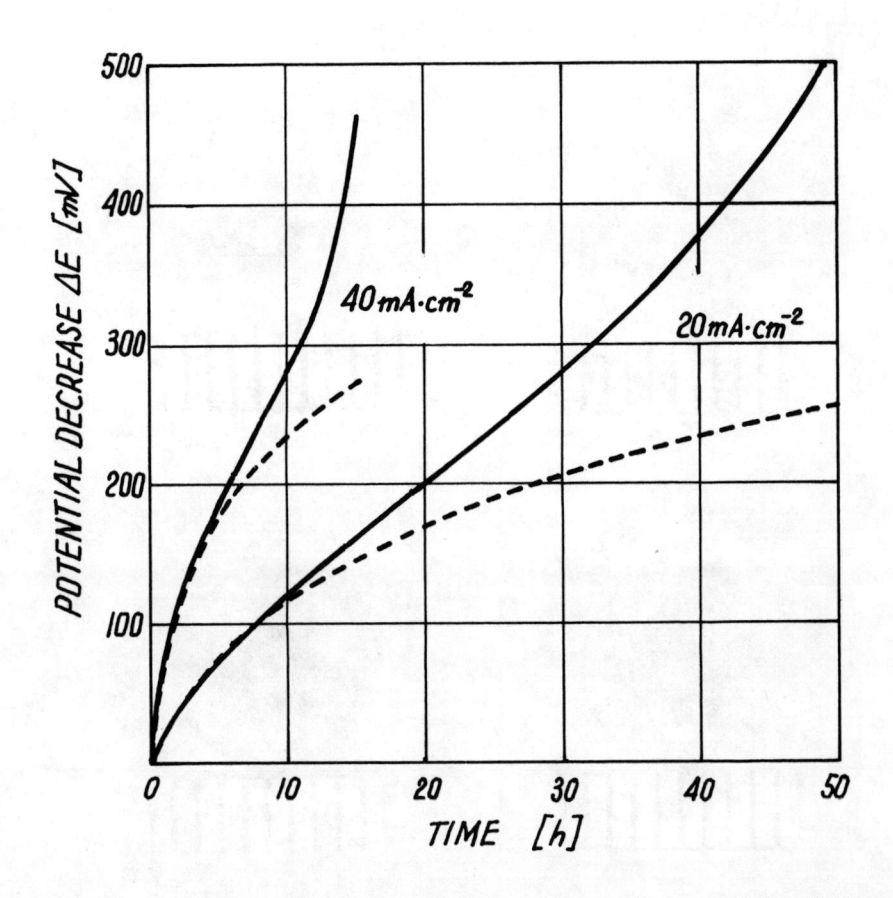

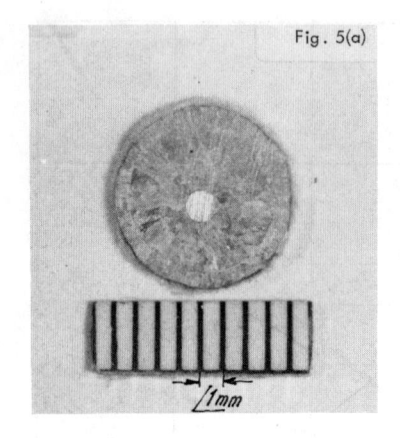

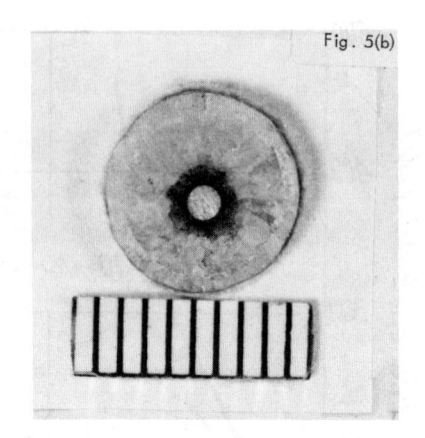

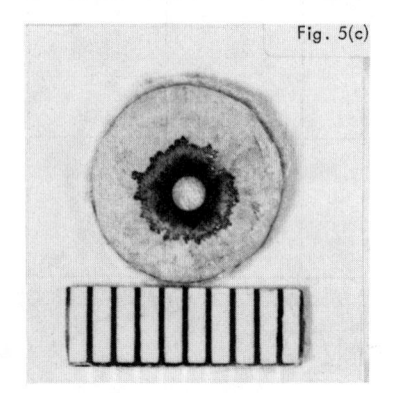

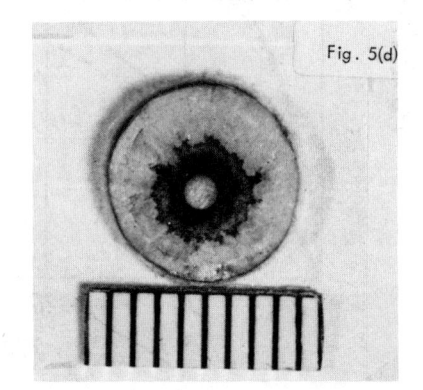

Surface of a lead chloride electrode at different periods of discharge with 5.2 mA.cm^{-2}.
(a) 0 min. (b) 10 min. (c) 30 min. (d) 110 min.

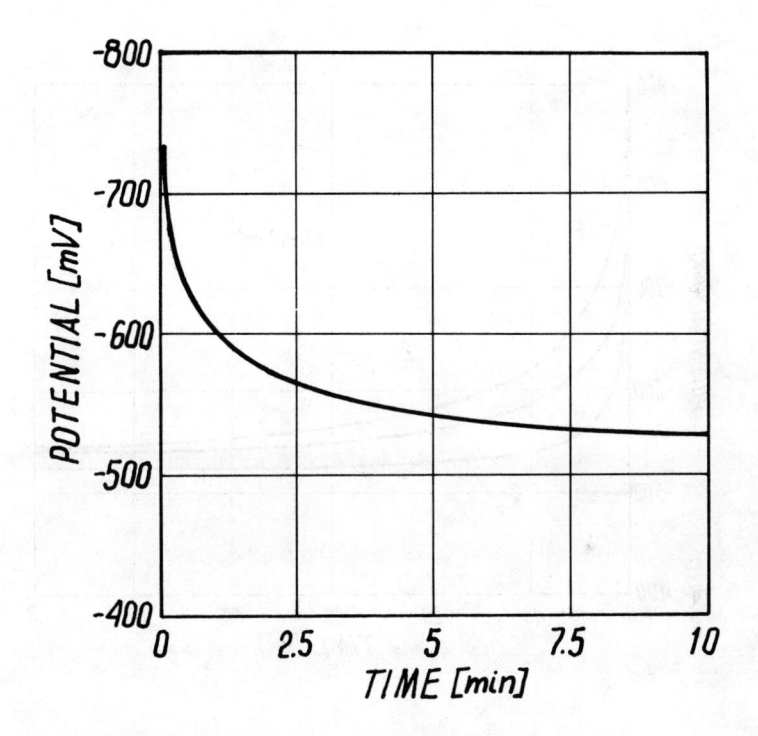

Fig. 6

Discharge curve of the electrode shown in Fig. 5.

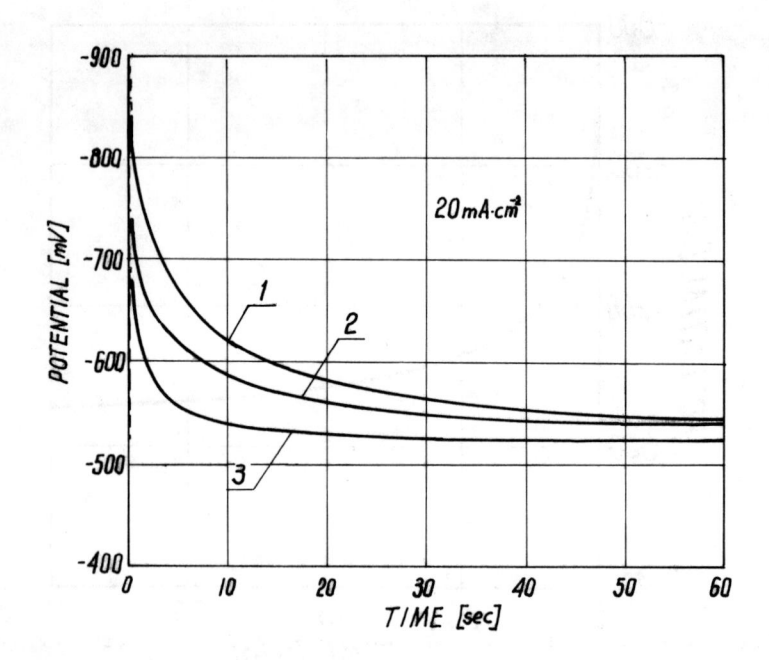

Fig. 7

First phase of reduction: $PbCl_2$ plates with screens of different mesh.

Fig. 8

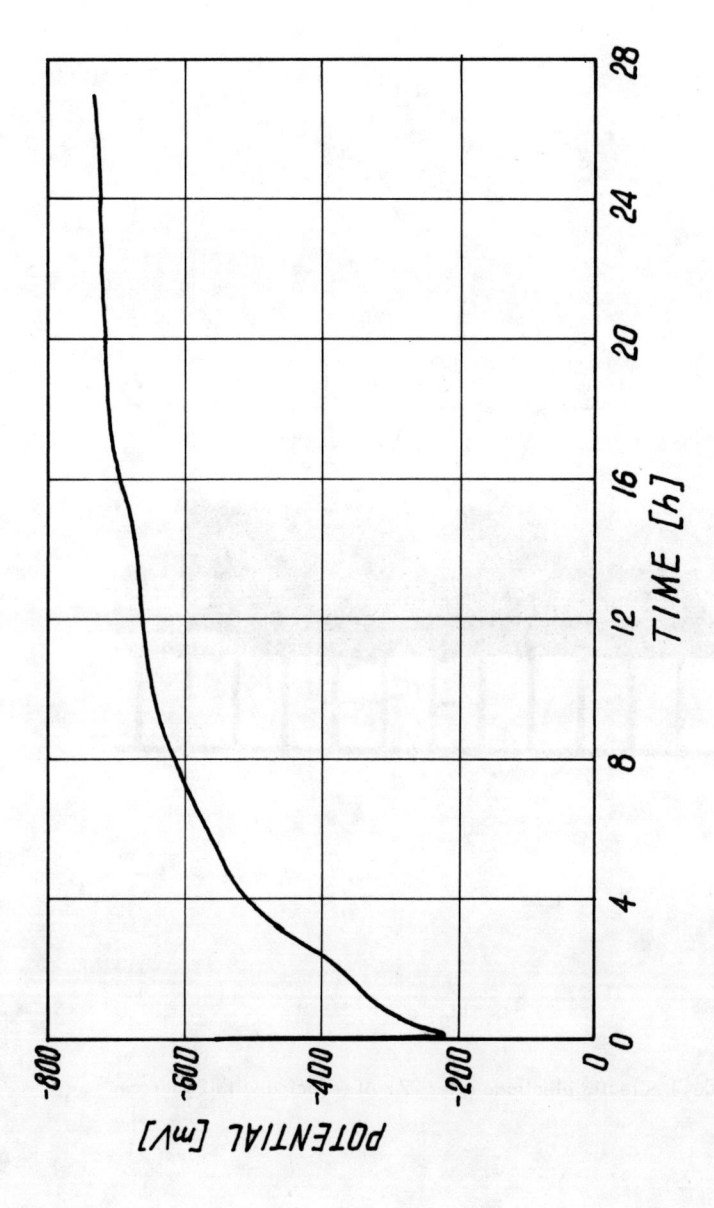

Galvanostatic reduction of $PbCl_2$–$CuCl$ eutectic (24 mA.cm^{-2}).

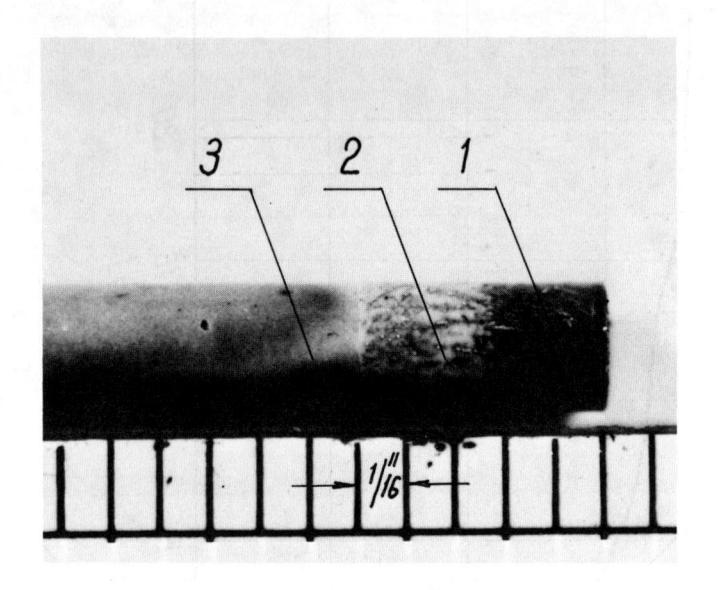

Fig. 9

PbCl$_2$-CuCl eutectic electrode after 27 h of reducing with 24 mA.cm^{-2}.

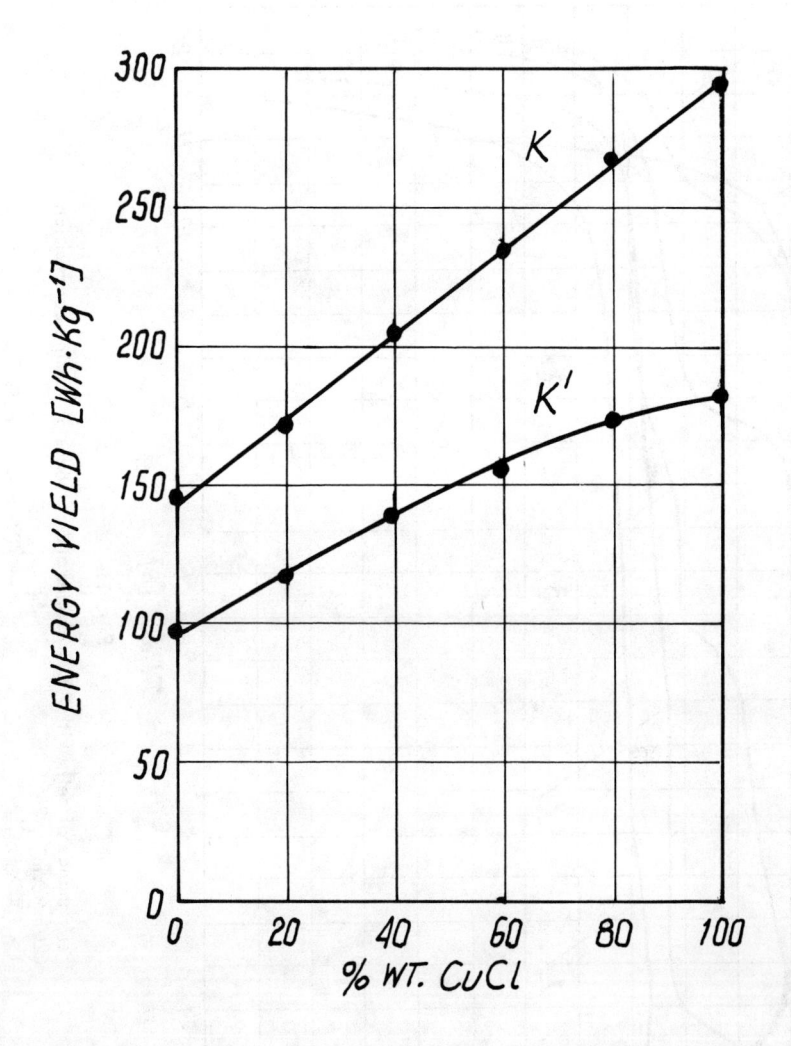

Fig. 10

Influence on cell's energy yield of the proportion of CuCl in the cathodes.

Fig. 11

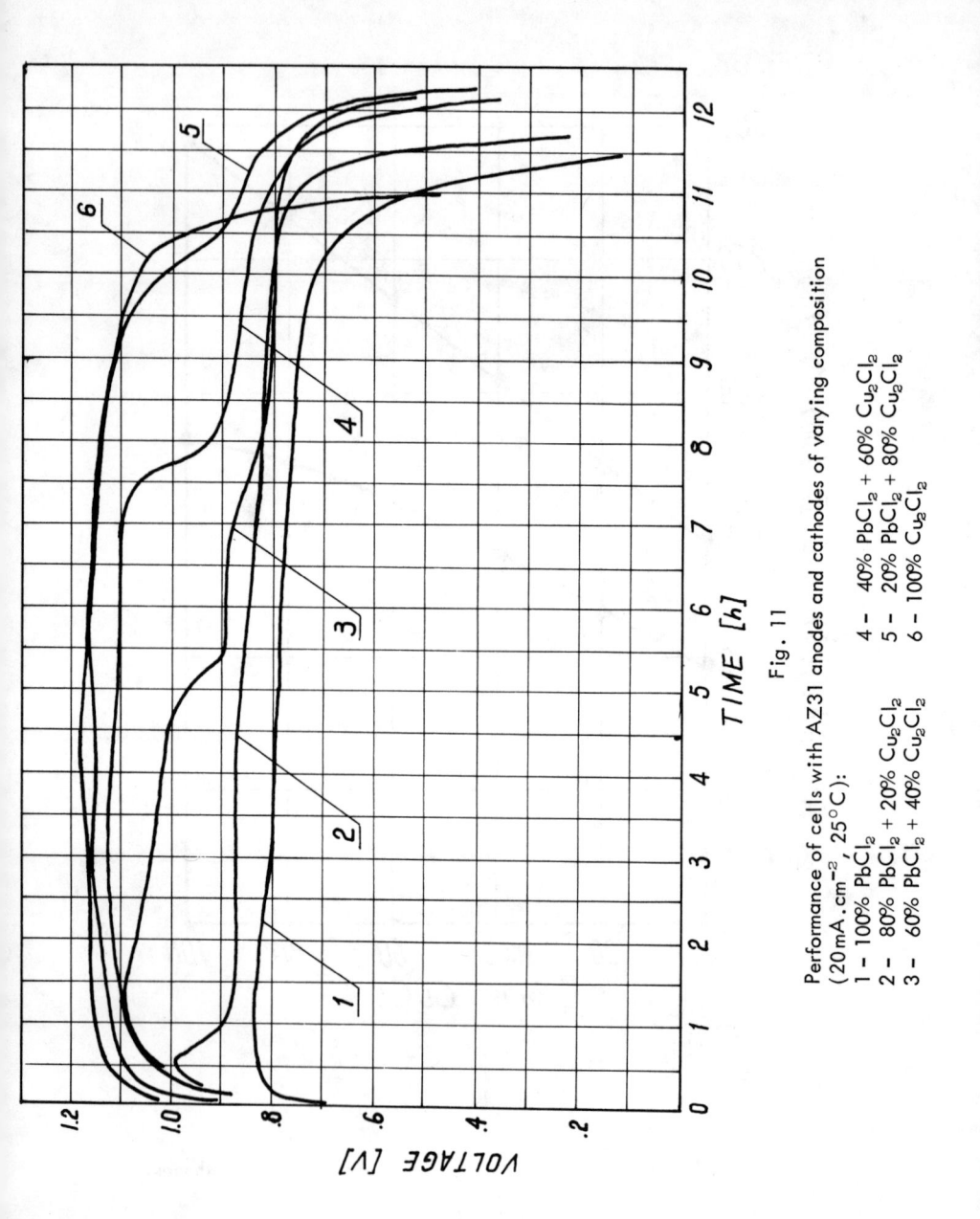

Performance of cells with AZ31 anodes and cathodes of varying composition
(20 mA.cm^{-2}, 25°C):
1 – 100% PbCl$_2$
2 – 80% PbCl$_2$ + 20% Cu$_2$Cl$_2$
3 – 60% PbCl$_2$ + 40% Cu$_2$Cl$_2$
4 – 40% PbCl$_2$ + 60% Cu$_2$Cl$_2$
5 – 20% PbCl$_2$ + 80% Cu$_2$Cl$_2$
6 – 100% Cu$_2$Cl$_2$

Fig. 12(a)

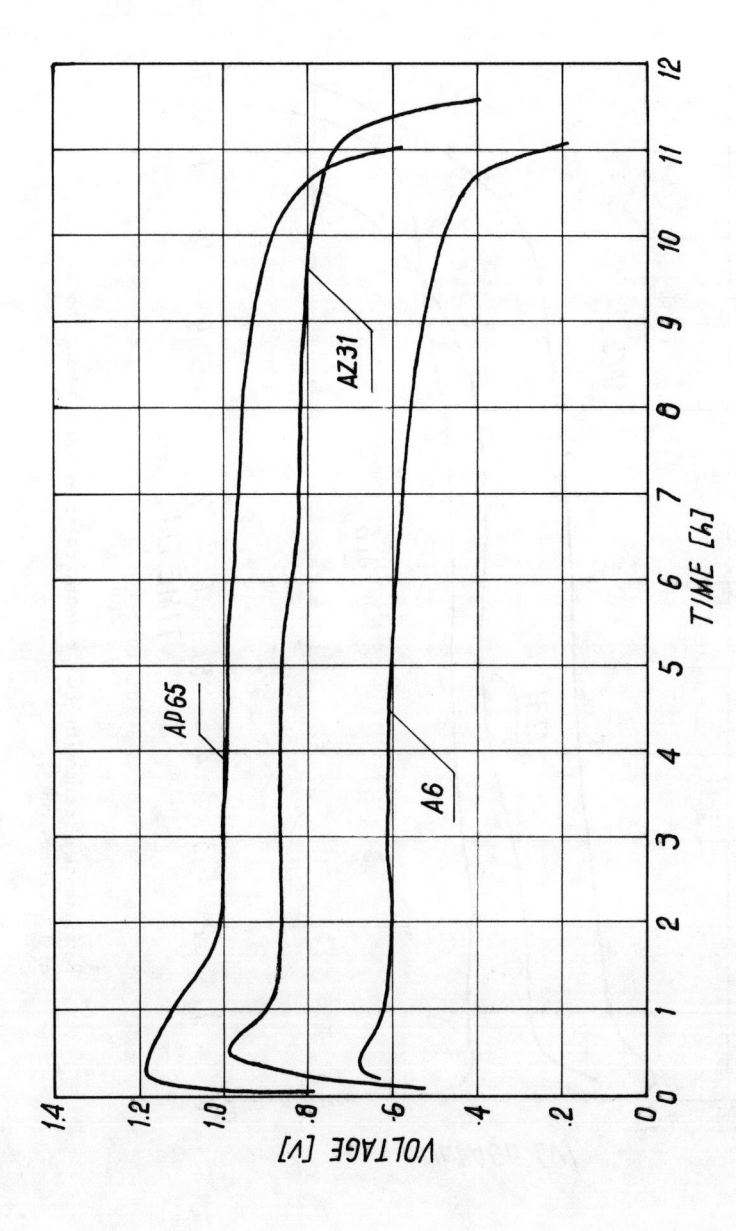

Performance of cells with $PbCl_2$–CuCl eutectic cathodes and different anodes at room temperature (20 mA.cm^{-2}).

Fig. 12(b)

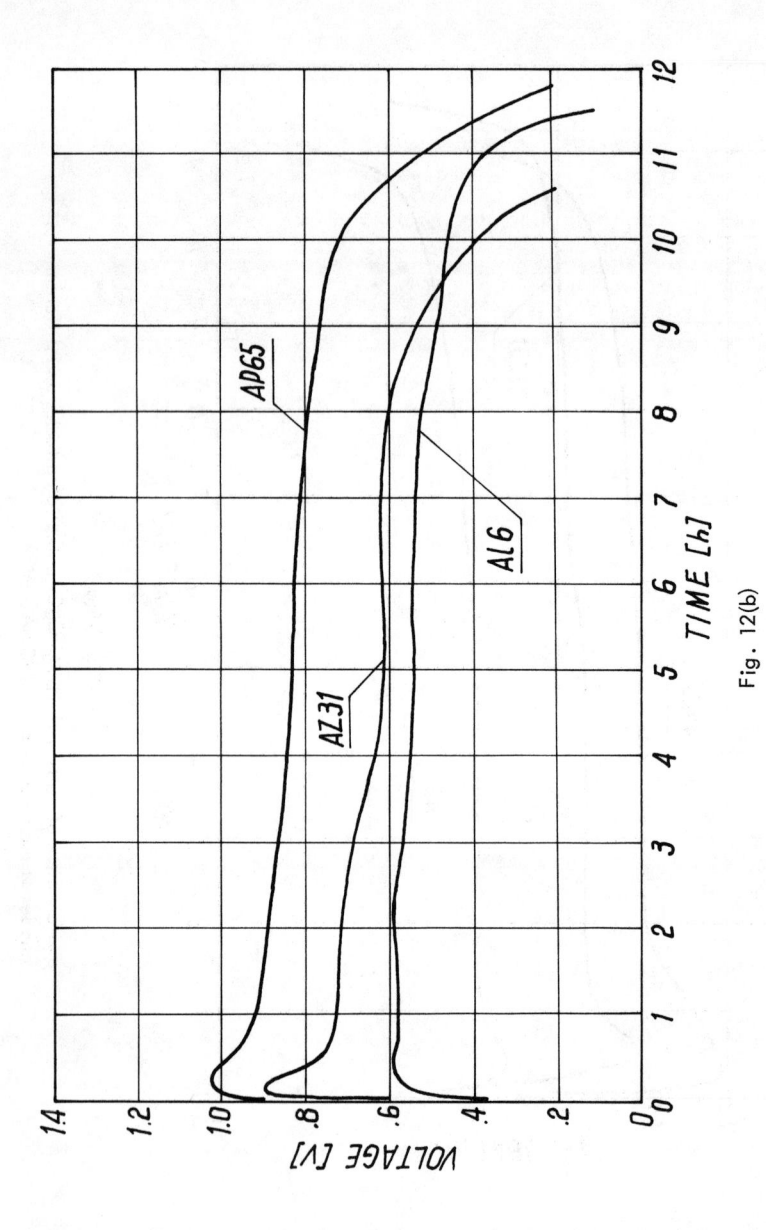

Performance of cells with $PbCl_2$ eutectic cathodes and different anodes at 0°C (20 mA.cm^{-2}).

DISSCUSSION

M. Eisenberg, Electrochimica Corporation, 2485 Charleston Road, Mount View, California 94040. Question: You used a mathematical formulation which is based on a one dimensional model of orthogonal [to electrolyte face] cylindrical pores with a further assumption of unchanging pore geometry. Obviously, none of these assumptions apply to the lead chloride electrode you used. Totuosity is considerable and pores constantly grow irregularly in all directions. Why then use this mathematical treatment at all for the case in point? Wouldn't the presentation of experimental data be sufficient by itself?

Dr. T.J. Gray, Atlantic Industrial Research Institute, Halifax, Nova Scotia, Canada. Answer: Because of the symmetry of the system with the field perpendicular to the plate there is not point in treating the model otherwise than as a one-dimensional continuum with "effective" rather than true transport coefficient. In spite of the face that we do not have at present any independently obtained (e.g. by means of permeability measurements) data concerning the correction factor β, it seems worthwhile to apply a general formula for $\bar{c}$ to the particular case under investigation. The only sensible step is to consider the limiting case of $\beta = 1$, especially in view of what has been said in note (*). Even if a more realistic value of β were available, a considerable degree of uncertainty would remain as far as $\bar{D}$ is concerned (D and t_- are concentration dependent, the local temperature in the pores is not known and it certainly is not constant). An accurate value for the effective transport coefficient or, what comes to the same thing, for the constant in the equation $J^2 t_* =$ const. cannot unambiguously be established. However, the relationship itself and the information on the significance of the factors on which the constant depends are instructive, and have already proved useful in the design of batteries of this type for operation during extended periods of time under power-programmed conditions.

There is considerable merit in introducing an appropriate model into a research project at the earliest possible stage of its development. Even if very crude, quite often through the exposition of its own deficiencies. The present case is no exception.

Consequently, our answer to Dr. Eisenberg's rhetorical question is negative.

CHARACTERISTICS OF LEAD ACID AND SILVER-ZINC BATTERIES IN THE DEEP OCEAN ENVIRONMENT

K. K. BERJU, E. L. DANIELS, E. M. STROHLEIN
E. S. B. INC., PHILA., PENNA. 19120

ABSTRACT

The utilization of lead acid and silver zinc batteries in oceanographic applications has led to the development of specialized pressure compensated systems. Studies of the charge/discharge characteristics, material compatibility and hardware development of pressure compensated batteries for the deep ocean environment are examined. Normal capacity reduction is encountered at low temperature, i.e. 30°F, in each couple; however, the lead acid system exhibits a beneficial effect at elevated pressures. Results obtained from charging both systems at high pressure and/ or cold temperature are discussed. A comparison of the power densities and related performance criteria of the lead and silver batteries is reviewed. A listing is presented of the many applications of both batteries for oceanographic use.

INTRODUCTION

Power requirements for propulsion, instrumentation and life support of oceanographic submersibles have grown by leaps and bounds over the past two decades. Where the power requirement was small or the vessel was large - such as a conventional submarine, the required battery power could be placed within the pressure hull. As the pressure hulls decreased in size because of cost and feasibility of diving to greater depth, space within the confines of the pressure hull become so limited that some other location for battery placement became mandatory. The placement of battery systems outside the pressure hull became feasible with the advent of the pressure compensated battery system. The very removal of the battery from the safe confines of the pressure hull, however, exposed the battery to the hostile environment of the deep ocean such as high pressures, salt water and 30°F temperatures. Therefore in addition to the development of the system itself, the evaluation of battery systems under the above conditions was dictated.

The principle of the pressure compensated battery system really differs very little from any other pressure compensated hydrodynamic system and is schematically illustrated in Figure 1 and pictorially in Figure 2. The battery of interconnected cells is contained in a tray or battery module fitted with a dome shaped cover. Each cell is filled with electrolyte. Hydraulically connected to the tray is a flexible compensating bladder. The volume of the module, including the space under the cell cover immediately above the electrolyte is flooded with the pressure transmitting fluid. It also fills the bladder and interconnecting tube. The electrolyte-oil interface is thus located under the cell cover. As the module is submerged, and is subjected to an increase in pressure, the fluid filled compensating bladder is compressed, forcing fluid into the battery module, thus causing an equal increase in the internal pressure of the module. This movement of the pressure transmitting fluid thus results in a zero or negligible pressure differential between the inside and outside of the module. On ascent, the decreasing pressure forces fluid from module to bladder. Evident in Figure 1 are also the special unique hardware items which are necessary to make the system operational.

HARDWARE DEVELOPMENT

Development of hardware components of the system centered about their compatibility with each other, the cell constituents and electrolyte, as well as the seawater that surrounds it. Of initial primary importance was the tray or module container which would structurally support the battery cells. Development activity of this structure included consideration of weight, strength, manufacturability and resistance of the external surface to seawater corrosion and internal surface to electrolyte attack. Trays have been constructed of #1010 Steel, #316L Stainless Steel, #6061 Aluminum and reinforced plastic. Each tray design excelled for the specific application, marrying minimum allowable weight and size with the economics in question. For surface protection both polyurethane and epoxy coatings were developed to be resistant to the seawater and electrolytes such as sulfuric acid and potassium hydroxide.

The requirements asked of the pressure compensating fluids were rather extensive. Of primary importance, of course, was the non-reactivity with any of the system components as well as a suitable viscosity over the broad range of temperatures and pressures encountered during operation and storage. As an example, one of the many evaluations of a pressure transmitting liquid was that of the effect of the liquid on the physical properties, such as, tensile strength of the cell container material. Figure 3 graphically depicts the results of this investigation. High impact rubber was subjected to continuous immersion in various fluids. As may be seen, the effect of Primol 207, a high grade white oil manufactured by Humble Oil and Refining Company gave significantly superior results. A repeat test program is presently being conducted to determine an alternate for this fluid, since the manufacturer has indicated that Primol 207 will no longer be available.

It was discovered very early in the system development that during cell gassing, electrolyte was carried with the gas through the oil causing considerable loss of electrolyte as well as ground voltage problems when the electrolyte settled on top of the cells and exposed electrical conducting components. This discovery led to the development of two important components of the system.

First, each cell of the system is equipped with an electrolyte deentrainment device commonly termed an electrolyte scrubber. It is mounted on top of each cell, replacing the commonly used vent cap, and provides a circuitous path for the gas - electrolyte bubble prior to its emergence from the cell. This device successfully separates the gas from the electrolyte permitting the electrolyte droplet to return into the cell and the gas to exit.

Secondly, should any electrolyte be carried past the scrubber and drop onto the cell tops, an encapsulation system now protects all exposed electrical conducting components. The encapsulation material is a specially developed polyurethane which is resistant to electrolyte, the pressure compensating fluid and salt water. Figure 4 illustrates the encapsulation of the battery intercell connectors. The specially designed encapsulation retainer is put in place prior to intercell connector placement and the encapsulating material is cast in place subsequent to assembly. Even though this is a permanent encapsulation the material can be mechanically removed should cell replacement be found necessary.

Gas evolution from the battery during operation while submerged can upset the pressure balance between inside and outside causing a pressure differential. This is most likely to occur during ascent especially if the ascent rate is too rapid which will result in a violent release of previously evolved compressed gas. To prevent this condition, a pressure relief valve is used which minimizes this pressure build-up. Reliability of this valve is naturally of the utmost importance since failure could result in a bulkhead explosion. The one-way valve operates on a 1 to 3 pound pressure differential venting the gasses which collect in the dome-shaped cover. Valve components were selected on the basis of their resistance to oil, electrolyte and salt water degradation.

Development of other hardware items such as, compensating bladders, electrical penetrators and special fittings followed the same scheme of compatibility, corrosion resistance and functional reliability.

CELL DESIGN

A lead acid cell design has evolved which includes an electrolyte-oil interface that permits pitch/roll angles to 45°. The rise and fall of electrolyte level which occurs during charge and discharge also makes the oil-electrolyte level critical since continued exposure of plates to the oil has a detrimental effect on cell performance.

With the selection of Primol 207, it was found that the conventional cell container - cover asphalt sealant was not compatible. This led to the development of an epoxy sealant which is inert to sulfuric acid, mineral oil and salt water.

The problems of adapting the silver zinc cell into a pressure compensated system differed somewhat from the lead-acid cell. Unlike the lead cell, the silver zinc cell is of the unflooded type, i.e., the electrolyte level is always below the top of the separator wrap. Under normal ambient conditions, the electrolyte level frequently drops considerably below the top of the electrodes which under pressure compensated conditions would cause the oil-electrolyte interface to drop below plate level and permit contact of the oil with the plates. Our previous investigations had determined that repeated or prolonged contact of oil with the plate surface would render the affected area inactive.

By increasing the separator envelope above the electrode top, the electrolyte variation could be restricted to this area. To determine the minimum increase required, the following contributing factors were considered:

1. Normal electrolyte variation during cycling. This value was obtained by observation of cells operating under numerous test conditions.

2. Pressure effect on compressibles. The major part of this value is due to trapped gases within the plates which was obtained by the alternate application of vacuum and pressure at various operating stages.

3. Compressibility of electrolyte and oils. These values were obtained from literature and manufacturers.

The resultant increase in separator envelope extension amounted to 25% of the plate height, a considerable increase over the approximately 5% used under normal conditions.

CELL PERFORMANCE

Performance of the lead acid battery under pressurized conditions has been extensively examined. Figure 6 illustrates the increase in discharge capability. The specific example indicates the ampere-hour capacity to have been improved 15% and the watt-hours by 16%. We have rationalized the improved discharge performance under pressure to be related to the pressure's forcing electrolyte into normally inaccessible pores of the plate - resulting in more efficient use of acid and active material. Also, the volume of entrapped gas at plate surfaces is reduced and this decreases cell internal resistance. All of these effects contribute to the general increase in discharge capability.

Conversely, the lead acid battery discharge performance at 80°F will be reduced by 10% to 30% when operating at 30°F. The example shown in Figure 6 represents an 18% reduction in ampere-hours @ the five hour rate @ 0 PSIG. The net effect of pressure and low temperature in this example is a loss of about 3% in ampere-hours.

Studies on the effect of charging the lead battery under elevated pressures indicates that a capacity decline will result. This condition is made worse by the incorporation of low temperature, i.e., 30°F. Figure 6 graphically illustrates this capacity decline with various charging regimes at room temperature. It is apparent that the slope of capacity degradation is similar for each charging procedure evaluated. The loss of capacity ranged from approximately 25% to 35%. The capacity loss attributed to charging under pressure is directly related to the lack of electrolyte mixing, normally present under ambient charging conditions but greatly reduced under pressure. This condition manifests itself as electrolyte stratification.

Another phase of the program to determine the effect of pressure was that of determining the point of optimum discharge performance. Figure 7 represents some of the data generated in this study and indicates that the major discharge effect occurs at about 1000 PSIG.

Evaluations of silver-zinc cell performance at elevated pressures did not yield increases in capacity such as were found in lead-acid cells. An explanation for this lack of increased performance is that the silver-zinc couple is inherently much more efficient than the lead-acid. The effect of cold temperature on cell performance is evident but not as great as it appears in lead-acid cells. Figure 8 depicts the effect of these environments on silver-zinc cell performance. Identical discharge curves for 0 PSIG and 10,000 PSIG are shown. Performance at 32°F is approximately 90% of 80°F performance.

It is well known that the charging of silver zinc cells differs somewhat from that of other cells. It is usual practice to charge cells at constant current until they reach a finite voltage - usually 2.10 volts. This prevents cell overcharge which can seriously limit the cycle life of the silver cell. Figure 9 shows the effect of temperature on the charge characteristics. As may be observed, the normal charge at 80°F is such that the voltage remains at the lower charge voltage during the initial 25% of the charge. The voltage then rises to the higher plateau charge voltage until the end of charge. Cold temperature (30°F) charging, however, seriously limits the charge acceptance causing the cell to reach the 2.10 cut-off voltage at approximately 60% of charge return.

Elevated pressures have not been found to have any effect on charge acceptance of silver zinc cells. Cycling regimes have been conducted which included repeated recharges at pressures up to 1000 p.s.i. which have shown no detrimental effect on the discharge capacity of subsequent cycles.

The power density of the lead acid battery varies from 10 to 15 watt hours per pound at ambient conditions. This value is reduced by approximately 20% when the lead cell is incorporated into a pressure compensated system. The silver zinc battery is inherently a much more efficient system exhibiting a power density of 35 to 45 watt hours per pound under normal conditions. This power density figure also is reduced by approximately 20% when pressure compensation is added. Figure 10 illustrates the comparative differences in tabular form of an identical installation using both silver zinc and lead acid pressure compensated power supplies. In the space that housed a 21.8 KWH lead acid power system, a 64 KWH silver-zinc power supply is possible. The weight and space advantage is obvious - approximately 4:1. The inherent life limitation of the silver-zinc couple, however, is also obvious - in this case approximately 10:1 in favor of lead acid.

APPLICATIONS

Numerous custom engineered Sea Space® Power Systems have been provided for the many oceanographic installations in existence and are providing reliable service. The table in Figure 11 is a partial list of installations of these systems - both pressure compensated and non-pressure compensated - using lead acid and silver zinc batteries in manned and unmanned vehicles, habitats and buoys.

Each installation is an example of mating the ideal electrochemical couple, module design and structure as well as materials and components to the individual application. Some of the applications include supplementary devices such as charger modules as well as individual cell voltage monitoring units. This latter device, so vital to the proper operation of the silver-zinc system, has been produced using electromechanical and static components depending on the specific application. Use of these devices had enabled the utilization of silver zinc cells in large multicell installations.

SUMMARY

The adaptation of electrochemical batteries to the ocean environment has been accomplished. During the past decade ESB has successfully designed and developed power systems for submersible applications. While much has been learned during these extensive investigations, there remain areas to be pursued to further optimize these pressure compensated systems. Charging under pressure and improved cold temperature performance are certainly two problems, the solutions to which will lead to a better source of underwater power.

As always, the demands of the future will require more and more intensive effort to satisfy the vehicle designers. Power density, performance and reliability will be in the fore of research and development effort.

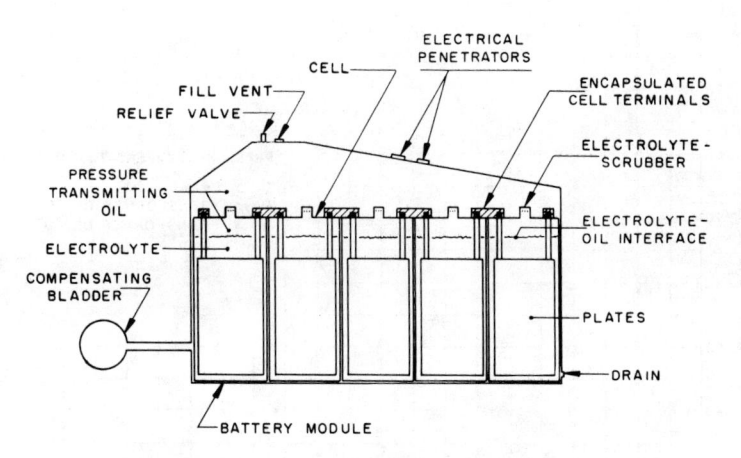

FIGURE 1
SCHEMATIC OF PRESSURE COMPENSATED BATTERY SYSTEM

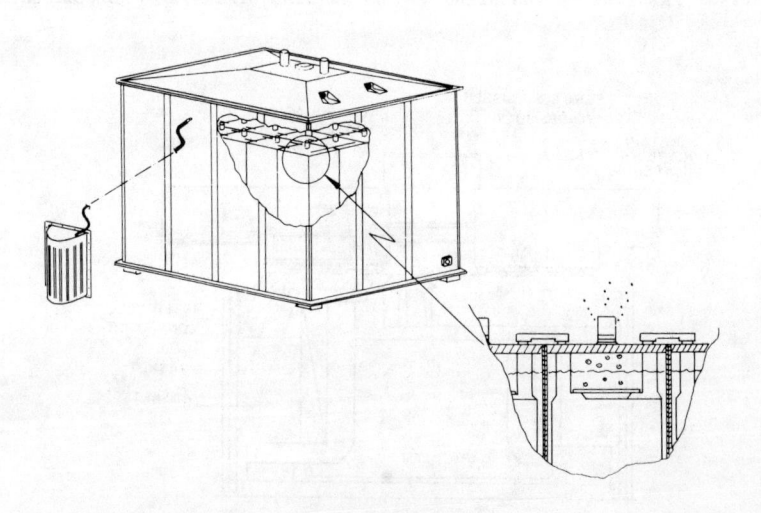

FIGURE 2
PICTORIAL VIEW OF PRESSURE COMPENSATED BATTERY SYSTEM

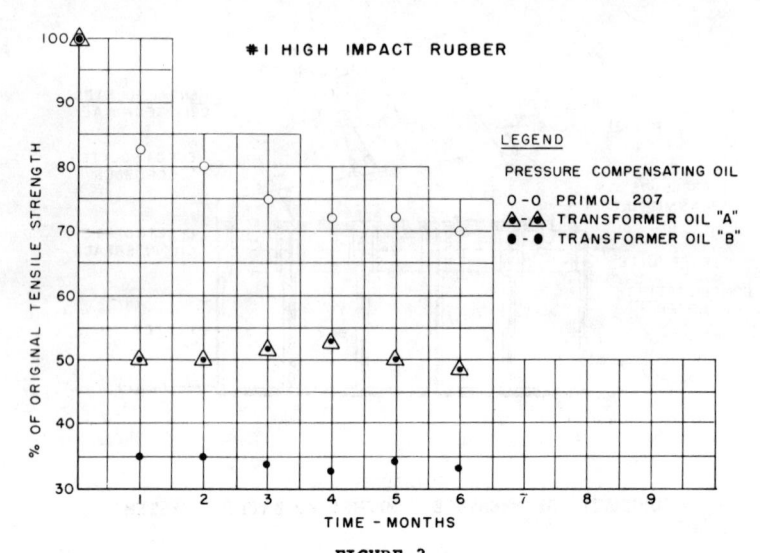

FIGURE 3
EFFECT OF PRESSURE COMPENSATING OIL ON TENSILE STRENGTH OF JAR MATERIAL

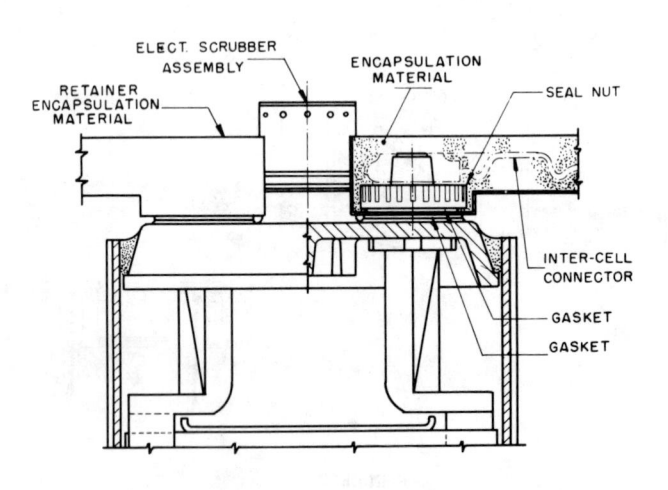

FIGURE 4
ENCAPSULATION OF INTER-CELL CONNECTORS

DISCHARGE RATE - 42.0 AMPS
FINAL VOLTS - 1.73

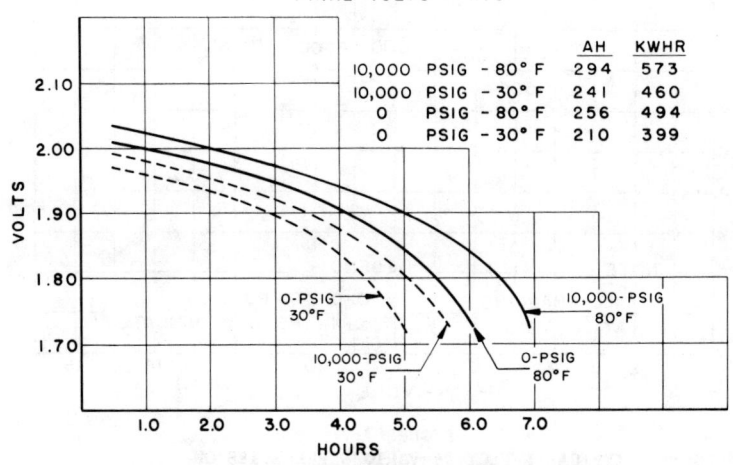

		AH	KWHR
10,000	PSIG - 80° F	294	573
10,000	PSIG - 30° F	241	460
0	PSIG - 80° F	256	494
0	PSIG - 30° F	210	399

FIGURE 5
TYPICAL EFFECT OF PRESSURE AND TEMPERATURE ON LEAD ACID DISCHARGE CHARACTERISTICS

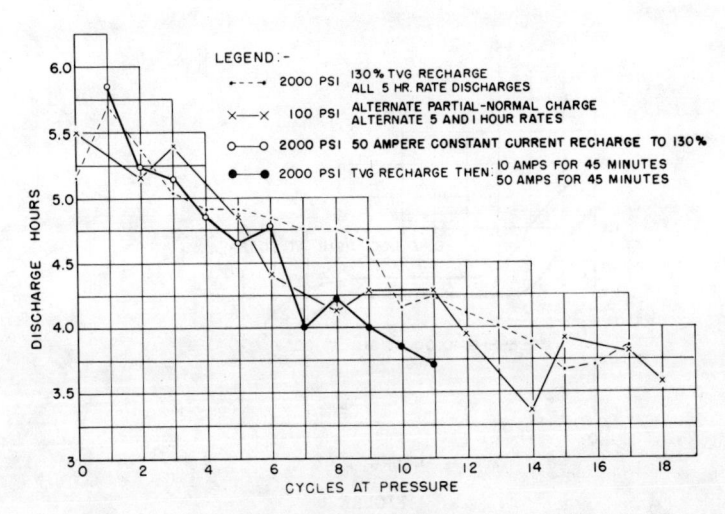

LEGEND:-
- - - - 2000 PSI 130% TVG RECHARGE ALL 5 HR. RATE DISCHARGES
×——× 100 PSI ALTERNATE PARTIAL-NORMAL CHARGE ALTERNATE 5 AND 1 HOUR RATES
○——○ 2000 PSI 50 AMPERE CONSTANT CURRENT RECHARGE TO 130%
●——● 2000 PSI TVG RECHARGE THEN: 10 AMPS FOR 45 MINUTES / 50 AMPS FOR 45 MINUTES

FIGURE 6
EFFECT OF CHARGING UNDER PRESSURE ON LEAD ACID
DISCHARGE PERFORMANCE

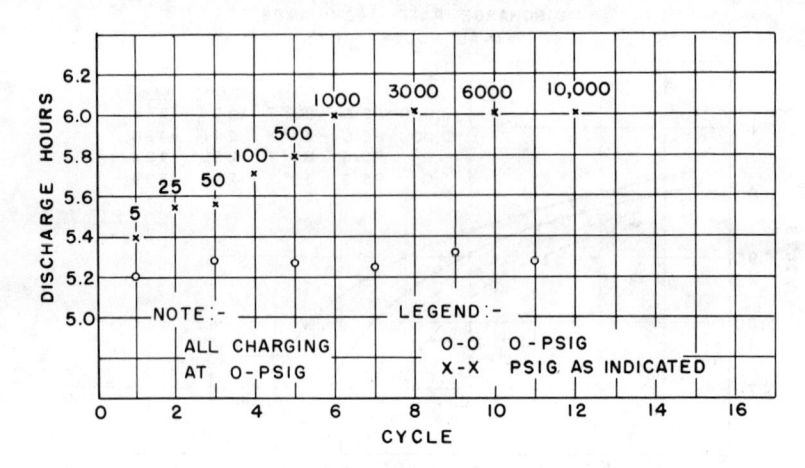

FIGURE 7
TYPICAL EFFECT OF VARIOUS PRESSURES ON
LEAD ACID DISCHARGE PERFORMANCE

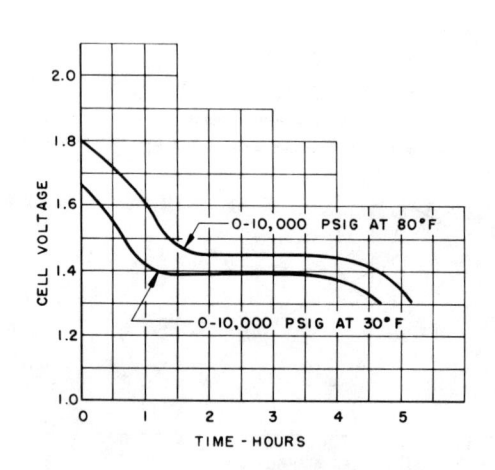

FIGURE 8
TYPICAL EFFECT OF PRESSURE AND TEMPERATURE
ON SILVER ZINC DISCHARGE PERFORMANCE

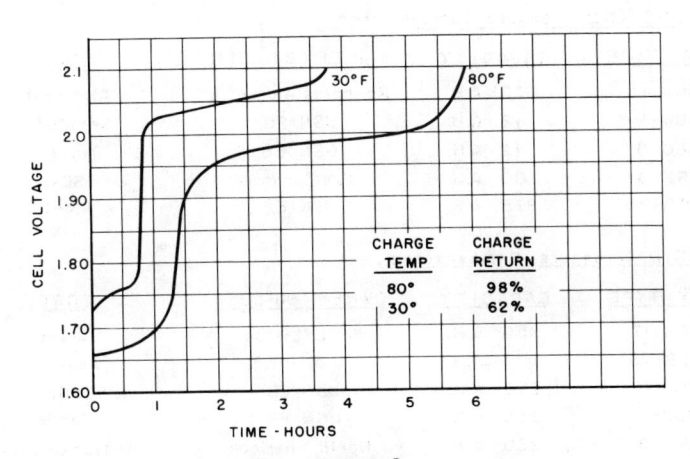

FIGURE 9
TYPICAL EFFECT OF TEMPERATURE ON
SILVER ZINC CHARGING CHARACTERISTICS

	LEAD ACID	SILVER ZINC
KILOWATT-HOURS @ 6 HOUR RATE @ 80°F	21.8	64
KILOWATT-HOURS / lb	8.38	32.0
KILOWATT-HOURS / cu. ft.	1.42	4.15
CYCLE LIFE	500-1000	50-100

FIGURE 10
COMPARISON OF A TYPICAL PRESSURE COMPENSATED
LEAD ACID vs SILVER ZINC SYSTEM

SILVER ZINC (Not Pressure Compensated)

BATTERY TYPE	CAPACITY	OWNER/BUILDER	BOAT
308 AGA-11	650 A.H.	Reynolds Alum/EB	Aluminaut
12 AGC-9	12 A.H.	USN/EB	Sea Cliff
12 AGC-9	12 A.H.	USN/EB	Turtle
AGE-31	100 A.H.	USN/Convair	SDV
150 AGB-39	875 A.H.	USN/EB	NR-1

LEAD ACID (Pressure Compensated)

BATTERY TYPE	CAPACITY	OWNER/BUILDER	BOAT
20 DFH-17	150 A.H.	WHOL/Litton	Alvin
12 DTSC-23	935 A.H.	U of P/EB	Asherah
90 DMSC-11	250 A.H.	USN/EB	Sea Cliff
90 DMSC-11	250 A.H.	USN/EB	Turtle
14 DRSC-9	260 A.H.	North American	Beaver (MK IV)
16 DMSC-11	250 A.H.	North American	Beaver (MK IV)
128 DRSC-11	780 A.H.	Lockheed	Deep Quest
56 DMSC-9	200 A.H.	USN/Lockheed	DSRV
54 DKCS-9	132 A.H.	Gen. Dyn.	Star II
60 DMSC-11	250 A.H.	Gen. Dyn.	Star III
14 DMEC-25	2040 A.H.	USN	Trieste
66 DRSC-17	520 A.H.	USN	Trieste
12 DMSC-13	300 A.H.	Cammel-Laird	Sea Bed Vehicle
24 DMEC-21	1700 A.H.	Makai Rainge	Makai Rainge
14 DMSC-11	250 A.H.	USN	Sea Lab III
63 DFH-17	150 A.H.	USN/SWRI	Nemo
15 DFH-17	150 A.H.	Lehigh U.	Lehigh Sea Probe
14 DTG-33	1152 A.H.	Smithsonian Inst/ Singer-Link	Deep Diver
30 DRSC-13	390 A.H.	USN/Marine Resources	CAV Vehicle
14 DEHC-9	568 A.H.	USN	Habitat

FIGURE 11
ESB SEASPACE POWER SYSTEMS

SILVER-ZINC BATTERY POWER FOR DEEP OCEAN VEHICLES

Guy W. Work
Yardney Electric Division
Pawcatuck, Connecticut 02891

ABSTRACT

Silver-zinc batteries are now a major power source for deep ocean vehicles. Two types are in use, those carried within the pressure-resisting hull and those pressure compensated and mounted directly in the sea. This paper discusses battery characteristics and problems in the pressure compensated applications only.

Units sized from 3KWH to 120KWH are operating four different vehicles to depths as great as 16,000 ft. The pressures and temperatures of the deep sea have relatively small effects on silver-zinc battery performance. Application related problems caused by electrolyte carryover, tilt angles, inaccessibility for maintenance and temperature differentials across large batteries still need work to bring this system to its full potential with the desired reliability.

BACKGROUND

The silver-zinc battery has the highest energy density of any field tested power system in the range required for present deep ocean vehicles. In concept and now in the field it is also showing that it meets most of the other generalized requirements for an energy supply for low cost submersible systems (1). Therefore, it has become a major power source for such vehicles.

Installations are of two types. Some are located within the pressure hull or sphere at near atmospheric pressure and temperature. Others are in the free flooding area of the vehicle at ambient sea pressure and temperature. They serve as the sole power source for propulsion and auxiliary loads, the standby power for a nuclear powered vehicle and as essential or emergency power for life support, instrumentation and control. Examples of these applications are shown in Table I. In each instance, weight and volume are critical.

TABLE I
SILVER-ZINC DEEP OCEAN BATTERIES

INSTALLATION	ENVIRONMENT	USE	EXAMPLES
Conventional		Propulsion	DOLPHIN
Inside	Atmospheric	Stand-By	NR-1
Pressure	Pressure and	Emergency	DSV-1, DSRV
Hull	Temperature	Essential	DSV-1, DOLPHIN
Compensated			DSRV, SDV
Outside	Sea	Propulsion	DSV-1 (80-cell)
Pressure	Pressure and	Auxiliary	DSRV, SDV
Hull	Temperature		DSV-1 (16-cell)

Small silver-zinc batteries have been used within the sphere of deep ocean vehicles for several years. These and other inside-the-hull applications may in some instances require larger cells and different hardware but are basically similar to many silver-zinc applications. Therefore, only pressure compensated batteries will be considered here, those filled with oil and mounted outside the pressure hull.

The first major installation of this type was in DSV-1 in 1968. Since that time at least three more vehicles have been designed around pressure compensated silver-zinc batteries.

DESCRIPTION

Pressure compensation is accomplished by installing the cells in a liquid tight box, filling all voids with a dielectric fluid such as oil and incorporating some variable volume device so that the pressure of the sea is distributed equally over every component in the battery. In one application, a swimmer delivery vehicle (SDV), the compensator is a neoprene bag within the battery box, opening only to the sea. It begins the descent completely collapsed due to the weight of the oil but as pressure increases with depth, sea water enters the bag and the bag serves as a diaphragm between sea water and oil. In another application, DS-20,000, the bag is in a free flooding compartment of the battery unit, the bag opening into the battery box. This begins the descent full of oil. The TRIESTE (DSV-1) compensator is a stainless steel cylinder open to the sea through a pipe entering at the base of the compensator. DSRV has a more sophisticated closed steel compensator which serves the two batteries as well as other components.

Pressure compensated batteries differ from other types also in that all electrical conductors are encapsulated. This includes terminals, intercell connectors, shunts, fuses and the like. This is done to reduce resistance-to-ground problems.

The corrosive nature and the enormous compressive forces of the deep sea are taken into account in battery container design and hardware selection (2). Both steel and epoxy-fiberglass battery boxes are being used successfully. These never see more than a 1 to 2 psig differential pressure inside and out. A pressure relief valve is mounted at the highest point in the cover for venting of gas. This is most likely to happen during an ascent.

Weight and power characteristics for six batteries are shown in Table II.

TABLE II
SILVER-ZINC DEEP OCEAN BATTERIES
PRESSURE COMPENSATED TYPE

BATTERY	WEIGHT LBS.	KWH	POWER WH/LB#	WH/IN CU	DEPTH METERS
8xLR250(DS)*	85 (44)	3	35.0	3.3	90
16xLR250(DS)*	156	6	37.8	3.0	90
74xLR525(DS)	2070 (918)	68	32.7	2.2	1830
74xLR735(DS)*	2500 (1272)	78	31.3	2.3	6100
80xLR952(DS)	3056**	117	38.3	2.5	6100
16xLR5000(DS)	3520**	122	34.7	2.2	6100

* Includes built-in compensator   ** Cells only
( )Weight in water   # Based on weight in air

The key to battery nomenclature can be illustrated from the first listing in Table II. An 8xLR250(DS) is an 8-cell battery designed for low rate (LR) operation with a nominal capacity of 250 ampere hours in a pressure compensated (DS) application. Note that the negative buoyancy is approximately half the air weight of most batteries. The power values are based on capacity in early cycles. These will be exceeded initially but will decline with cycling at a rate depending on the type of usage. They usually level off for a good part of life in the 75 to 90% of nominal capacity range. The batteries under discussion are in research vehicles or vehicles being "de-bugged" so usage has generally been light. Capacity data on a 8xLR250(DS) cycled for life in oil is shown in Figure 1.

The effects of temperature and pressure on the electrical characteristics of two deep sea type cells were discussed elsewhere (3). Some of these data are reviewed here to introduce additional observations. Charge and discharge voltage curves for a LR735(DS) cell in early cycles are shown in Figure 2. While polarization is less at higher pressures, it is greater at lower temperatures so that the net result in the deep ocean environment is a somewhat lower discharge voltage.

The DSRV rescue mission requires a minimum recharge time and these batteries are charged at rates not normally associated with silver-zinc batteries. Figure 3 shows charges of a LR525(DS) cell at 300 amperes. As used in the rescue mission seven high rate cycles might be run without interruption. Some preliminary testing of an old 74xLR525(DS) equilibrated in water at 45 to 50°F at the Battery Laboratory, Mare Island Naval Shipyard, showed that the first recharge at 300 amperes only put back 75 percent of the capacity discharged but that by the seventh cycle, the recharge was nearly 100 percent. This was attributable to the fact that charge acceptance is better at higher temperatures. In fact, the major concern with this type operation was the size of the heat build up toward the center of the battery. Heat transfer measurements on DSV-1 batteries in steel boxes with steel "egg crating" between cells did not show a large temperature gradient across the battery (4). The DSRV batteries, however, had epoxy-fiberglass boxes and only .030 steel along the heat transferring surfaces of the cells for heat conduction. The temperature rise measured in the electrolyte of four cells at different distances from the center of the battery are shown in Figure 4.

PROBLEMS

The problems encountered are those characteristic of the state-of-the-art of the silver-zinc battery in general plus those introduced by the application. In the first category are capacity maintenance and life to short. These are still the principal failure modes. The third set of batteries have been powering DSV-1 the past 17 months. The first set was replaced when slow shorts began to develop. The second set was in service two years and capacity was still good on both sets when they were removed. As the DSRV vehicle evolved the batteries were marginal on both voltage and capacity. The first pair for DSRV was replaced after nine months with a redesigned set having more capacity. These are still operable, after 28 months. The second set for DSRV I was replaced recently after 17 months with capacity in the 350 to 425 AH range. The first set on DSRV II has seen 14 months and capacity is about 80% of nominal.

One of the major problems introduced by the nature of the application is development of low resistance-to-ground conditions. This is determined from a voltage measurement between one current conductor and any other part of the vehicle. Translated to resistance, a clean battery will show several megohm and a vehicle will not be permitted to operate if the value is below 50,000 ohms. The conductive bridge which develops is normally sea water or electrolyte. Every pressure compensated battery to date has had sea water in it at one time or another but this is not the major problem. Electrolyte gets out of the cells due to spillage or carryout with gas. Individual cells will take angles to 60° from vertical without spilling. However, it is very difficult to adjust and maintain the electrolyte level to the optimum height in every cell of a large multicell battery. Experience to date shows that vehicles have gone to angles of 80° to 90° on occasion.

Although the gassing rate of a silver-zinc cell is relatively low, each bubble of gas leaving a plate carries a film of electrolyte. This is dropped at the electrolyte-air interface in most applications but where the head space of the cell is filled with oil, as in pressure compensated batteries, it tends to continue on out of the cell with the gas bubble and ultimately deposits on top of the cells. Entrainment devices in the form of chimneys containing nylon net are installed on each cell vent to minimize electrolyte carryover. Notwithstanding, all conductors are encapsulated with polyurethane.

CONCLUSIONS

The pressure compensated silver-zinc battery has demonstrated its capabilities in powering several small deep ocean vehicles. Additional work should bring it to its full potential and produce the desired reliability.

ACKNOWLEDGEMENTS

The author wishes to thank J. Delorey, Mare Island Naval Shipyard, for permission to use his data on temperature rise in a DSRV battery. Thanks is also extended to SUBDEVGRU-1 and operating personnel of the vehicles which use these batteries for their complete cooperation.

REFERENCES

1. Work, G.W., "Batteries for Deep Ocean Applications", Application of Marine Technology to Human Needs, MTS Preprints 8th Annual Technology Conference, 1972, pp 429-433.

2. Work, G.W., "Effects of the Deep-Sea Environment on Battery Materials and Characteristics", Material Performance and the Deep Sea, ASTM STP445, ASTM, 1969, pp 31-40.

3. Work, G.W. and Fried, S., "Silver-Zinc Battery Power for Deep Ocean Vehicles,"SAE 690728, Society of Automotive Engineers, 1969.

4. Momsen, F., and Clerici, J.C., "Evaluation of First Silver-Zinc Batteries Used in Deep Submergence", Marine Technology Society Journal, Vol. 5, No. 2, 1971.

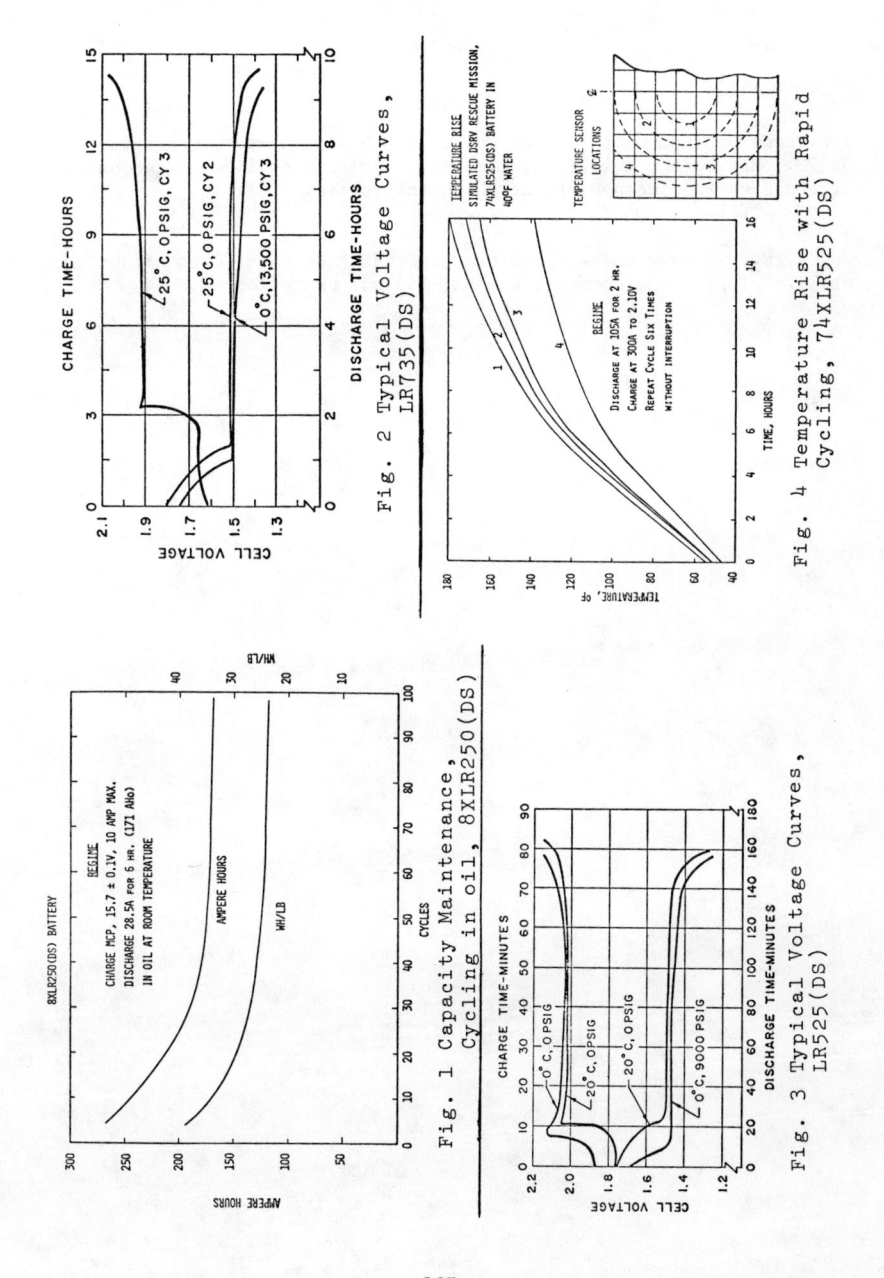

Fig. 1 Capacity Maintenance, Cycling in oil, 8XLR250(DS)

Fig. 2 Typical Voltage Curves, LR735(DS)

Fig. 3 Typical Voltage Curves, LR525(DS)

Fig. 4 Temperature Rise with Rapid Cycling, 74XLR525(DS)

DISCUSSION

J.P. Pemsler, Kennecott Copper Corp., 128 Spring Street, Lexington, Mass. 02173: How is the relative reliability and service life of the battery affected by the ocean depth at which they operate?

Guy Work: We have no evidence that operating depth affects either reliability or life of pressure compensated (oil filled) batteries.

ENERGY DENSITY, SERVICE LIFE, AND COST OF DEEP-OCEAN VEHICLE POWER

Donald O. Newton
Mare Island Naval Shipyard
Vallejo, California 94592

Abysmal depth power sources generally are exterior to crew's pressure hull. Fuel cell systems could be installed in separate pressure hulls, but almost all known operating deep-ocean vessels use batteries in oil-filled ambient-pressure tanks. Undisplaced weight is critical, so maximizing energy density with high-priced silver-zinc batteries usually provides overall operating economy. Attained kilowatthours/kilogram undisplaced weight have been 0.13 for silver-zinc, and 0.037 for lead-acid systems.

Minimum acceptable levels of energy and power density for deep-ocean use cannot be set for every application. However, as a yardstick[1], we might look at the goals set by the Environmental Protection Agency for a satisfactory substitute for the internal combustion engine when used for private transportation. These are:

A power level of 0.22 kilowatts per kilogram
An energy density of 0.22 kilowatthours per kilogram
A cost of not over $10 per kilowatthour

Meeting the above power, energy, and cost objectives would enable a family-size sedan to have a 300-Km range at a speed of 100 kilometers per hour, and at a cost comparable to our modern automotive engine. Other EPA goals were:

Sealed construction, for safety and zero environment contamination
High charge-discharge efficiency
Rapid charge capability

All of the above goals, taken together, may be more stringent than necessary for a deep-ocean vessel, but there is no doubt that, if the EPA meets its needs, the result would be eminently satisfactory for ocean use. Let us see how close we can come with the present technology.

Every known energy source has been proposed for underwater use, including biological, electrical, mechanical, chemical, and nuclear devices. The first underwater vehicle was built early in the 17th century by Van Drebbel[2], who could think only in terms of manpower. His submersible used men, operating oars; it looked something like an inverted rowboat. The only modern suggestion for using biological power centered around something called the "Bug Battery," a device for using chemical energy via bacteria. No reports were made public, but it depended on investigations of others, such as Allen and Januszeski[3].

At present, there is no electrical energy storage device which can store more than a few wattminutes of energy per kilogram. This does not approach a useful energy density, so capacitors, electrets, and similar devices are not suitable. There is one special application of the direct use of electrical energy which cannot be excelled: the umbilical-cord connection to shore or ship power. Where the application makes this possible, it cannot be improved on for power level, simplicity, and cost. However, it is hardly suitable for deep-ocean use, which is the subject of this discussion.

Mechanical energy storage is also not being considered seriously, but several investigators have proposed developing such devices[4,5]. Energy in an amount almost equal to the lead-acid battery can be stored in a high-speed flywheel. Twenty-five watt-hours can be stored in a one-kilogram flywheel rim with a radius of 13 centimeters rotating at 30,000 RPM. This energy is clean and non-contaminating (except for sound?). The great mechanical difficulty of (a) designing an economical system, (b) extracting the energy in an efficient manner, and (c) coping with the powerful gyroscopic effect make the flywheel a very unlikely energy source. However, if a flywheel technology is ever developed for surface-vehicle use, it would certainly be applicable to underwater vessels.

As a matter of fact, since the formula for the energy stored in a flywheel is $E = \frac{Wv^2}{2g}$, where E is energy, W is weight of flywheel rim, v is rim velocity, and g is the acceleration of gravity -- all in consistent units, it is obvious that the energy stored increases as the square of the rim velocity. From this, it is also obvious that any technological advance which would allow four times the present maximum rim velocity would enable us to store enough energy to meet the EPA requirements. In the example above where an energy density of 25 watthours per kilogram was mentioned, it was assumed that the flywheel material had a tensile strength of

180,000 lbs per square inch (HY 180), so a major increase in the tensile strength of the rim metal is all that is required! Unfortunately, the gyroscopic effect also increases at the same rate as the amount of energy stored.

Chemical and nuclear power are the only two types of "energy bucket" being given serious consideration for deep-ocean use. In this meeting we will not discuss the application of a nuclear reactor to deep-ocean vessels.

Most underwater vessels operate from stored chemical energy - the internal combustion engine, the fuel cell, or the battery.

The internal combustion engine is still the primary source of power for many Naval submarines which use the diesel-electric motive system. When submerged to any depth, they are actually operating on battery power. However, a number of proposals have been made to use various heat engines when operating at depth. A recent SAE Symposium reviewed the Sterling engine[6], the Brayton-cycle engine[7], and the Rankine engine[8] as possible deep-ocean power sources. All of the above operate with the same efficiency disadvantage as the Otto engine; none can exceed the efficiency limits set by the Carnot cycle of any heat engine. At presently attainable compression ratios, this is only about 25-30%.

It has been apparent from the beginning of the modern deep-ocean exploration effort that fuel cells could provide several times the fuel and oxidant utilization efficiency of any available heat engine. Also, it was expected that a fuel cell could provide from 5 to 30 times the energy density of any battery. However, costs have been out of bounds and operating life unpredictable. Fortunately, potential users of fuel cells can take advantage of the extensive and expensive work done for the Space Program. Many hundreds of millions of dollars have gone into research, development, and the procurement of Space fuel cells. It is only thus that they have been brought to their present somewhat less than perfect, state. A large percentage of this Space effort can be translated directly to the deep-ocean "innerspace" program. References 9, 10, 11, and 12 are recent surveys of the deep-ocean use of fuel cells.

In an undersea environment the fuel cell has several unique advantages over batteries. Mainly, of course, it is the much higher energy density of the reactants. The fuel cell stack can be installed in a separate pressure hull, and operated at a pressure of several thousand psi. This chamber, and the cylinders of compressed fuel and oxidant, can have a net neutral buoyancy. If hydrogen and oxygen are used the buoyancy effect is maximized, and this hydrogen-oxygen system has the highest possible energy density per unit weight of reactants. It is also the most reliable, thoroughly proven fuel-cell system. Many engineering difficulties must be overcome; one example is the hydrogen embrittlement of metals when exposed to a hydrogen-rich atmosphere at several thousand psi.

Inasmuch as the fuel cell is not a heat engine, and therefore is not bound by the thermodynamic constraints of the Carnot cycle, its efficiency can approach a much higher limit. A fuel cell could provide over three times the useful work from the energy in the same amount of fuel and oxidant as an internal combustion engine, but this level has seldom been reached.

Unfortunately, because of limited experience, it is almost impossible to calculate the real operating cost of the simplest fuel cell. Several manufacturers will be happy to sell you their model, but the purchase price is meaningless for a device which is usually so complex that it requires the services of top quality engineers and scientists to keep it working.

In spite of the above, I do believe that, except for the small, low-budget operation where the lead-acid battery cannot be improved upon, most deep-ocean vessels are eventually going to be powered by fuel cells.

However, at the present time operators of deep-ocean vessels are limited to battery power[13,14,15]. This is illustrated in the recent book by Shenton[16], which listed all known submersibles, their dimensions, and their power source. Of the 64 vessels listed, 48 use lead-acid batteries, 8 use silver-zinc, 4 use shore or ship power via an umbilical cord, 3 were unknown, and there was 1 each which used: nickel-cadmium battery, silver-cadmium battery, and fuel cell. (These add up to more than 64 because one vessel used two power sources: fuel cell, lead-acid.)

Operators concerned with cost effectiveness want an answer to the question of which battery is best. The answer depends on the use and the user. If capital investment must be minimal, if operating and support crews are small or "free" (such as college students), if maximum battery life is required, and if missions are normally short - - the lead-acid battery is best. On the other hand, adequate funding, large, expensive crews and support vessels, and a need for the maximum attainable submerged operating time restrict the operator to the silver-zinc battery.

A tabulation of the properties of these two batteries, and of the fuel cell, per present goals, makes the above conclusions clearer. The wide ranges given for the various characteristics reflect a wide gamut of opinions, and the usual dichotomy between cost and quality. The data on the fuel cell is the design goal of a team presently planning on using a fuel cell.

Table I

ENERGY STORAGE

	LEAD-ACID BATTERY	SILVER-ZINC BATTERY	FUEL CELL (Goals)
Requirement, per Kilowatt-Hour			
Volume, m^3	0.02-0.04	0.005-0.008	0.0026
Air Weight, Kg	40-70	10-20	1.6
Undisplaced Wt, Kg	20-40	5-12	0
Cost, $	25-75	300-600	Unknown
Anticipated Life			
Cycles	200-600	20-100	Unknown
Years	3-6	1-2	3-10

The troublesome decision arises when there is the contradictory requirement of minimum cost and maximum life vs. deep dives of maximum duration. There is one obvious pitfall to avoid; there is no point providing an energy system capable of submerging for 60 hours if the vessel is so small, cramped, cold, and uncomfortable that 15 hours represents the limit of human endurance. However, the Navy never seems to be overpowered; the electronics systems designers are always anxious to hang on one more piece of power-consuming equipment. In any case, if the basic mission of the vessel is to go down 3000 fathoms, search for eight hours, obtain and lift a heavy load, the only presently thoroughly proven power source is a silver-zinc battery.

In addition to the decision as to the kind of battery to use, there is the question of how it will be used, where will it be located. Almost all deep-diving submersibles use oil-filled, ambient-pressure-compensated battery tanks exterior to the crew's hull. Very fortunately, high pressure clearly improves the performance of lead-acid batteries[16] and, to a somewhat lesser extent, that of silver-zinc batteries[17]. There are many factors involved, but the most obvious is the suppression of gas bubble size. Unfortunately, the performance of all batteries is seriously downgraded by low temperatures, but waste heat from charge-discharge operations usually helps alleviate the effect of the cold of the abyss.

Locating a battery in an oil-filled tank leads to a number of special maintenance problems. It is difficult to check the gravity of lead-acid cells. It is very time consuming to replace one poor cell. It is generally very difficult to guess the cause of poor battery performance from the voltages of the individual cells. Other difficulties which have been experienced and, by various means, partially overcome include: excessive blowout of electrolyte due to release of gases during rapid ascent, contamination of the electrolyte with seawater, contamination of the battery active material with compensating oil, loose voltage connections, oil degradation of busbar insulation, oil degradation of cell sealants, faulty operation of cell voltage scanner and other battery-monitoring devices.

Partial solutions to most of these problems have been found. Better scanning systems and procedures are being developed so the crew can know what is going on in their big black box. One item which would be most welcome would be an economical, effective specific gravity measuring device with a remote electrical readout. Even though the silver-zinc battery is more sensitive to some operating troubles, particularly to poor voltage control during charging, it can reduce other problems due to its very low gassing rate and to its unconcern with electrolyte specific gravity. A low gassing rate translates into very little water loss, smaller gas removal devices, and reduced carryover of electrolyte into compensating oil. One easily contrived solution to the problem of charge control was to use larger wire for the voltage-scanning line to each cell so that those lines could be used to equalize-charge individual cells when necessary.

Early in the undersea program, considerable effort went into designing complex baffles to be installed in the "bubble breakers" on top of each cell for the purpose of preventing electrolyte carryover. However, one of the major battery manufacturers developed a simple, effective device which incorporated a small plastic open-ended jar filled with plastic gauze. Almost everyone now uses a similar device with, at least moderate, success. An additional aid in reducing electrolyte carryover is to make the battery jar oversize, allowing enough head space in the cell for an excess of compensating oil and electrolyte. This space imposes only a small weight penalty in battery case material because of the buoyancy effect of the oil. This reduces the likelihood of the contamination of the active material with compensating oil.

The plastics manufacturers have finally learned which insulation materials to use on the busbar, cell case covers, and other components. The use of Teflon-coated wire for voltage-indicating lines is a poor practice; it is almost impossible to seal a plastic penetrator material around Teflon wire.

Finally, most submersibles are now being designed with battery boxes located so that the entire box can be readily removed and a new, freshly-reconditioned battery installed with a minimum of effort.

In conclusion, an enormous amount of effort is being invested in developing new energy buckets for military and civilian uses. Any significant breakthrough will help solve the energy problem of the operator of deep-ocean submersibles.

REFERENCES:

1. Transportation II, Two For The Road, J.T. Salihi, Spectrum, v9 n7 p43-47, July 1972

2. *Encyclopedia Britannica*, "Diving Bells"

3. Electrochemical Aspects of Some Biochemical Systems, M.J. Allen, R.L. Januszeski, Electrochem Acta v9 n11 Nov 1964, p1423-7, 1429-32

4. Utilization of Flywheel Energy, R.C. Clark, SAE paper 711A, June 10-14, 1963, 35 pp

5. Storing Energy in Springs, J.T. Gwinn, Jr., Machine Design v36 n8 Mar 26, 1964, p166-72

6. Sterling Engine for Underwater Vehicle Applications, J.N. Matavi, F.E. Heffner, A.A. Miklos; SAE paper 690731, meeting of October 27-29, 1969, 25 pp

7. Closed Brayton Power System for Deep Ocean Technology, R.A. Rackley; SAE paper 690733, meeting of October 27-29, 1969, 8 pp

8. Rankine Cycle Power Plant with Boron Slurry Fuel, C.V. Burkland; SAE paper 690732, meeting of October 27-29, 1969, 7 pp

9. Fuel Cell Systems for Undersea Work, P. Shimoji; Marine Technology Society Journal v3 n4 Jul 1969, p37-38

10. Hydrazine-Hydrogen Peroxide Fuel Cell Systems for Deep Submergence Vessels, G.E. Rich; SAE paper 690730, meeting of October 27-29, 1969, 6 pp

11. Hydrospace Fuel Cell Power Systems, J.F. McCartney, Marine Technology Society 6th Annual Conference and Exposition, June 29-July 1, 1970, v2 p879-907

12. Development of Fuel Cells for Commercial Power, R.A. Sanderson, M.B. Landan, Marine Technology Society 6th Annual Conference and Exposition, June 29-July 1, 1970, v2 p865-877

13. Selection and Utilization of Batteries for Deep Submergence Vehicles, N. Kuska, J.A. Cronander, J. Hydronautics, v2 n1 Jan 1968, p20-25

14. Batteries for Undersea Applications, A. Himy, Energy Conversion, v8 n3 Nov 1968, p117-119

15. The Large Seawater Battery - A New Submersible Power Source, A.E. Ketler, Marine Technology, v5 n6 Nov-Dec 1971, p52-54

16. <u>Diving for Science</u>, E.H. Shenton

17. Operation of Batteries at Great Ocean Depths, R.A. Horne, Undersea Technology, v4 n7 July 1963, P16-18

18. Effect of High Pressure on the Voltage and Current Output of Silver Oxide-Zinc and Mercury Oxide-Zinc Miniature Batteries, T.N. Anderson, B.A. Miner, M.H. Chandehan, R.J. Brodd, H. Eyring; J. Electrochem Soc. v116 n10 October 1969, p1342-7

A FUTURE LOOK AT SMALL MANNED SUBMERSIBLE
ENERGY STORAGE SYSTEMS

A. W. Petrocelli
E. S. Dennison
A. S. Berchielli

Yardney Electric Division
Pawcatuck, Connecticut 02891

ABSTRACT

The ZENO energy system is a completely self-contained zinc/oxygen refuelable battery system of modular design.
The basic unit consists of a series of "monocells" utilizing zinc as the anode (fuel) and pressurized oxygen, which acts both as the cathode depolarizer and as a coolent.
The self-enclosed construction of the battery and the use of pressurized oxygen eliminate the well known problems associated with zinc/air batteries, and enable the use of this battery system in underwater applications. The modular construction provides ease of maintenance.
These performance characteristics make the Yardney ZENO System a viable contender as a power source for submersible applications.

The need for the development and deployment of low cost manned submersible systems capable of satisfying the requirements related to: underwater surveillance, deep sea rescue, tactical deterrence, and deep ocean exploration is recognized as a critical consideration in the overall development of the Navy's role in the current decade.

The current operating fleet of small, manned submersibles, DSRV, NR-1, Trieste, Dolphin, has served to prove the usefulness of their potential. In addition, significant strides have been made in the state-of-the-art related to: sonar, hydrodynamics, hull material and fabrication, life support systems, lock out-lock in technology, mating, and nuclear technology.

The system will provide an on-board energy source consisting of a compact renewable battery, a sub-system to store and supply oxygen under pressure, and a battery cooling circuit utilizing oxygen as the heat transfer medium. Variations are possible, especially in the sub-systems, depending on the size of the unit and the applications. A mother ship, or a shore-based station, is fitted with facilities for replacing the battery with a fresh unit, and for re-supply of oxygen. Both tasks can be arranged for expeditious handling, and the power plant is then ready for a new mission. Turn-around time is held to a minimum.

The on-board energy source, together with the necessary mother ship installation, comprise the ZENO energy system.

ELECTROCHEMICAL CONSIDERATIONS

The primary zinc/oxygen electrochemical system became of interest because it can provide energy densities which exceed present primary batteries at high rates of discharge and is competitive with power sources such as fuel cells for periods of operation approaching several hundred hours. This system is presently being studied at Yardney and elsewhere (1, 2, 3,).

A zinc/oxygen monocell, the basic unit in the Yardney system, consists of a porous zinc electrode, oxygen cathode and alkaline electolyte (potassium hydroxide). The individual electrode reactions for this couple are as follows:

$$\text{Zinc Anode:} \quad Zn + 2OH^- \longrightarrow Zn(OH)_2 + 2e \qquad 1.245 \text{ v}$$
$$\text{Oxygen Cathode:} \quad \tfrac{1}{2} O_2 + H_2O + 2e \longrightarrow 2OH^- \qquad .401 \text{ v}$$
$$\text{Cell:} \quad Zn + \tfrac{1}{2}O_2 \longrightarrow Zn(OH)_2 \qquad 1.646 \text{ v}$$

While the virtues of the zinc/oxygen (air) couple have long been recognized, the cells utilizing this couple were typically low current density, long life units used in inaccessible locations. Their prime virtues were a fairly high weight energy density, long life after activation, stable discharge voltage, and low cost (4). These cells could support high current drains for only short periods, but this was not a disadvantage in their limited applications. With the advent of fuel cell research, however, the technology requisite to improving the zinc/oxygen cell current density capabilities for long duration discharges became available. The oxygen depolarized electrode had been the limiting electrode during the early development of the zinc/air cell, but the oxygen cathodes

developed for fuel cells removed this limitation. Coupled with the
development of high current density oxygen cathodes were the improvements in high current density zinc electrodes resulting from the
developments in the systems such as the silver/zinc and mercury
oxide/zinc cells. The new technology for oxygen cathodes and zinc
anodes which enables the construction of high current density cells
capable of sustaining the high current for long discharge times plus
the extremely high theoretical weight energy density of the zinc/
oxygen cell (541 W-hr/lb), forms the foundation for Yardney's commitment to the ZENO system.

While much of the prior and current research and development work
on the zinc/oxygen cell uses air as the oxygen source, the ZENO
system uses pure oxygen under pressure as the cathode depolarizer.
The use of pure oxygen in zinc/oxygen cells has primarily been
limited to electrically rechargeable secondary cells, (1, 5) and
these for the most part operated at current densities in the 100-
200 ma/in^2 current density range. The ZENO system is designed for
operation at current densities above 1.0 amp per square inch and is
adaptable to either a primary or a mechanically rechargeable mode.

The basic advantage of using pure oxygen under pressure in the ZENO
system becomes evident from the equations relating current density
and cathode potential in alkaline solutions,

$$\varepsilon = A' + \frac{RT}{1.5F} \ln \left[O_2\right] + \frac{RT}{1.5F} \ln \left[H+\right] - \frac{RT}{1.5F} \ln |i| , \quad (2.1)$$

where ε is the electrode potential with respect to a reference electrode, A' is a constant which depends on the reference electrode
used, $\left[O_2\right]$ is the concentration of dissolved oxygen, $|i|$ is the
absolute value of the current, and all the other symbols have their
usual meaning.

Krasilshchikov (6) found this equation to hold good for a 0.21 to
70 atmosphere oxygen pressure range with silver as the catalyst.
The significant feature of this equation is that at a constant potential the current is a linear function of the oxygen concentration. Another study on the same subject by Hartner et al. (7)
found that while the equipotential current is linearly dependent
on the oxygen concentration in solution, the relationship becomes,

$$\frac{i_1}{i_2} = \left(\frac{p(O_2)_1}{p(O_2)_2}\right)^{1/2} , \quad (2.2)$$

when the oxygen was introduced as gas which first had to dissolve
before being reduced at the catalyst surface.

In Equation 2.2, i_n is the current density associated with gaseous oxygen pressure $p(O_2)_n$ over the solution under equipotential conditions. The oxygen dissolution rate was found to be the factor which reduced the linear relationship between current density and the concentration of dissolved oxygen, to a relationship involving current density and the square root of the oxygen pressure over the solution.

Experimental results similar to those from which Equations 2.1 and 2.2 were derived, have been obtained in our laboratories. It has been clearly demonstrated that besides the gains in current density at a fixed potential when the pressure of oxygen is increased, a marked improvement is achieved in the limiting current region.

The benefits derived from using pure oxygen under pressure go beyond an increase in current density at a given potential and the elimination of oxygen gas concentration overpotential. In addition, there is an increase in the useful life of the cathode, more stable operation of the cell, and reduced power losses.

In using pure oxygen in the ZENO system, not only is the inert nitrogen eliminated, but also the carbon dioxide which air contains. Carbon dioxide is detrimental to power cells with alkaline electrolyte because of carbonate formation. This carbonate has two profound effects on the cell formation: a pH gradient is set up across the cell, and eventually a carbonate precipitate can form within the electrode structure (8). A contributory factor in both of these effects is the necessity of a separator which slows the processes of diffusion and convection within the electrolyte, both of which processes are needed to reduce concentration gradients within the electrolyte. The reaction of carbon dioxide with the potassium hydroxide at the cathode to form potassium bicarbonate and ultimately potassium carbonate decreases the pH at the cathode. This decrease in pH is another form of concentration polarization within the cell. Furthermore, if the carbon dioxide reaction is allowed to proceed for some time, the formation of solid potassium bicarbonate precipitates within the cathode structure can occur because the solubility of the bicarbonate is less than the potassium hydroxide solubility. This bicarbonate precipitation, which is more likely to occur with cells operating at ambient temperature, eventually can cause electrode failure, as the electrode pores become clogged (8).

In this context of concentration changes within the cell, another definite advantage of the ZENO system is pertinent. Because the proposed ZENO system is sealed, water losses from the electrolyte are minimized. If the oxygen is presaturated with water, there would only be the evaporation associated with any increase of the electrolyte vapor pressure. The net result of this system stability is that dry-out is no problem for the ZENO system.

The ZENO system also removes two causes of increases in the electrolyte vapor pressure. One of the side effects of carbonate formation is an increase in solution vapor pressure at a given temperature. With no carbonate formation, no electrolyte solvent evaporation will occur from this source. Secondly, the decreases in polarization associated with the use of pure oxygen at elevated pressure mean that there is a commensurate decrease in internal power losses in the cell. These power losses are primarily in the form of heat, so that electrolyte is heated and its vapor pressure increased as a result of this polarization. Decreasing polarization by using oxygen at elevated pressures also means that there is a lessened problem with heat dissipation if the cell is used in a confined space.

While the oxygen reduction reaction requires a catalyst in order to proceed at the rates required for an oxygen cathode to support a current density of 1 A/in^2 (155 ma/cm^2), the zinc anode with its high exchange current density (12-400 ma/cm^2, depending on temperature and hydroxide ion concentration (9)), has no such requirement. The discharge behavior of the zinc anode does not completely conform to what would be expected of a material with such a high exchange current density, however, because of passivation films which form on the surface of the zinc anode during discharge (10). Some of the factors affecting the rate at which this passivating film forms on the surface of the zinc anode are the current density, real surface area (as opposed to the geometrical or apparent surface area usually used in determining current density), hydroxide concentration, and temperature. Although a thorough discussion of this phenomenon is not possible in the present paper, two pertinent observations are: The zinc anode has shown a maximum ability to withstand passivation in approximately 7-8 M KOH (10), and electrodes with porosities of approximately 65 to 80% were found capable of sustaining current densities well above 1.0 amp per square inch with moderate polarization and high zinc utilization at ambient temperatures. (11)

THE BATTERY

The objective of the ZENO system is to utilize the zinc/oxygen reaction as an energy source, in much the same way that the hydrogen/oxygen reaction is utilized in that type of fuel cell. "Hydrogen logistics" is to be replaced by "Zinc logistics." The basis of the ZENO system is pressurized primary zinc/oxygen battery, designed for simplified replacement and renewal, and achieving a maximum of avialability and service continuity.

The individual cell is illustrated by Figure 1 and the assembled cell stack, enclosed in its canister, by Figure 2. On-board auxiliaries, also shown schematically by Figure 2, will be discussed at a later point.

In contrast to most existing zinc/air and zinc/oxygen batteries, the ZENO employs a "mono-cell" design, a single cathode being paired with each anode. Several advantages are gained thereby, the most obvious being that cell-to-cell jumper connections are eliminated, and that the voltage gain per unit height of stack is essentially doubled. For motor application, power at low amperage and relatively high voltage is preferred.

Referring to Figure 1, the cell is of annular disc form, the central opening being about 1/3 the outer diameter. The active materials are contained within a frame consisting of inner and outer rings, with intermediate spokes to provide structural integrity. The frame is molded from a high-temperature plastic, and is non-conductive.

The diaphragm is a light sheet of copper, which acts as a partition and conductive path between successive cells. In combination with the frame, which is seated on it with O-ring seals, it forms a shallow dish serving to contain the fuel slugs. These are of known type, such as Zn-KOH mixtures, commonly used in existing zinc/air batteries. Fuel slugs are molded in advance, conforming to the spaces in the frame, and are stored in moisture and gas-tight containers. They are eventually activated by addition of water, as will later be explained.

The cathode mat, like the fuel mixture, is of a type employed in zinc/air cells. It normally consists of separating, catalytic, and water-repellant layers and is backed by a light metallic screen. The cathode mat is attached to the frame and may be molded integrally with it in manufacture.

The exposed upper faces of the frame rings are formed with intermittent projections. Open gas passages, or ports, are left between the lands. The diaphragm, next above, rests on these lands. The crevice thus formed is filled by the grid, an open elastic screen formed of wire or expended metal, which makes distributed metallic contact with both the cathodic terminal below it, and the diaphragm which is the anodic terminal of the cell next above.

Three distinct functions served by the grid can be identified:

1) It maintains a clear space over the cathode face, giving access for gaseous O_2 to supply the cell.

2) It provides a path for forced circulation of O_2 to cool the cell, as will later be described. By acting as "extended surface," it further helps to promote heat transfer from the diaphragm.

3) By reason of multiple metallic contacts, as noted above, it acts as the means of series connection between successive cells.

Alignment of cells, to form the stack or pile, is fixed by the "spine" shown in section by Figure 1. This is a light non-conductive structural member, made of glass-reinforce plastic or similar material. It is designed to provide ample open cross-section, while avoiding obstruction of gas flow from ports in the inner frame rings. The result is a central duct extending continuously through the pile.

Referring now to Figure 2, the cells are stacked vertically in a canister, with outside clearance sufficient to allow circulation of oxygen. The stack is completed with structural heads, top and bottom, and a central bolt passing through an opening in the spine, clamping the cells into a firm assembly. The central duct is closed at its lower end, while its upper end opens directly into the suction of a jet pump, or ejector. Action of the ejector serves to force the circulation of O_2 downward through the clearance space surrounding the stack, radially inward between the cells, and back to the ejector suction. Typically, the working conditions of the battery would include: Oxygen pressure, 100 psia; temperature, 150°F. Electrical characteristics will be discussed in a later section.

THE ON-BOARD AUXILIARIES

The battery as described, and as illustrated in the exploded view in Figure 3, is arranged for expeditious withdrawal and replacement with a fresh unit. In operation, it is served by a set of auxiliaries, permanently installed in the vessel, which are represented schematically in Figure 2.

Services to be provided include: Make-up supply of O_2; compression of a secondary stream of gas, to energize the ejector; and removal of excess heat under conditions of high load. Commercially available equipment can be adapted to meet each of these requirements.

Liquefied gas storage systems, both supercritical and subcritical, have been developed for aircraft and aerospace purposes (1). Space and weight requirements for a LOX system of this kind are compatible with submersible vehicle limitations. A single LOX system may be made to serve both battery and life-support purposes.

The ejector system is activated by drawing a fractional part, such as 1/6, of the desired circulation from the canister. This stream is compressed by a ratio of 1.03 to 1.05, and returned to form the jet. The natural thermal circulation through the battery is thus positively supplemented. Small oil-free compressors are available to meet this requirement. Power demand is estimated at 3% of battery output. Drive by a spark-free (AC) motor is preferred, allowing the entire unit to be enclosed and shaft seal problems avoided. Silicone lubricants are available commercially for service in an oxygen atmosphere, up to 1000 psig (12).

The branch stream is also cooled, through a rather wide temperature range, to effect all needed heat removal. In the example cited, a $30°$ drop in the branch stream will result in a $5°$ drop in the main gas stream.

It may be noted, in passing, that the make-up oxygen stream, which comes from a relatively high-pressure source (150 psig) can be employed to supply some or all of the jet energy, as well as part of the required cooling.

As indicated by Figure 2, parasitic heat extracted from the battery may be utilized for space heating if required, or may instead be released outboard.

Combining the auxiliaries as a fixed package serves to relieve the battery proper of internal complications. Piping connections are provided with quick disconnect and isolation valves. Similarly the electrical connections, for power and instrument circuits, are arranged to be readily broken in preparation for battery removal and replacement.

THE MOTHER SHIP INSTALLATION

The ZENO being a primary battery (or zinc/oxygen fuel cell), periodic renewal and replenishment functions are assigned to the mother ship supporting the minisub operation. These functions are forseen to include the following:

1) Lift-out and replacement of the spent battery. With the vehicle surfaced, the outer hatch cover is removed and the battery freed of operating connections. It is then removed by direct lift. Normally a replacement has been prepared in advance, and can be inserted without time-consuming preliminaries.

2) Replenishment of oxygen supply. The mother ship is provided with ample LOX storage, which in most cases will be loaded from a land-based plant. In an isolated situation, the mother ship may instead be fitted with air liquefaction and separation equipment. Transfer of LOX to the minisub flask, by insulated hose, is routine.

3) Battery overhaul and renewal. The mother ship is provided with a stock of fully-assembled batteries, factory produced, which are available as replacements. But in addition, renewal of spent batteries is a feasible ship-board function. Fuel slugs and cathode films are expendable but all other battery components can be reused, subject to cleanup and inspection.

Fresh fuel elements, or complete cells, may be factory supplied; battery assembly requires a small trained group and relatively simple equipment. This possibility can prove attractive for a remote and active operation.

Battery capacity for a minisub installation would be selected to match the maximum endurance of the crew. This might be, for example, 72 or 96 hours. Battery replacement is then a normal step in preparing for a new mission. In emergency, time allowance for the exchange operation should not exceed one hour.

BATTERY ACTIVATION

Activation requires that a pre-determined quantity of water be supplied to the battery, with uniformity of distribution between cells, and covering the entire active area of each cell individually. The usual method of activating a zinc/air battery, open to the atmosphere, is to fill by hand a small receptacle attached to each cell. The water is then expected to permeate the entire anode compact. This method is not applicable to the enclosed and isolated ZENO battery. Also, it would not be practicable in a system consisting of 100 or more cells.

A novel procedure of "vapor activation" has been selected for the ZENO application, and has been successfully demonstrated on a laboratory scale. The process is based on the strongly hygroscopic character of KOH, which in dry condition is mixed with Zn to make up the fuel slugs. The rate of activity is determined mainly by the respective water-vapor pressures.

The canister is first evacuated to a level of 1 mm Hg or less. This produces two desirable effects. First, inert gases are removed from the voids in the fuel; they cannot obstruct the penetration of liquid throughout the mass. Second, when steam is admitted, it will not be stratified or diluted, and a uniform H_2O vapor pressure will prevail at all points.

Conditions are then as represented by Figure 4. The vapor pressure exerted by solid KOH is near zero. Hence initially the full saturation pressure of steam, at a given temperature, is available to promote the absorption process. The steam must penetrate the layers of the cathode mat in order to make contact with the KOH; the separating film has a retarding effect. But the controlling factor is the degree of dilution in the fuel pack itself, where an opposing pressure develops. The process tends to be self-regulating. Dilution and thermal effects combine to locally retard or accelerate condensation, helping to promote equal distribution over the active area. The thermal quantities involved, both of condensation and solution, are substantial, and provision must be made for them in a full-scale operation. Condensation and solution release about 1000 Btu/lb. of water vapor, plus 400 Btu/lb. of KOH dissolved.

The preferred method of introducing activation steam is by means of the ejector nozzle connection. Circulation throughout the battery assembly is thus assured, helping to equalize system temperatures. Similarly, external cooling can be applied if found necessary for heat removal.

PRELIMINARY RESULTS

There has not been opportunity for thorough investigation of the special factors influencing ZENO cell performance. The unfamiliar operating conditions have given rise to problems of material selection and cell design details. It has still been possible to obtain results that give an indication of future potential.

Figure 5 pictures the parts of a model test cell as employed in recent work. These parts are to be assembled in the order suggested by the overlaps. At the left is the fuel compact, bonded to the diaphragm in this model. At the center is the frame, with cathode mat and grid in place. At the right is the covering diaphragm, with copper exmet attached, ready to receive the next fuel compact.

The active diameter is 6 in. Gas passages are visible on the frame rings. The center opening is relatively small, but adequate for a group of a few cells. No spokes are used in this model.

Figure 6 illustrates the test chamber designed to receive these model cells. Both heating and internal cooling can be applied. Means of circulating the gas is being prepared but was not yet fitted to the chamber as shown.

In this exploratory work, a test commonly applied is to load the cell at 1.0 amp/sq. in., and to adjust resistance as the cell becomes gradually depleted. It is a short-term test, which provides a fair measure of the cell's overload response. Figures 7, 8, and 9 represent tests of this type.

Figure 7 gives the results of a series in which three similar cells were tested, one in the atmosphere, the others in oxygen at 20 psia and 45 psia respectively. The improvement in a pressurized O_2 atmosphere was expected, but the extent of gain has not been consistent in all tests. In the case of Figure 7, the cathode was doped with Pt at 9 mg/cm^2. Figure 8 represents another series of tests, with somewhat different pressures. In this case the cathode was doped at 45 mg/cm^2, an excessive amount from the economic standpoint. The effect was to lift performance substantially; also to cut down the spread due to pressure. These results, taken together, suggest that one effect of O_2 pressure is to compensate for low activity of the catalyst.

Figure 9 presents the results of three tests in which an improved type of carbon cathode was tested in oxygen at 30, 65, and 115 psia, respectively. As is evident here and in other data, the gain due to oxygen pressure is less pronounced at higher levels. Any increase much beyond 115 psia may prove uneconomic.

Comparison of cell performances can be made on a 2-hour discharge basis. The second and third tests were cut off at 0.60 volts, too low a figure for most applications. However, the results indicate the present state of this line of development, and they are summarized in Table I.

Figure 10 is an example of a polarization curve, in this case for a 6 in. disc cell, tested in O_2 at 115 psia and 180°F.

At this point it appears that the ZENO system can achieve viable status provided the following cell performance criteria are met:

1) 0.90 volts, average, at 1.00 amp/in^2 of active area.

2) 3.25 watt-hrs/in^2 available energy.

3) 70% coulomb efficiency of zinc utilization.

It is important that these results be secured without dependence on a cathode doped with Pt or other noble metal. The inexpensive carbon cathode appears capable of meeting the requirements.

TYPICAL POWER PACKAGE

The requirements specified for the DSRV silver/zinc battery are typical of a fairly large unit of this kind. Table II makes a numerical comparison between the actual Ag/Zn battery, and the estimated characteristics of a ZENO designed to meet the same requirements.

It is assumed that the specified power (KW) is to be met at both the 8-hour and 2-hour rates. The 2-hour case determines both frontal area and required number of cells, even though the energy demand (WH/in^2) is not excessive. The volt/ampere relation in the 8-hour case is adjusted to the cell polarization curve.

The estimates for this table are not based on the criteria previously stated, but on a coulomb efficiency of 61%, and in most respects conforming to capability already existing.

Were the selected criteria to be confirmed, the 8-hour KWH would rise from 60.8 to 82. The 8-hour period would be extended to 10.8 hours and the 2-hour period to about 2.75 hours. There would be a small increase in Zn and O_2 demand, totaling about 50 lbs., resulting in an increase to 1100 lb. total installed weight.

Based on these estimates, comparison of the two battery systems can now be summarized:

```
Ag/Zn Battery complete    - (2000/60.8) = 33.0 lb/KWH
ZENO Battery as installed -
        Present technology (1050/60.8) = 17.3 lb/KWH
        Projected criteria (1100/82)   = 13.4 lb/KWH
```

The comparison as stated is not entirely valid, since the pressure-compensated Ag/Zn battery is designed for direct exposure to sea pressure, while the ZENO must be enclosed and protected. This aspect will be considered in the following section.

INFLUENCE ON VEHICLE DESIGN

Probably the optimum minisub energy system, currently available, is the compensated silver/zinc battery, mounted external to the pressure hull. When immersed, this package is about 50 percent buoyant. The remaining 50 percent must be made up by excess buoyancy of the hull, usually supplemented by blocks of low-density syntactic foam. A heavier battery, such as the lead/acid type, has the same problem except that the negative buoyancy to be offset is increased by a factor of 2 or 3.

For stability, the secondary battery must usually be placed low in the vehicle profile. The space above it is then occupied by lighter outboard equipment and by buoyant material. The net effect of the imposed conditions is that access to the battery is limited and inconvenient. Inspection and servicing are difficult. Installation or removal require partial dismantling of the external plating and auxiliaries.

The weight of the ZENO is about one-half that of the Ag/Zn battery. Hence it requires hull buoyancy nearly equal to the supplementary requirement of the Ag/Zn. This may be provided by extension of the main hull, or by separate enclosure.

A true comparison of submerged displacements due to the two energy systems is not possible, since numerous factors are involved. A rough estimate was attempted for a vehicle designed for a working depth of 6500 ft. and a crush depth of 10,000 ft., using the data of Ref. (14) for steel structures. The basis is the two units previously in Table I. The ZENO case assumes a steel sphere 52 in. O.D.

    For ZENO: Weight of Battery 1,100 lb.; displacement 2,700 lb.
    For Ag/Zn: Weight of Battery 2,000 lb.; displacement 3,500 lb.

The difference in weight is partly offset by the buoyancy of the compensated Ag/Zn battery. (Comparable figures for the lead/acid battery would be: Weight of Battery, 5,000 lb.; displacement 11,000 lb.)

Neutral buoyancy of the ZENO battery package allows freedom of placement in the vehicle structure; specifically, direct access for liftout and replacement as is desired in this instance. Figures 11 and 12 represent two alternatives which would be applicable to an Autec type of vehicle. Figure 11 shows the preferred arrangement, opening up accessible space for equipment which would might otherwise crowd the crew compartment. Figure 12 is better adapted to existing Autec designs, since the central longitudinal frame member is cleared, but the battery space is inaccessible. These arrangements indicate the feasibility of installing two units of a selected size. Each arrangement provides adequate space for LOX storage and other battery accessories.

A vehicle of the Autec type, equipped as suggested by Figures 11 or 12, could be provided with 120 KWH or more of available energy, if so desired. This contrasts with 45 KWH now furnished by lead/acid batteries.

Installations such as those of Figures 11 and 12 require an external hatch cover. Connections are made accessible with the cover removed. In other cases, such as that of sub-surface coupling with a mother submarine, modules would be selected to allow clear passage through the hatchway, then to be relocated in suitable cradles.

CONCLUSION

The ZENO system is presented as a concept, not as an accomplished fact. Preliminary tests have been encouraging, and improvement over present results can be expected.

If proved feasible, the ZENO would occupy a place between the state-of-the-art silver/zinc battery and the advanced H_2-O_2 fuel cell. It requires no exotic materials, and would be less costly than either. It offers higher energy density than the Ag/Zn, and fewer problems of logistics and operational control than does the fuel cell.

In the opinion of the authors, expansion of development effort on the ZENO system is justified.

ACKNOWLEDGMENTS

The authors wish to acknowledge the valuable contributions of Mr. Ronald Cercone, who was the original proponent of "vapor activation," and who has been an important factor in advancing the experimental program.

The authors also wish to express their appreciation to Mr. Roland Chireau and Dr. Ronald Gunther for their cooperation and assistance in editing this manuscript.

REFERENCES

(1) Powers R. A., Final Report: Secondary Zinc Oxygen Cell For Space-craft Applications. Contract No. NAS-5-10247, April, 1971.

(2) Powers R. A., Bennett R. J., High Rate Zinc Oxygen Batteries, Power Sources #2: Res. Develop. Non-Mech. Elec. Power Sources,

(3) McCormick R. J., A Self Contained Primary Zinc Oxygen Battery For Oceanographic Application, Intersociety Energy Conversion Engineering Conference PP 536, (1971).

(4) Roberts, Ralph, The Primary Battery, G. W. Heise and N. C. Cahoon, Ed. John Wiley and Sons, Inc., New York (1971).

(5) Klein M., Sealed Zinc Oxygen Batteries, Research and Development Technical Report, ECOM-0249-1, Contract DAAB-07-71-C-0249 (1972).

(6) Krasilshchikov, A. I., Zh. Fiz. Khim., 26, 216 (1952).

(7) Hartner A. J., Vertes, M. A., Medina, V. E., and Oswin, H. G., Fuel Cell Systems, G. J. Young and H. R. Linden, Ed. American Chemical Society, Washington, D. C. (1965).

(8) Kunz, H. R. and Katz, M. J., Electrochemical Society, Atlantic City Meeting, Sept. 1970.

(9) Dirkse, T. P. and Hampson, N. A., Electrochimica Acta, 17, 135 (1972).

(10) Dirkse, T. P. and Hampson, N. A., Electrochimica Acta, 16, 2049 (1971).

(11) Shepherd, C. M. and Langelan, H. C., J. Electrochemical Society, 114, 8 (1967).

(12) Dow Corning Corporation, Correspondence (1972).

(13) "Power Plants for Deep Ocean Vehicles" by T. D. Morrison, J. F. McCartney and J. F. Blose, SNAME, (1966).

(14) Feasibility of Pressure Hulls for Ultra Deep Running Submarines, by E. Wenk, Jr. ASME, Paper 61-WA-187.

TABLE I

Oxygen Pressure Effects

O$_2$ pressure, psia	30	65	115
Current, amp/in^2	1.00	1.00	1.00
2-hour performance:			
Volts at cutoff	*(0.68)	0.75	0.75
Volts, 2-hr average	(0.76)	0.83	0.84
Amp-hr/in^2	(2.00)	2.00	2.00
Watt-hrs/in^2	(1.52)	1.66	1.68
Cut-off at 0.60 volts:			
Hours duration	-	2.58	3.00
Amp-hrs/in^2	-	2.58	3.00
Volts, average	-	0.825	0.785
Watt-hrs/in^2	-	2.12	2.35

* Data extrapolated beyond 1.83 hours

TABLE II

ZENO BATTERY AS DSRV REPLACEMENT

Original: Ag/Zn Secondary Battery, 74 cells, weight 2000 lbs, incl. tank and fluid.

Discharge rate, per spec., hrs.	8	2
Amperes	68	233
Volts	112	101
KW	7.62	23.5
KWH	60.8	47.0

Replacement: Canister: 21 in. O.D. x 52 in. high
Active cell diameter, in.   18
Number of cells   112
Active area per cell, in.   227
TOTAL Active Area, in.   25,400

Hours	8	2
Amperes	62	233
Volts	123	101
KW	7.62	23.5
KWH	60.8	47.0
Amp/in.	0.27	0.98
Volts/cell	1.10	0.90
HW/in.	2.40	1.85

Weights:
Zinc per cell   2.20 lb.
Cell complete, activated   6.50 lb.
TOTAL Cells (112)   730. lb.
Cell stack ass'y, complete   800. lb.
Canister   100. lb.
TOTAL Life-Out Package   900. lb.

On-Board Auxiliaries:

LOX 45 lb., incl. flask   65 lb.
Compressor plus motor   30 lb.
Heat Exchangers and Piping   40 lb.

Auxiliaries   135 lb.

TOTAL Installation, approx. 1050 lb.

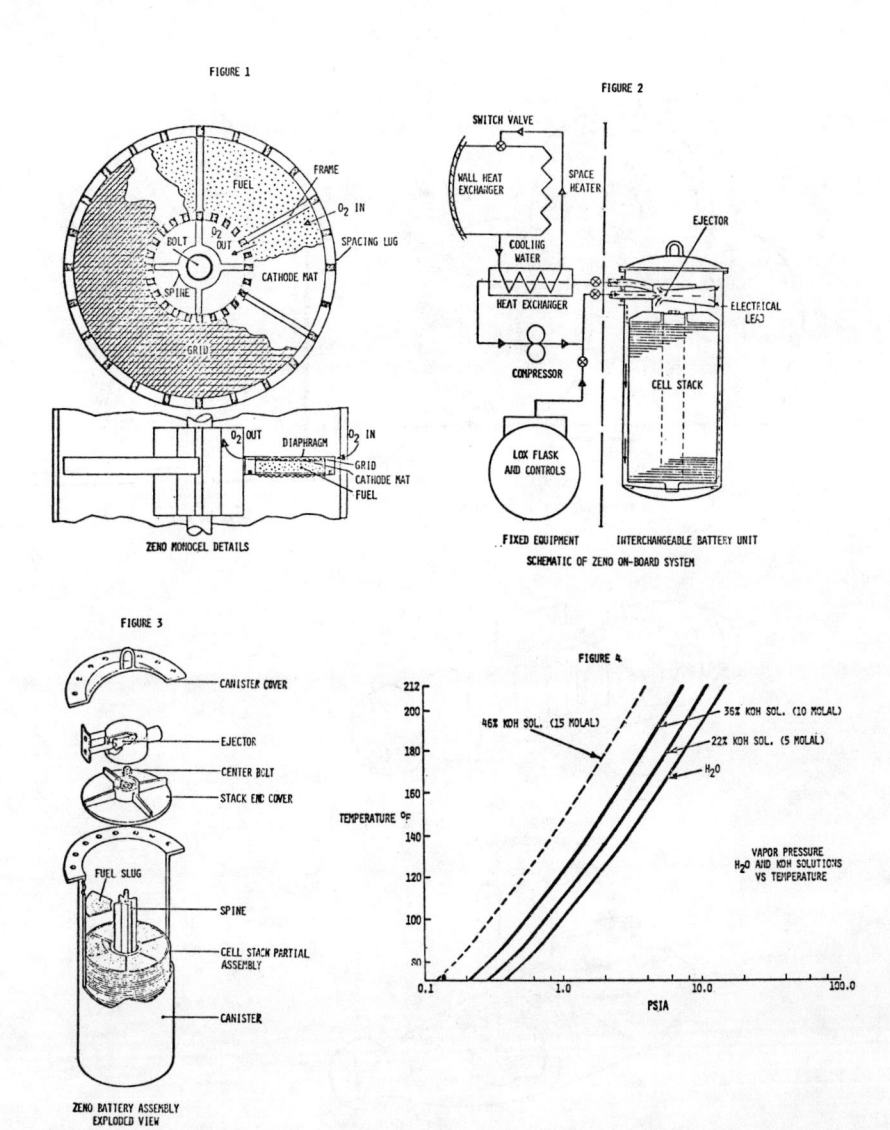

FIGURE 1

ZENO MONOCEL DETAILS

FIGURE 2

SCHEMATIC OF ZENO ON-BOARD SYSTEM

FIGURE 3

ZENO BATTERY ASSEMBLY
EXPLODED VIEW

FIGURE 4

VAPOR PRESSURE
H_2O AND KOH SOLUTIONS
VS TEMPERATURE

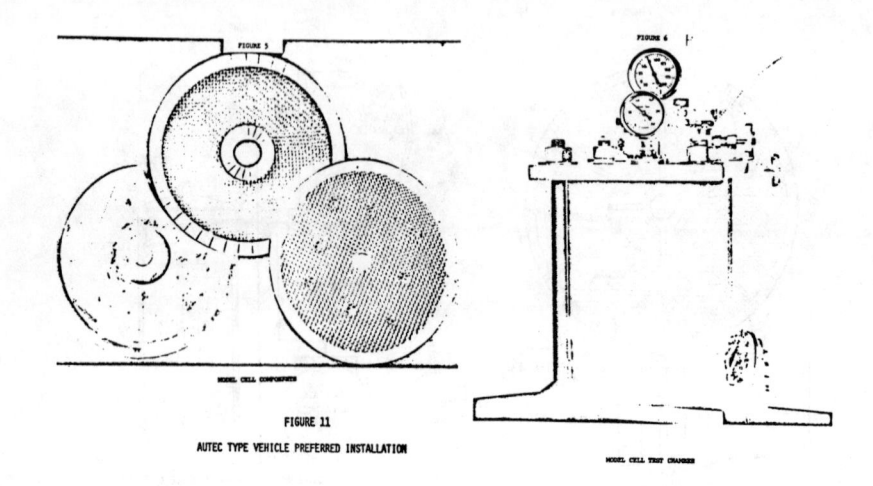

FIGURE 11
AUTEC TYPE VEHICLE PREFERRED INSTALLATION

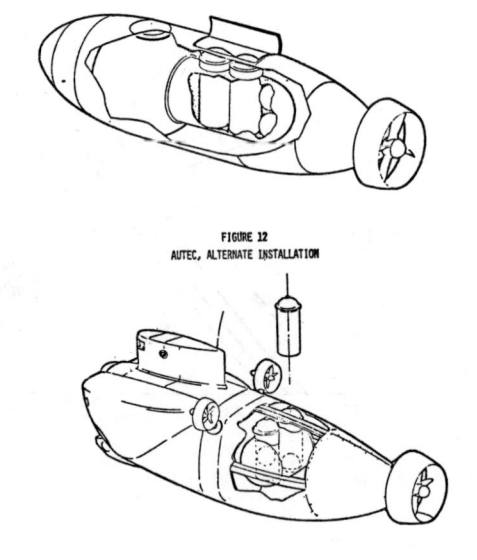

FIGURE 12
AUTEC, ALTERNATE INSTALLATION

214

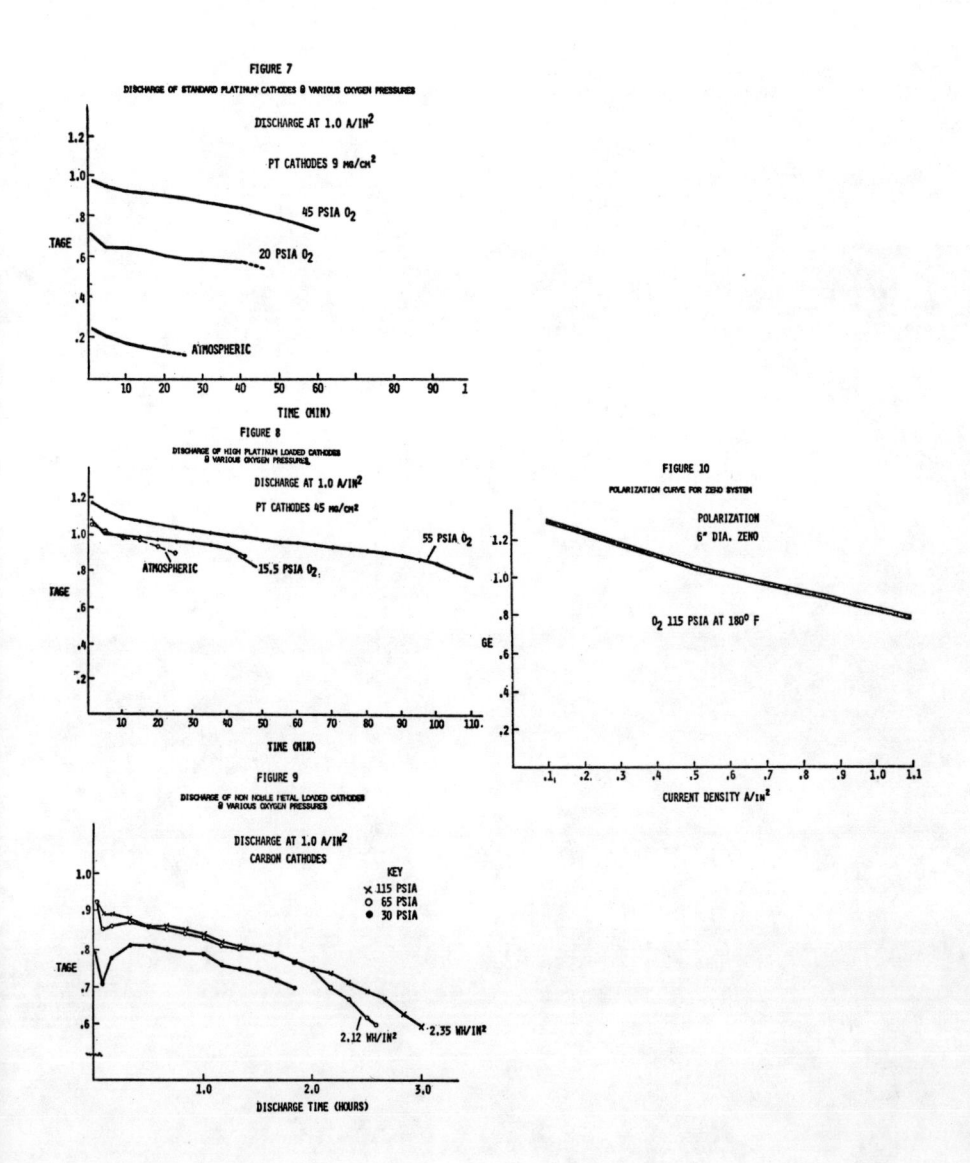

FIGURE 7
DISCHARGE OF STANDARD PLATINUM CATHODES @ VARIOUS OXYGEN PRESSURES

DISCHARGE AT 1.0 A/IN^2
PT CATHODES 9 mg/cm^2

45 PSIA O$_2$
20 PSIA O$_2$
ATMOSPHERIC

TIME (MIN)

FIGURE 8
DISCHARGE OF HIGH PLATINUM LOADED CATHODES @ VARIOUS OXYGEN PRESSURES

DISCHARGE AT 1.0 A/IN^2
PT CATHODES 45 mg/cm^2

55 PSIA O$_2$
ATMOSPHERIC
15.5 PSIA O$_2$

TIME (MIN)

FIGURE 9
DISCHARGE OF NON NOBLE METAL LOADED CATHODES @ VARIOUS OXYGEN PRESSURES

DISCHARGE AT 1.0 A/IN^2
CARBON CATHODES

KEY
× 115 PSIA
○ 65 PSIA
● 30 PSIA

2.12 WH/IN^2 2.35 WH/IN^2

DISCHARGE TIME (HOURS)

FIGURE 10
POLARIZATION CURVE FOR ZENO SYSTEM

POLARIZATION
6" DIA. ZENO

O$_2$ 115 PSIA AT 180° F

CURRENT DENSITY A/IN^2

MARINE CORROSION

Chairmen:
M.J. Pryor
Z.A. Foroulis
J.A. Ford

CORROSION IN THE OCEAN

F. L. LA QUE

Vice President (retired)
International Nickel Co. Inc.
1 New York Plaza, New York, N.Y. 10004

There is a need for precautions in devising, conducting and interpreting the results of corrosion tests in sea water and sea air. This is illustrated by reference to tests used for evaluating resistance to erosive effects, behavior in the tidal zone, salt spray tests and atmospheric tests at different elevations and degrees of shelter and effects of pollution.

Special attention is given to studies of cathodic protection and a demonstration that the important potential for control of cathodic protection is the potential of the cathodic reactions achieved rather than the potential of the original anodes.

Concern with how to deal with the corrosive effects of sea water and salt air has occupied the attention of engineers and scientists from the time man began to use metals on ocean highways and sea water itself as a cooling medium in heat exchangers on board ships and in oceanside plants.

To supplement guidance provided by practical experience and to predict the probable performance of new materials, various means have been used to test materials either under the natural conditions of exposure or under conditions designed to simulate such natural conditions. The latter approach has often extended to attempts to shorten the time required to get results by the use of what are called "accelerated" corrosion tests. Usually the acceleration is achieved by distorting one or more of the factors that control the rates of corrosion under natural conditions so as to subject the test material to what is hoped will be equivalent damage in a shorter time.

The object of this paper will be to draw attention to precautions that should be taken in devising and carrying out such corrosion tests and in appraising the significance of the results.

Since, unfortunately, many investigators of sea water corrosion must make their tests in inland laboratories with no access to continuous supplies of natural sea water they become forced to substitute either synthetic sea water or even simple solutions of sodium chloride in concentrations approximating the chloride content of natural sea water.

It is possible to make up a solution containing in proper proportions all the inorganic constituents of natural sea water that might reasonably be expected to have a significant effect on corrosion. Such a formula for synthetic sea water for use in corrosion tests was described by T. P. May and C. P. Black[1] and it is covered also in ASTM Special Publication 148-A[2].

Even though it is possible to reproduce the chemical composition of sea water this does not mean that there is an equal reproduction of its corrosive characteristics. Something, possibly living organisms, or their remains, seems to be missing. This was demonstrated by erosion tests on copper alloys as illustrated by the data covered by Figure 1.

The data in Figure 1 also show that full strength natural sea water has a greater erosive effect than brackish water from the Severn River at Annapolis, Maryland.

There is also a big difference in the erosive effect of full strength sea water when it is passed thru the British Non-Ferrous Metals Research Association jet test apparatus once as compared with similar water recirculated thru the same apparatus for the several days required for the test. The once thru water is much more aggressive, especially towards copper alloys, such as aluminum brass, more resistant to impingement attack than admiralty brass.

Jet tests with recirculated water used in the BNFMRA laboratories in London showed, properly, the advantage of aluminum brass over admiralty brass in resisting impingement attack. But when similar tests were made with once thru sea water at Kure Beach, N.C. the aluminum brass was attacked much more severely than by the recirculated water, did not demonstrate as great an advantage over Admiralty Brass and was shown, properly, to be inferior to the iron modified 70:30 cupro nickel alloy. This had not been disclosed by tests with recirculated water in London.

When the sea water at Kure Beach was recirculated it was found to be much less aggressive than water from the same source passed thru the apparatus only once.

In the studies with recirculated water in London there appeared to be a strong effect of the presence of air bubbles entrained in the water. In fact, there was evidence that air bubbles having a critical size were more damaging than larger or smaller bubbles[3].

In the tests with the once thru water at Kure Beach there appeared to be no effect of air bubbles in the jet since about the same damage occurred in the absence of air bubbles as when they were present[4].

Up to now there has been no explanation of why the once thru sea water is more aggressive than recirculated water. Efforts to filter out any fine solids in suspension did not make the once thru water less aggressive.

There is a **continuing problem** in finding some way to evaluate quantitatively the corrosive characteristics of polluted sea water except by tests in each specific locality with results applicable only to that test location.

Dr. T. Howard Rogers[5] endeavoured to classify sulfide polluted waters in terms of what he called a Copper Corrosion Index based on the rate of corrosion of copper in a sample of the water under specified test conditions. It was found, however, that the results were affected strongly by the length of time between when the sample was taken and the tests were made.

It has been found also, that jet test specimens conditioned by prior exposure to sulfide polluted water so as to develop a sulfide corrosion product film will suffer severely aggravated impingement attack in subsequent tests with clean, unpolluted sea water. This is illustrated by Figure 2 and the data in Table 1.

These results suggest that the corrosion product films first formed on copper alloys can have a strong influence on their subsequent behavior. Sulfide films are likely to be particularly damaging and can continue to have harmful effects even after sulfide pollution has been eliminated unless they are removed at the same time.

On the other hand, favorable films formed during early exposure to clean sea water might be expected to enable alloys to withstand at least short or intermittent exposure to sulfide polluted water.

The notorious inadequacy[6] of the usual salt spray tests in measuring the resistance of metals to attack by sea water and sea air has been attributed incorrectly to the failure to use either natural sea water or synthetic sea water in place of straight sodium chloride brines in the salt spray test.

This was dealt with very thoroughly in a program described by T. P. May and A. Alexander[7]. Typical results for zinc and steel are shown in Figures 3 and 4.

As a matter of fact there is no reason to expect that metal surfaces exposed to a salt atmosphere near the ocean will encounter particles having the exact salt concentration of the sea water in which they originate. On hot, dry days the water particles carried by the wind will become concentrated by evaporation while they are in the air and will concentrate further soon after they land. On rainy days there will be considerable dilution. This also will occur on surfaces kept wet by fog or dew.

Effects such as these will account for the considerable influence of altitude and distance from the ocean on the corrosivity of what are loosely called "marine atmospheres". This is illustrated by the data in Figure 5.

The influence of orientation relative to the ocean and of partial shelter was demonstrated dramatically by tests carried out by the late C.P. Larrabe, using the test set up shown in Figure 6 with results shown in Figure 7.

Special precautions must be taken in tests covering behavior in the tidal zone. Specimens must be large enough to span the distance from well below low tide to well above high tide.

Exposure of isolated specimens at various elevations within this zone will not take into account the effects of differential aeration cells involving cathodic areas in the alternately wet and dry areas within the tidal zone and anodic areas under the fouling organisms below low tide. A potential difference in excess of 100 millivolts and a current density of more than 30 ma per sq. ft. (3 ma per sq.dm) have been measured.

The gross difference in behavior between isolated and continuous steel specimens is shown by Figure 8..

The mechanism of erosion or impingement attack of copper alloys generally involves corrosion cells in which surfaces that are able to retain corrosion product films act as cathodes to the usually relatively small areas over which films do not form or are frequently removed. As a result, tests to measure resistance to such erosive attack are likely to be affected seriously by the dimensions of the test pieces. With larger specimens the areas that are covered with cathodic films will be relatively larger than will be the case with smaller specimens. This is true of specimens used for jet tests. It is also true with tests using specimens in the form of spinning discs. It has been found that, irrespective of the diameter of a disc specimen, there will always be an anodic area suffering accelerated attack near the periphery of the disc. Consequently, no quantitative significance can be attached to the depth of attack and the velocity of movement calculated from the radius at which the attack is measured and the number of revolutions per minute. This is illustrated by Figure 9.

There is a similar effect of specimen size with iron discs except that, here, the anodic areas are towards the center of the test discs which become anodic to the peripheral areas to which there is a relatively greater supply of passivating oxygen as a result of the faster movement.

Tests of this sort can be used to establish an order of merit in resisting erosion provided the size of the test specimens and the rate of movement are appropriate and are kept constant.

There is also a powerful effect of test specimen size on results of crevice corrosion tests. It has been shown [8] for example, that the extent of crevice corrosion on stainless steel is directly proportional to the area of the freely exposed surfaces outside the crevices. See Figure 10. Consequently no quantitative significance can be attached to results of crevice corrosion tests where the ratio of freely exposed areas to areas within crevices in the service application is different from that in the tests.

Accelerated tests to measure resistance to cavitation erosion generally tend to accentuate the mechanical forces relative to the chemical forces causing cavitation damage. Results, therefore, tend to indicate greater advantage of hardness than of corrosion resistance. This is particularly true of tests in which cavitation is induced by high frequency vibration of test specimens. A modification introduced by Plesset[9] corrects a good deal of this deficiency by providing a pulsed vibration in which the contribution of corrosion resistance is given a chance to manifest itself during the rest periods between each interval of high frequency vibration.

Due to lack of a sufficient supply of sea water in most laboratories, tests to observe the effects of electrical currents applied to specimens as anodes or cathodes are usually carried out by applying a range of current densities or potentials to single specimens.

Studies using sufficient sea water to enable a separate specimen to be used to study the effect of each current density within a range of cathodic current density have yielded results which suggest a quite different mechanism of cathodic protection in sea water than the mechanism based on tests with single specimens.

The commonly proposed mechanism is that the effect of an applied current is to achieve an equipotential surface by polarizing cathodic areas to the potential of the anodic areas.

Such an explanation calls for protection to be achieved at a potential equal to, or greater than, that of the original anodes. This is no doubt the case with galvanic couples of dissimilar metals where the anodic and cathodic areas are well defined and do not change in dimensions or position as a result of the application of current.

The usually proposed mechanism of cathodic protection has apparently been related to what happens with such galvanic couples. It does not necessarily apply to the effects of applied currents on a single metal surface where the anodic and cathodic areas are not so well defined and do not have the same relative areas after the application of current as they had when only local action currents were involved.

This difference in behavior was demonstrated dramatically by tests in flowing aerated sea water, using a series of polished steel specimens each of which was subjected to the effects of applied cathodic currents in a range from zero to enough current to achieve complete protection[10].

This technique permitted observations of the relationships between applied current densities and weight losses, corrosion rates, potentials and distribution of corrosion.

As shown by Figures 11 and 12, the principal effect of increasing current density was to reduce the corroded area acting as an anode relative to the easily identified unattacked cathodic area.

As shown by Figure 13, the effect of increasing current density on the rates of corrosion of the smaller anodic areas that continued to corrode was very much less clear cut.

Potential measurements indicated, as might have been expected, that the potential did not change substantially and might even approach the potential of the cathodic surfaces rather than that of the anodic surfaces as the relative cathodic area increased with increasing applied current density and complete protection was being approached.

This suggests that the mechanism of cathodic protection, at least of steel in moving sea water and probably other media, is the creation of a cathodic reaction over the whole of the metal surface by furnishing sufficient electrons from an appropriate source to accomodate a cathodic reaction over the whole of the surface and thereby to extinguish all anodic reactions.

It follows from this that the important potential when cathodic protection is achieved is that of the cathodic reaction that is occurring. This has nothing to do with the potential of the original anode which no longer exists when the potential of interest is being measured.

In aerated sea water the most important cathodic reaction is oxygen reduction which occurs on steel in moving sea water at a potential of about -0.6 volt relative to a saturated calomel half cell. In laboratory tests with the potential of steel maintained at this value by application of a cathodic current, the steel was found to be practically completely protected from corrosion, as anticipated.

With increasing current density beyond that required to accomodate the oxygen reduction cathodic reaction this reaction is supplemented by a hydrogen evolution reaction occurring at a higher potential. This change occurs over a small range of added current as shown by Figure 14.

The fact that the potential finally associated with cathodic protection and some hydrogen ion discharge is higher than the anodic potential of steel in sea water appears to be a matter of coincidence between the potential of the final cathodic reaction and the potential of steel anodes rather than support for the notion that original and persistent cathodic and anodic areas continue to occupy their original areas and have been brought to the same potential by the applied current.

Observations such as this tend to justify questions as to the adequacy of studying corrosion by techniques which apply current to a specimen acting as either an anode or a cathode without concern for the other half of the overall corrosion reaction in either case. The experiments just described also raise a question as to the propriety of using a single specimen subjected to a range of increasing cathodic current densities. If there are deficiencies in this approach in the case of cathodic polarization phenomena there may well be deficiencies in the even more popular practice of using single specimens subjected to a range of applied currents to observe anodic polarization phenomena in an effort to characterize the corrosion resistance of metals.

In drawing attention to precautions that should be taken in undertaking and interpreting corrosion tests it is not intended to negate the value of such tests. Considerable progress has been made and will continue to be made with the help of better corrosion tests in understanding corrosion in the ocean, improving materials and applying them properly.

No doubt the several papers to be presented in the symposium will contribute further to this desirable result.

TABLE 1

Effect of Prior Exposure in Polluted Water
on Impingement Attack of Alloys
by Clean Sea Water at Kure Beach, N.C.

Maximum Depth of Impingement Attack in Jet Test

Inches

Alloy	Exposed Previously at New Haven	Exposed only at Kure Beach	Exposed Previously at Los Angeles
90:10 Cu Ni	0.004	0.002	0.002
70:30 Cu Ni	0.004	0.003	0.005
Admiralty Brass	0.009	0.005	0.018
Aluminum Brass	0.005	0.005	Perf.(0.031)

REFERENCES

1. May T. P. and Black C. E.
   Report p 2909 Naval Research Lab.
   Washington, D.C. 1946

2. ASTM Manual of Industrial Water
   Special Technical Publication # 148A
   1954

3. Bengough G. D. and May R.
   J. Inst. Metals 32,81, 1924
   and May R.
   J. Inst. Metals 40,141, 1928

4. Gilbert P.T. and La Que F. L.
   J. Electrochem. Soc. 101, 9,
   p 448 -455 1954

5. Rogers T.H.
   J. Inst. Metals 76, pt 6, 1950
   p 597 - 611

6. La Que F. L.
   Materials and Methods 35, 2,
   1952 p 77-81

7. May T. P. and Alexander A. L.
   Proc. Am. Soc. Test & Mat. 50, 1950

8. Ellis O.B. and La Que F. L.
   Corrosion 7, 11, 1951 p 362-364

9. Plesset M.S.
   Corrosion, 18, 5, 1962

10. La Que F. L. and May T. P.
    2ND Int. Congress on Metallic Corrosion
    NACE Houston 1966 p 789 - 794

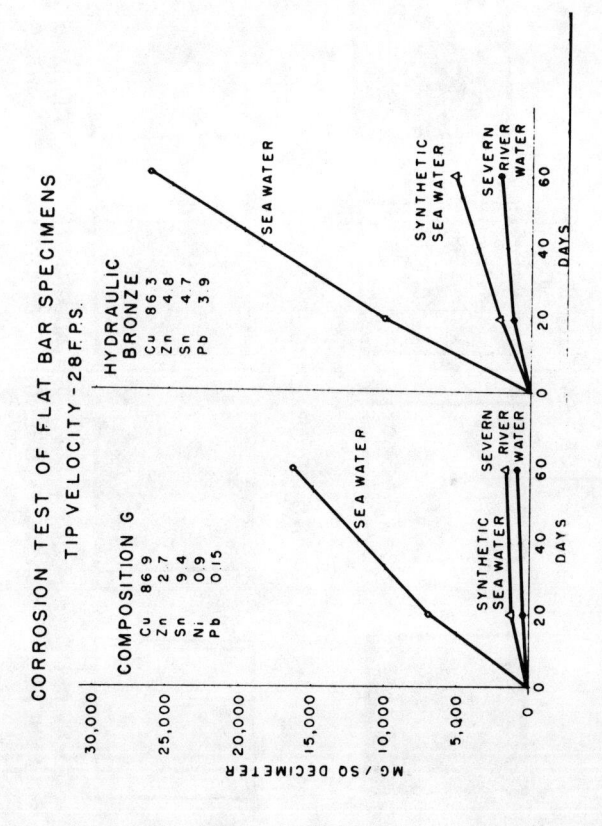

Fig. 1. Results of erosion tests on copper alloys in natural sea water, brackish water, and synthetic sea water.

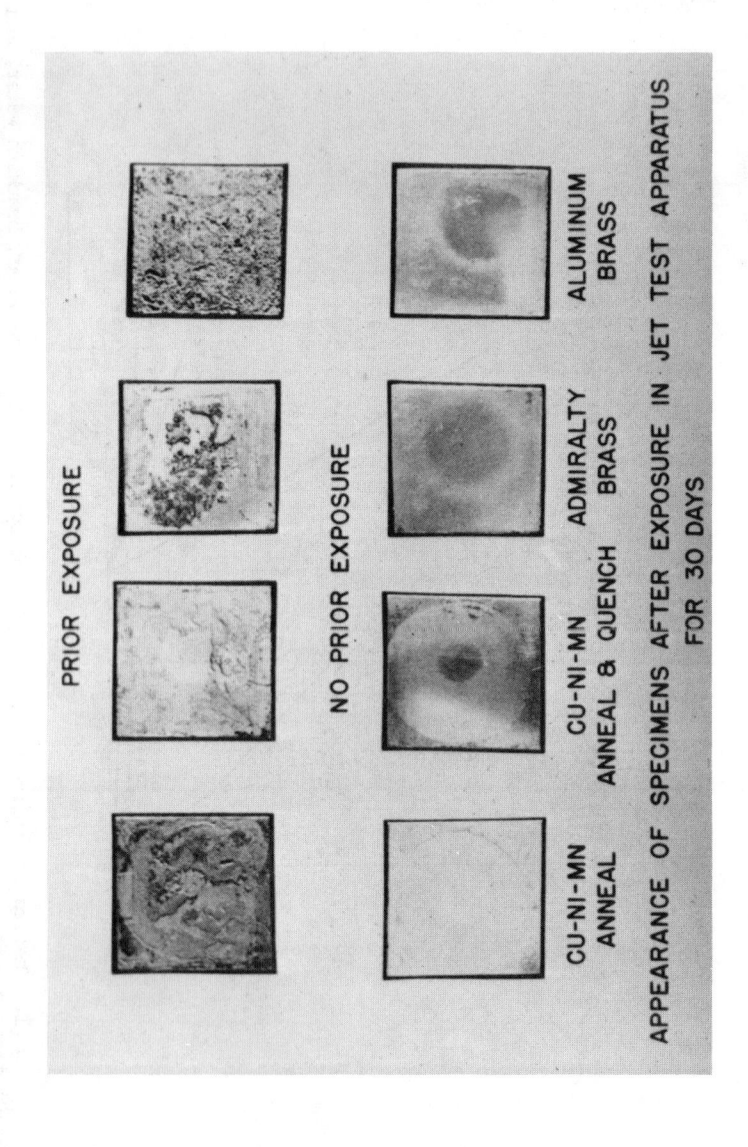

Fig. 2. Results of jet tests on specimens after prior exposure to sulfide polluted water.

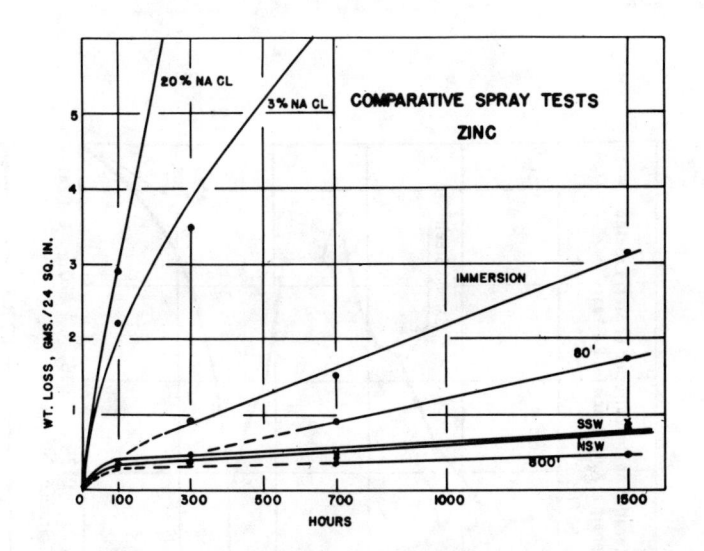

Fig. 3. Results of salt spray test on zinc with different brines.

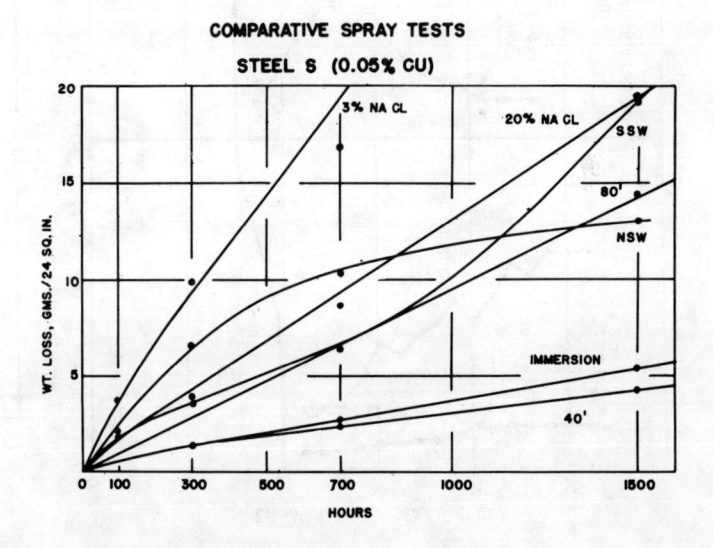

Fig. 4. Results of slat spray tests on steel with different brines.

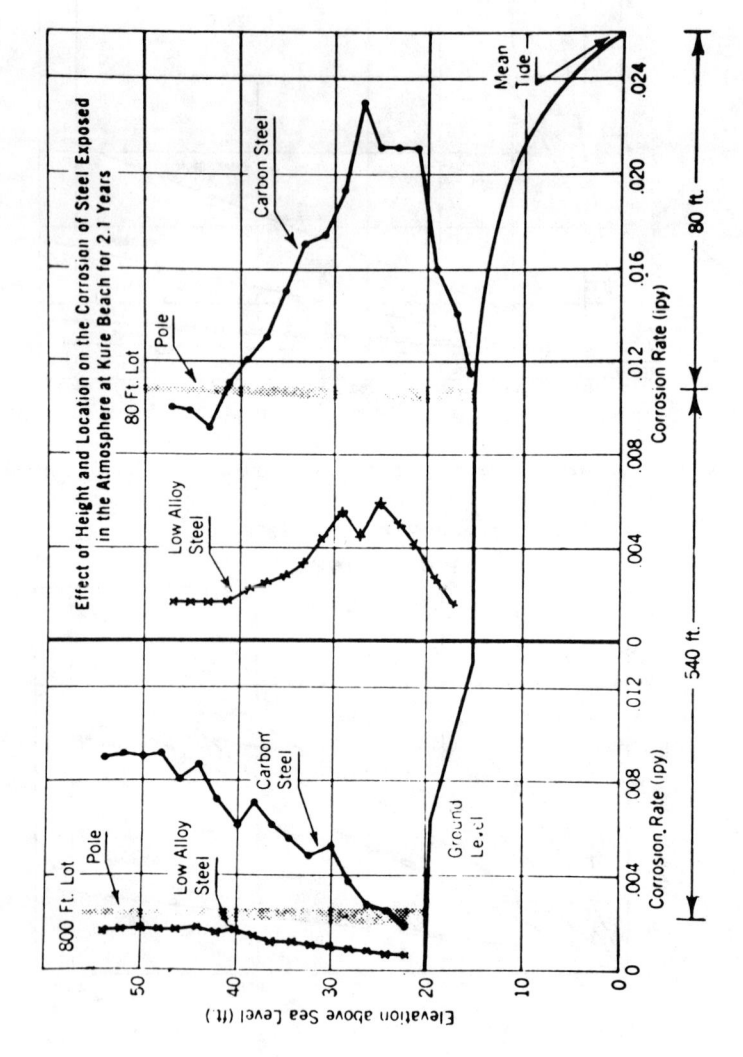

Fig. 5. Effects of altitude and distance from ocean on atmospheric corrosion of steels.

Fig. 6. Test racks for studying effects of orientation and partial shelter on atmospheric corrosion.

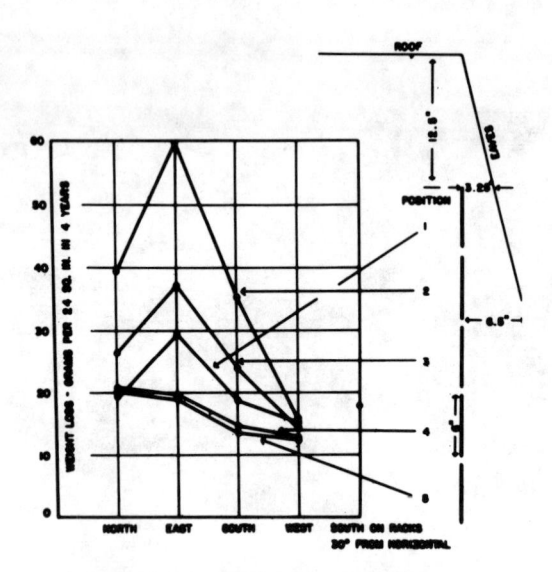

EFFECT OF DEGREE OF SHELTER AND ORIENTATION ON CORROSION OF CARBON STEEL (0.04 CU) SPECIMENS EXPOSED AT KURE BEACH, N.C.

Figure 7

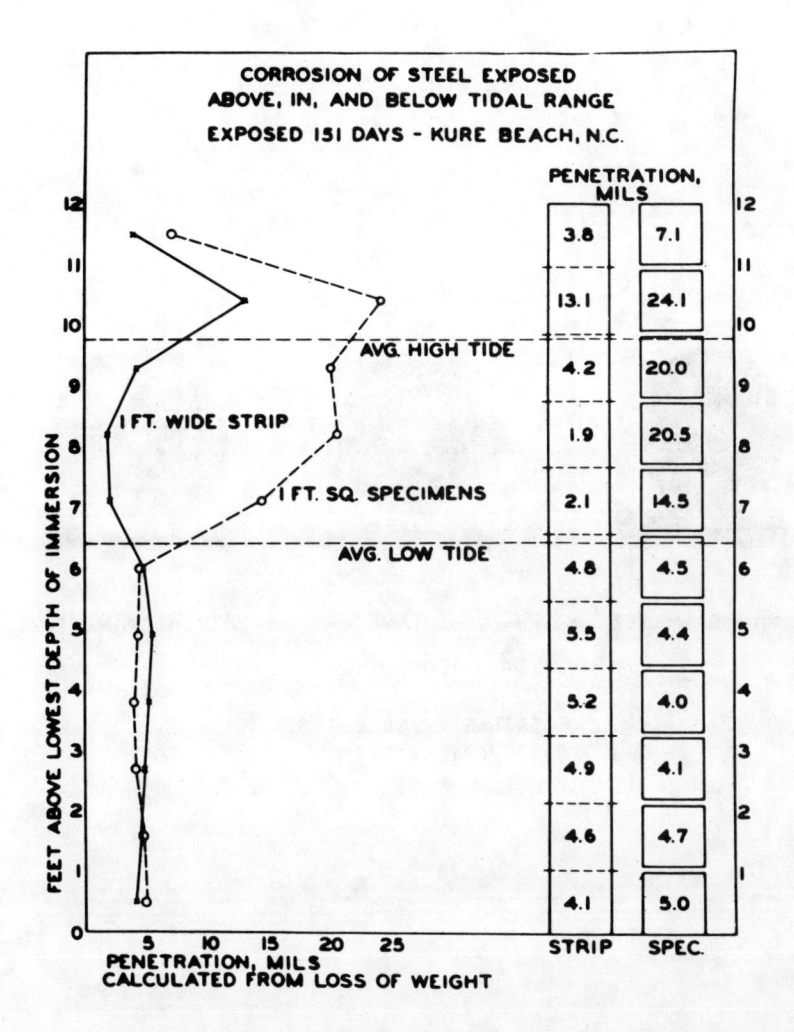

Fig. 8. Comparison of continuous and isolated specimens in tidal zone.

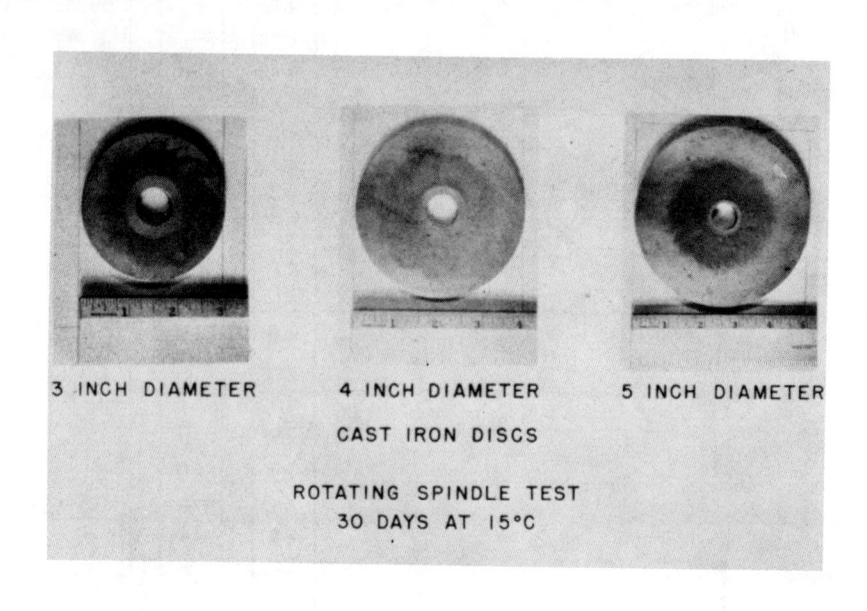

Fig. 9. Distribution of corrosion on iron discs of different sizes.

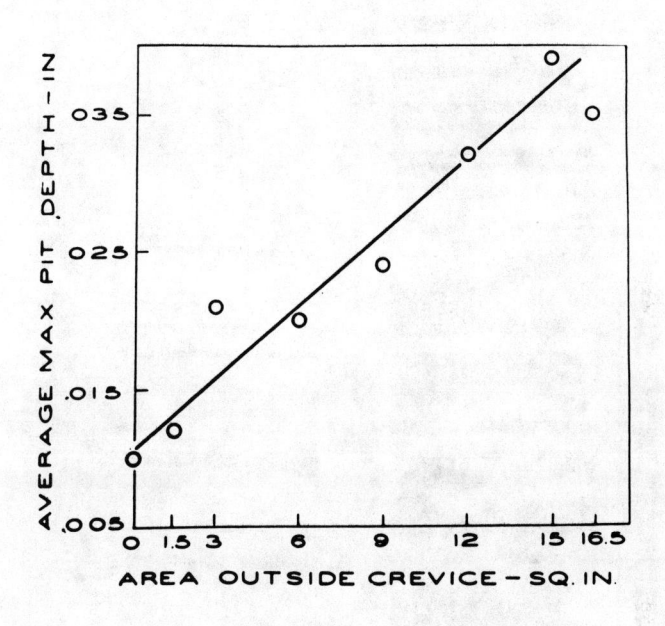

Fig. 10. Effect of area outside crevice on depth of crevice corrosion.

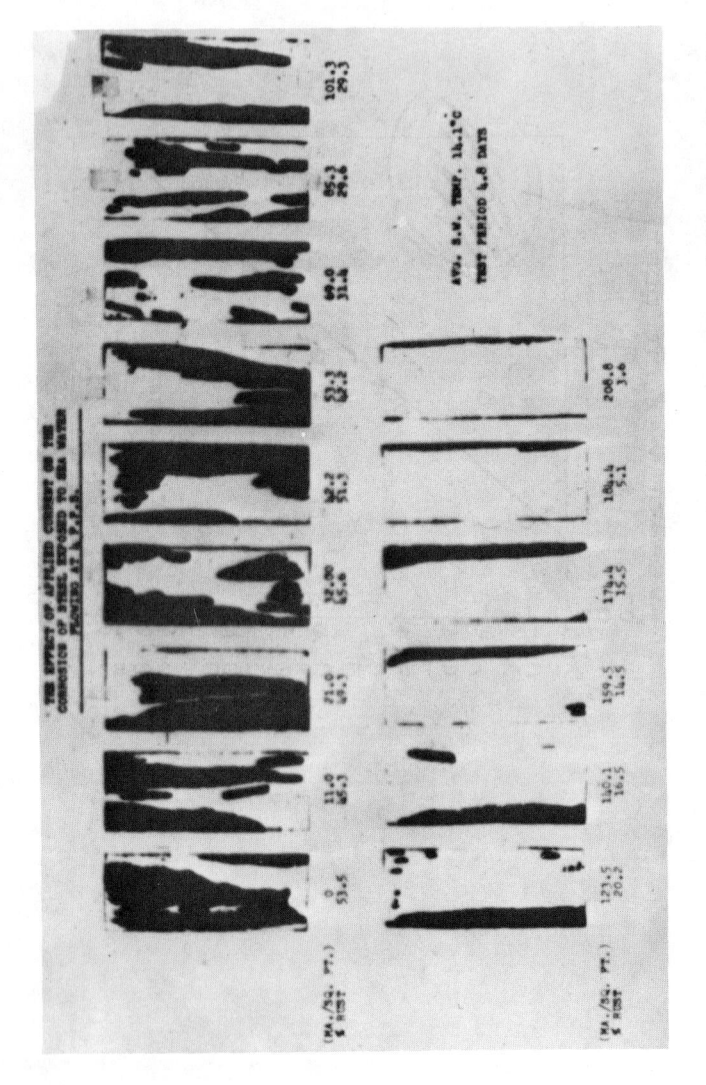

Fig. 11. Effect of current density on area corroded.

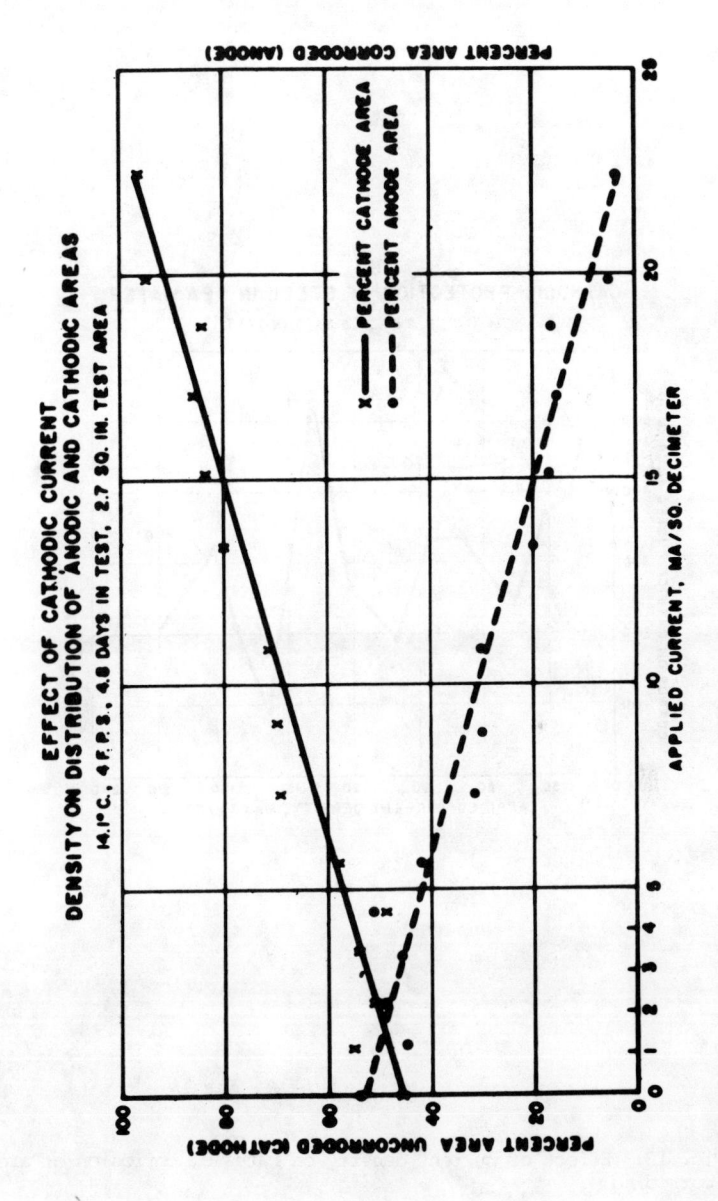

Fig. 12. Effect of current density on per cent area corroded.

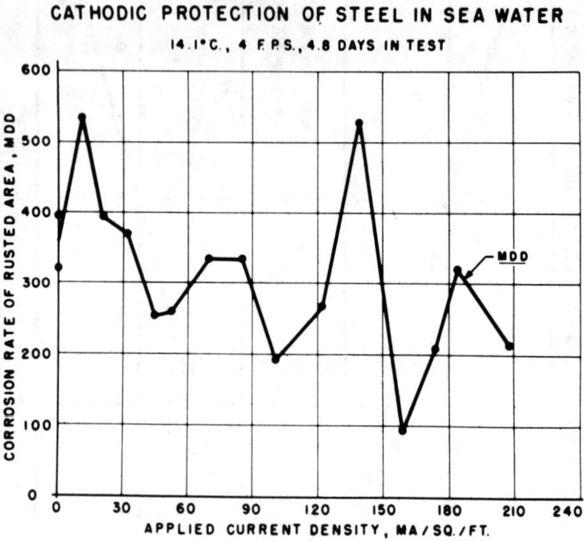

Fig. 13. Effect of current density on rates of corrosion of areas still corroding.

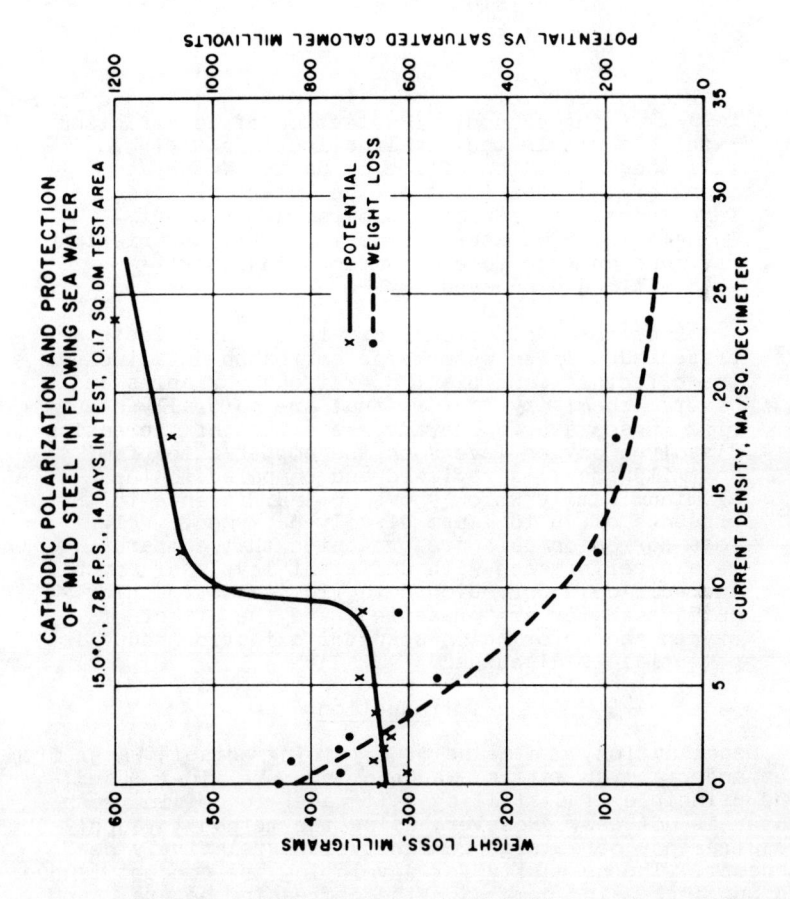

Figure 14

THE INFLUENCE OF WATER CHEMISTRY
ON THE BEHAVIOR OF METALS IN HIGH-TEMPERATURE SEAWATER

Oliver Osborn and Alan L. Whitted
The Dow Chemical Company
Texas Division, Freeport, Texas 77541

ABSTRACT

Lack of sufficient corrosion data has caused problems for the economical selection of construction materials within the desalination industry. To fill this void, the Office of Saline Water (an agent of the Department of the Interior) is funding a corrosion research program at Freeport, Texas. The OSW materials test program is briefly reviewed and the role of various industry associations discussed.

"Baseline" data from short-term coupon tests are presented. These water-side corrosion data include the effect of some plant operations variables and several chemistry factors that are naturally occurring in seawater. The adverse effect of increasing dissolved oxygen levels on the copper alloy family and mild steel is reviewed and compared to the aluminum family which shows an indifference to the presence of up to 1 ppm dissolved oxygen. Weight loss corrosion data are presented that compare metals performance with different levels of ammonia and sulfide. Corrosion data from tests in chlorinated seawater are presented, and the effect of oxygen and chlorine on seawater oxidation reduction potential is discussed.

Introduction

Desalination, as a means of obtaining potable water from sea and brackish waters, is becoming more widely used. The principle of distilling salt water to obtain fresh water is not new. However, using the method to supply the requirements of a large community is a relatively new concept. The necessity for developing the concept is due to our increasing demand on the decreasing natural supply.

Population growth and industrial expansion have put greater demands on the potable water supply in this country. Evidence indicated an ever-increasing problem of supplying this demand. To meet the problem in a positive manner, the United States Government established the Office of Saline Water as an agent within the Department of the Interior. The purpose of the Office of Saline Water is to conduct a research and engineering program for the development of new or improved processes for low-cost desalination of seawater and brackish waters.

Distillation is presently the most widely used method of desalination. Considerable work has been done in the design and engineering of plants of this type; however, economical material selection has been a problem. Many corrosion studies have been performed by LaQue and others in the seawater environment. However, there were very few data available regarding the performance of the various metal families in the specific environment of interest-- treated and concentrated seawater in heat exchanger tubing, flash evaporators, etc. Experience has shown that hot seawater is extremely corrosive, and unless it is carefully treated with chemicals, it will severely attack most of the metal families. Because of the economic importance of material selection for desalination plants, the Office of Saline Water has sponsored extensive work to investigate the economics of various metals and the water treatments which must be employed for their use.

The OSW Materials program has two basic, long-range goals: (1) Reduce cost of water from large-scale plants of the future through selection of more economic materials of construction; and (2) Reduce cost of water in present and future plants by demonstrating proper control of the water chemistry factors which relate to materials performance.

Materials Test Center

A site was provided by the Office of Saline Water for corrosion testing. The OSW Materials Test Center is located at Freeport, Texas, next door to their Test Bed Plant No. 1. (Figure 1)

The Materials Test Center has a fivefold objective:

1. To provide a site and service in which:
   a. Joint OSW-Industry materials programs can be conducted.
   b. OSW-sponsored materials programs can be carried out.
   c. Unexplained materials failures in operating plants can be studied under simulated conditions.
   d. New candidate materials can be evaluated.

2. To provide and demonstrate techniques for control of water chemistry.

3. To demonstrate new techniques for materials evaluation within an operating plant.

4. To provide technical service in the materials field to OSW and concerned metal industries.

5. To implement technical findings.

This Test Center provides a site not only for the OSW-sponsored materials programs but also for industry. The industry metal-producing associations (American Iron and Steel Institute, Copper Development Association, Aluminum Association) are involved. Each association has its own test plant for testing the materials of its particular interest. The block diagram (Figure 2) shows the general layout of the Materials Test Center. The Test Center is built in a T-shape. In one wing of the T, there are offices for the technical staff. The other wing encompasses the Analytical and the Metallurgical Laboratories. The leg of the T is the control room which houses the control panels for the test plants that are situated on the outside slab.

All of the test units at the Center are operated by Dow personnel under contracts with the associations and under a joint agreement with the Office of Saline Water. The Office of Saline Water furnishes each association a site for its test units, utilities, and analytical and metallurgical backup. Each association is responsible for its own program and funds it accordingly.

MTC Corrosion Test Program

Corrosion testing at the Materials Test Center was begun in a single loop funded by the Office of Saline Water. The original work was performed to determine the performance of six alloys of the copper family. The environment was treated seawater with no evaporation. The controlled variables were temperature, pH, dissolved oxygen content, and velocity.

Since the initiation of the program, selected alloys of all the metal families have been tested in an expanded OSW miniplant. The results of this testing are considered as baseline data because the tests are short-term and performed with coupons. This baseline information forms the substance for determining which alloys to consider for further testing and what environmental variables to investigate further.

The Test Center has grown with the addition of the three association corrosion test units and an additional OSW test loop. These units are actually miniature desalting plants. Testing is done on tubular products in actual service heat exchange conditions. Performance of nonmetallics for desalting service is being tested in a separate test unit.

Source of Test Water

All seawater for testing comes from the Gulf of Mexico off the coast of Texas at Freeport. The water comes through the Dow Chemical Company intake system. It is pumped from the intake pit on the facility site to the 200-gpm treating plant of the Test Center. Here, the seawater receives its chemical treatment. Figure 3 is a line diagram of the treating plant. The pretreatment is designed to remove carbonates and oxygen. The carbonates are removed because they promote low-temperature scaling when seawater is heated to 140°F. Oxygen is removed as it is known to be responsible for much of the corrosivity of seawater. (However, it will be shown that the oxygen corrosivity is not all-inclusive for all metal families).

The incoming seawater is acidified to 4 pH in an open vessel for removal of carbonates and bicarbonates. It is then passed into a packed tower where the combination of vacuum and stripping steam serve to sweep out the non-condensible gases. This water is forwarded into a 10,000-gallon storage vessel that is padded with nitrogen. At this point the seawater is at a pH of 4 with less than 3 parts per million dissolved carbon dioxide and less than 5 parts per billion dissolved oxygen. This water is pumped through a closed, recirculating header for use at the various test units. The chemistry of the water is modified at the test units to fit the requirements of any particular test.

In addition to the 200-gpm treating plant, there is a softening unit with its own pretreating equipment. This softening plant removes magnesium and calcium from treated water by an ion-exchange method. Testing will be conducted at temperatures up to 325°F using the treated, softened seawater. The present temperature limit is 250°F.

Chemistry Variables

The controllable seawater chemistry factors are temperature, pH, oxygen, and carbonates. From a scaling and heat transfer aspect, we know that it is essential to have the carbonates as low as possible. This is accomplished in the treating plant. The part that oxygen plays in enhancing the corrosivity of seawater is also well documented. However, even in the low part-per-billion range, it is detrimental to most metals while a few seem to react favorably. The level of oxygen entering a desalination plant is dependent on the efficiency of the treating facility.

Due to the characteristics of seawater, many uncontrollable chemistry variables enter the picture. Seawater composition varies with locale and time of year as well as with the variance of pollutants that our society is dumping into it. Variables that have been studied to date are: chlorine as a biological growth inhibitor; ammonium ion as a natural element due to nitrate breakdown; and hydrogen sulfide as it is present from natural biological decay and pollution sources.

Corrosion Test Results

Corrosivity of Oxygen. It is well established that oxygen increases the corrosivity of most metals and alloys in water. Dissolved oxygen acts as a depolarizer by accelerating the cathodic reaction. Therefore, corrosion can be reduced by deoxygenating the water by means of deaerating systems or oxygen-scavenging chemicals. Deaerating systems seldom remove 100 percent of the oxygen. Therefore, corrosion tests were performed to determine the effect of different levels of dissolved oxygen in the parts-per-billion range. Figure 4 shows the corrosion rates experienced by mild steel and two copper alloys in 250°F, treated seawater with different dissolved oxygen levels up to 500 ppb. In the short-term tests, mild steel corroded at a rate of 10 to 15 mils per year even at zero oxygen. (The zero oxygen level was obtained by adding a scavenger to deoxygenated water.) The aluminum alloys exhibited much more tolerance to the presence of oxygen. The data plotted in Figure 5 illustrate this fact. With all other conditions held constant, the aluminum alloys tested experienced the same rate of metal loss over the range of 5 ppb to 1 ppm dissolved oxygen.

Aluminum is normally considered a pitting metal. Therefore, can the coupon weight loss data have any meaning? We believe it does. The coupon surfaces did show evidence of scattered pitting. However, all the pits were flat-bottomed. This seemed to be indicative of dormant pits. Long-term tests at the facility further supported this supposition. Aluminum tubes tested in heat exchanger service also exhibited the flat-bottomed characteristic as shown in Figure 6. Depth measurements at intervals during the test proved that the pits were not increasing in depth up to a period of 24 months. Most cases actually demonstrated a decrease in pit depth with longer exposure times, thus indicating passivating pits with uniform surface corrosion.

Interrelationship of Oxygen and Temperature. Due to the nature of the distillation process, it is generally desirable to operate at the highest temperature practical. As a general rule, chemical reaction rates increase with a temperature increase. However, this is dependent on the amount of species present and the rate-controlling steps in the reaction. The bar graph shown as Figure 7 illustrates the interrelation of dissolved oxygen level with temperature. At less than 5 ppb dissolved oxygen, the copper alloys tested experience only a slight increase in corrosion rates at the higher temperature. However, the mild steel corrosion rate at 250°F is nearly double the rate experienced at 180°F. The bottom set of data on the graph shows the increasing deleterious effect of oxygen at higher temperatures. At 180°F CDA alloy 715 (70/30 Cu/Ni) experiences essentially the same corrosion rate with both levels of oxygen. When exposed to 250°F seawater with 500 ppb dissolved oxygen, the rate is increased, and the performance of all the other alloys exposed is drastically worsened. Aluminum alloys exposed under these same conditions experience only slightly increased corrosion rates in the high-oxygen, high-temperature environment.

Chlorination for Antifouling. In whatever part of the world a seawater operation is established, some form of antifouling measure must be taken. The most popular is the procedure of chlorination. Some work had been done previously by others to determine the corrosive effect of chlorinating raw seawater. However, no reference was found relating the corrosivity of chlorinated seawater after it passed through a treating facility.

To perform this series of experiments the system was set up as shown in the line diagram of Figure 8. Chlorine gas was bubbled into a mix tank fed from the 200-gpm treating plant. This chlorinated water was passed through an

auxiliary deaerator before passing over the test coupons in
the OSW corrosion test units. It was found that when the
chlorine residual before the deaerator was less than 1 part
per million, there would be no residual after the deaerator.

Corrosion tests in this system, with the chlorine
residual in the range of 1 ppm before the deaerator, showed
little evidence of any increase in mild steel corrosion
rates. Figure 9 is a bar graph showing the different corrosion rates experienced by four copper alloys. A 21-day
test was performed on two sets of coupons. One set received
the chlorinated seawater while the other set, exposed at
the same time, had no chlorine added. It is obvious from
data analysis that the chlorinated waters increase the rate
of metal loss on the copper alloys. Subsequent to this
test, the coupons that had been exposed to the nonchlorinated water were exposed to chlorinated water. The photo
in Figure 10 illustrates the detrimental effect of the
chlorinated water. The protective film that had been
formed in the initial, nonchlorinated exposure was removed
during the exposure to the chlorinated water.

Seawater Oxidation Potential--Oxygen vs. Chlorine. Since
there was no residual chlorine in the water reaching the
test coupons, the question arises as to what is causing the
increased corrosion rates. It was proposed that the
chlorine acts as an oxidizing species. It perhaps reacts
with the heavy metal species in the seawater raising their
oxidation state; such as cuprous to cupric. If the oxidation state of these heavy ions is indeed changed, the
oxidation potential of the seawater should, in turn, be
changed. This would be reflected in the redox potential
of the seawater as measured between a platinum versus
saturated calomel electrode. To experimentally determine
if any change in the seawater redox potential was occurring,
a micarta weeping electrode bridge was placed in the
dynamic test loop. The weeping bridge is shown in
Figure 11. The chlorine injection was changed so controlled amounts of chlorine could be injected directly
into the test loop header. Other electrodes were also
installed in the loop. Along with the platinum electrode
the other electrodes included alloy 20 stainless, CDA
copper alloys 122 and 706, and mild steel.

The initial steps of this test determined the potentials
of the platinum and other alloys versus calomel with varied
amounts of dissolved oxygen. These results are shown in
bar graph form in Figure 12. With the addition of dissolved oxygen up to the 1.5-ppm level, the seawater redox
potential increases. The potentials of the other alloys,

except for mild steel, also are raised. The hot seawater was again reduced to a level of less than 5 ppb dissolved oxygen, and chlorine addition was initiated. Figure 13 shows the resultant potentials with 2.8 and 5.6 ppm chlorine injected. At these injection levels, there is no detectable chlorine residual. Both levels of chlorine produce the same potentials indicating a maximum redox is reached where addition chlorine has no further effect on the redox potential. The potentials attained with the chlorine are about what is experienced with 100 ppb dissolved oxygen. This indicates that addition of chlorine to a deoxygenated system will increase the oxidation-reduction potential of hot seawater to the same level as that of seawater with 100 ppb dissolved oxygen.

Ammonium Ion in Hot Seawater. Ammonium ion is a naturally occurring component in seawater. It arises from nitrate breakdown during biological decay. It also is found in heavy concentration in polluted waters and areas where erosion washes fertilizers from soils. During a close examination of the raw seawater received at the Test Center, it was found that the ammonium concentration would fluctuate from less than 100 ppb to 1 ppm in a week's time. To examine the effect that the ammonium ion may have on water-side corrosion, a 30-day test was performed. Ammonia gas was used to get ammonia into the system. It was used in place of the sodium hydroxide to control the seawater pH. Controlling the treated seawater pH at 7 gave an ammonium concentration of approximately 4 ppm. Figure 14 compares the corrosion rates of the alloys tested in ammoniated water to those tested without ammonia added. There was essentially no difference in the corrosion characteristics between the two systems except for the aluminum. The aluminum coupons in the ammoniated seawater corroded at half the rate of those in the nonammoniated environment. This test was only short-term and observed the effect on the water side. It is proposed that a detrimental effect would be found on the vapor side of brass since samples of vapor in another unit showed evidence of ammonia carryover. Therefore, more extensive work is needed in this area.

Hydrogen Sulfide Pollution. Many coastal areas where water intakes may be found are contaminated with sulfide. This sulfide results from the decay of organic materials--naturally occurring from normal biological decay--and from the pollution produced from the wastes created by civilization. To determine if sulfide-bearing hot seawater had any detrimental effect on several common alloys, a comparative test was performed in two separate test loops. One loop received normally treated seawater at 225°F with less than 5 ppb dissolved oxygen. The other loop received the same water with approximately 1.4 ppm hydrogen sulfide added. Figure 15 shows the results of this test in bar graph form for the mild steel and copper alloys tested. Mild steel was essentially not adversely affected by the hydrogen sulfide addition, the rates being 10 to 10 1/2 mils per year. Of the copper alloys tested, only the aluminum brass displayed an indifference to the presence of the sulfide. The phosphorous-deoxidized copper and the two copper-nickels each had higher rates in the hot seawater with sulfide.

Corrosion films that formed on the aluminum brass (CDA alloy 687) in both environments were thin and tenacious and had the same dark appearance. The film on the copper-nickels exposed to the sulfide-bearing water was much thicker and more brittle than the film formed in the non-sulfide environment. X-ray diffraction of scrapings from a 70/30 copper-nickel coupon showed it to be composed of Ni_3S_2, NiS, and $Cu_{1.96}S$.

Conclusions

1. It is essential to remove oxygen from treated seawater to the lowest level possible except for aluminum alloys. Aluminum is not adversely affected by oxygen levels up to 1 ppm in neutral seawater.

2. When pits appear in aluminum exposed to treated seawater, they appear to be self-passivating.

3. As operating temperatures are increased, the presence of oxygen becomes more detrimental to the corrosion performance of mild steel and the copper alloys.

4. Chlorination of seawater increases the oxidation potential of seawater to the equivalant of 100 ppb dissolved oxygen.

5. Ammonium in hot seawater appears to have no detrimental effect on water-side corrosion.

6. Mild steel and aluminum brass are not adversely affected in short-term corrosion tests with sulfide present. Copper and the copper-nickels, however, exhibit an extreme sensitivity to sulfide.

Acknowledgment

The work discussed was funded by the Office of Saline Water, U. S. Department of the Interior, under contract No. 14-01-0001-2150, Materials Division, Dr. F. H. Coley, Division Chief. The authors appreciate the many seawater analysis performed by B. Paul Webb and the many hours of helpful discussion with him. Our thanks go also to Steve Hill who operated the test plant and performed the major portion of the corrosion tests.

Bibliography

Schrieber, C. F., Billy D. Oakes, and Alan L. Whitted, "Behavior of Metals in Desalination Environments: Fifth Progress Report", paper presented Corrosion '72, NACE, St. Louis. Mo. (March, 1972).

Verink, Ellis D., Jr., "Performance of Aluminum Alloys in Desalination Service", paper presented at Corrosion '72, NACE, St. Louis, Mo. (March, 1972).

Whitted, A. L. and C. F. Schrieber, "Some Effects of Altered Seawater Chemistry on Materials Performances in Desalination Plants", *Water, 1972*, AIChE Symposium Series.

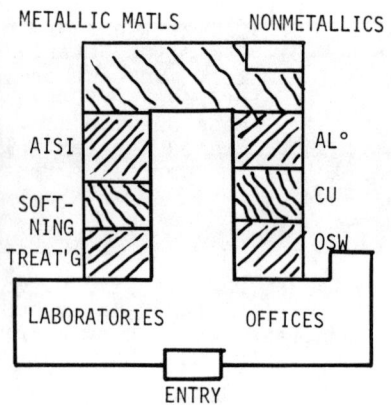

Figure 1 - OSW MATERIALS TEST CENTER

Figure 2 - FLOOR PLAN AND TEST AREA OF MATERIALS TEST CENTER

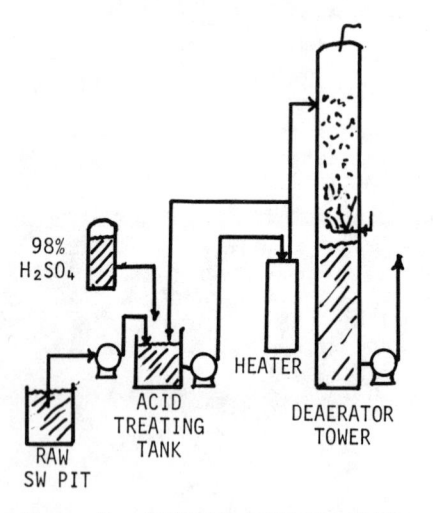

Figure 3 - SEAWATER TREATING PLANT

Figure 4 - CORROSION RATE VERSUS DISSOLVED OXYGEN (Ferrous and Copper Alloys)

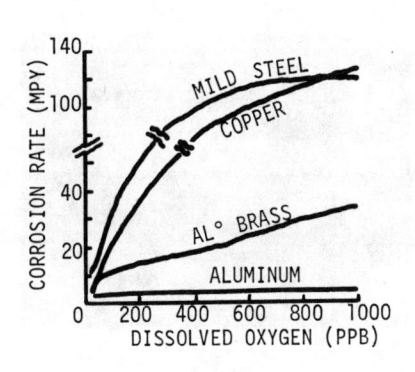

Figure 5 - COMPARING CORROSION RATES OF ALUMINUM TO MILD STEEL AND COPPER AS AFFECTED BY DISSOLVED OXYGEN

Figure 6
PHOTOMICROGRAPH SHOWING FLAT-BOTTOMED CHARACTER OF AL° PIT

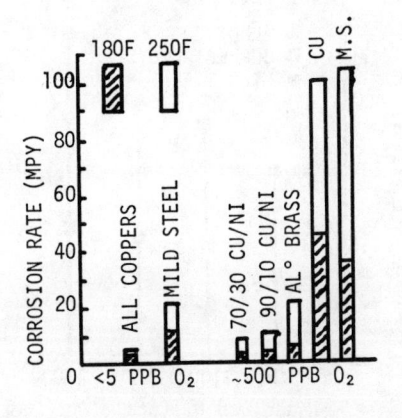

Figure 7 - MILD STEEL AND COPPER CORROSION RATES AT TWO OXYGEN LEVELS AND TWO TEMPERATURES

Figure 8 - FLOW SHEET OF CHLORINE INJECTION

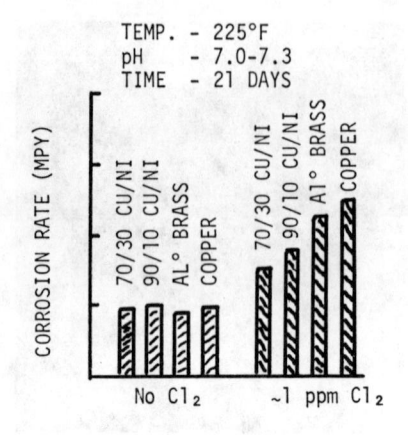

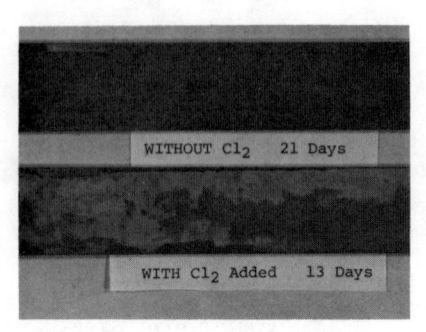

Figure 9 - COPPER ALLOY PERFORMANCE IN HOT, CHLORINATED SEAWATER

Figure 10 - PHOTO SHOWING FILM REMOVAL EFFECT OF HOT, CHLORINATED SEAWATER ON PRE-EXPOSED ALUMINUM BRASS COUPON

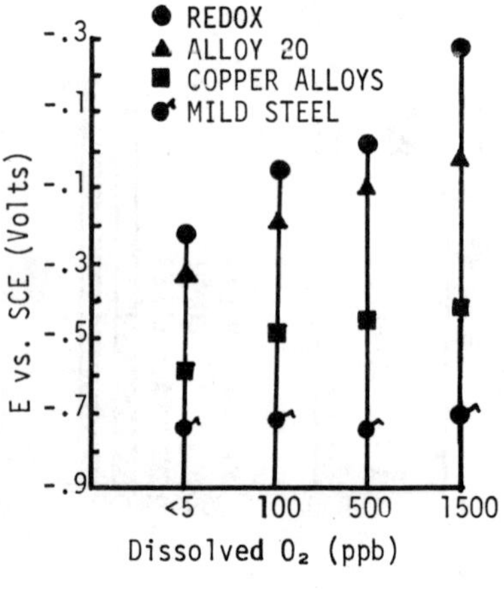

Figure 11 - MICARTA WEEPING BRIDGE USED FOR POTENTIAL MEASUREMENTS

Figure 12 - ELECTRODE POTENTIALS VS. DISSOLVED OXYGEN CONCENTRATION IN HOT TREATED SEAWATER

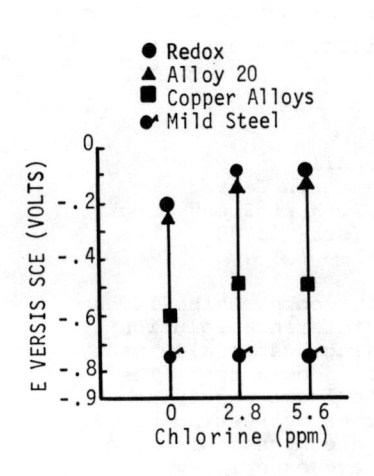

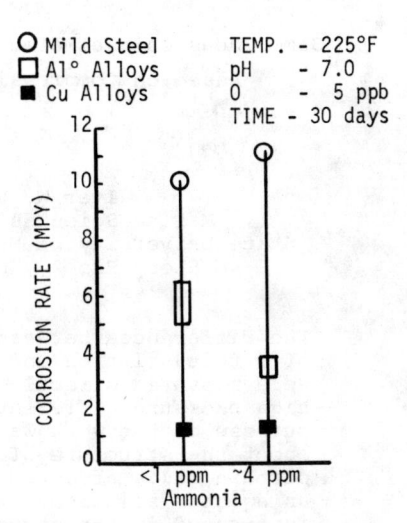

Figure 13 - ELECTRODE POTENTIALS VERSUS CHLORINE IN HOT TREATED SEAWATER

Figure 14 - METAL CORROSION RATES IN SEAWATER WITH HIGH AND LOW AMMONIA LEVELS

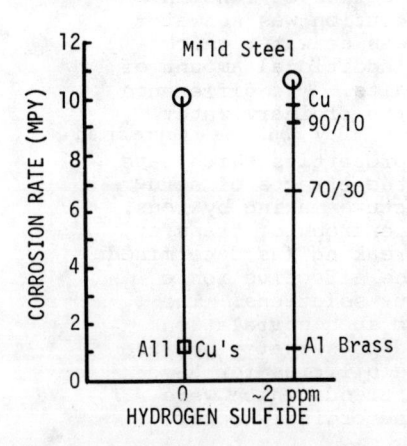

Figure 15 - MILD STEEL AND COPPER CORROSION RATES IN TREATED SEAWATER WITH AND WITHOUT H_2S ADDED

Some Ideas on the Molecular Structure of Seawater Based on Compressibility Measurements

Iver W. Duedall
Marine Sciences Research Center
State University of New York at Stony Brook
Stony Brook, New York 11790

The differences between the compressibilities of a test solution and a reference solution were measured directly using a specially made high pressure differential densimeter. The purpose of the work was to learn something about the structure of water by testing how certain salts affected the compressibility of an aqueous solution. We are especially interested in the structure of water in seawater. Two series of runs were made: A series of runs in which the reference solution was ordinary water and the test solution was a single salt solution containing one of the major sea salts, and a series of runs in which the reference solution was seawater and the test solution was seawater which had been doped with an additional amount of one of the major sea salts. The difference compressibilities for the ordinary water system were found so that they could be correlated with physico-chemical properties which have been used to estimate the effects of structure breaking and structure making by ions. These properties were: entropy of transfer, excess hydrogen bond breaking (as determined by NMR studies), and the effective ionic radii of ions in aqueous solutions. The information gained from such correlations can probably be applied to seawater because the difference compressibilities for a seawater system showed trends which were similar to those for the ordinary water data.

INTRODUCTION

In the past, chemical oceanographers have mainly been concerned with the analytical composition of seawater while neglecting the more fundamental aspects. For instance, prior to 1960 virtually no attention was given to the role played by the water in seawater, despite the obvious fact that seawater is comprised mainly of water. Only in the past few years have chemical oceanographers begun the task of defining seawater in terms of ion-ion interactions and ion-water interactions. Indeed, I find it very significant that the first chemical oceanography textbook to deal with the molecular structure of seawater and to show how water structure affects marine chemistry was published in 1969 (Horne, 1969).

The problem of devising a structure for seawater is complicated for the following reasons: 1) there is no universally accepted theory for the structure of "pure" water; 2) the effect of ion-water interaction on the water structure of simple two-component systems is not fully understood; and, 3) the combined effects of ion-ion interactions and the multitude of ion-water interactions on the structure of water in a multicomponent system such as seawater is unknown and is perhaps beyond our present level of comprehension. In spite of this outlook, it is possible to learn something about the structure of seawater by discovering how the addition of certain salts affects its physical properties. In this approach the added salt acts as the agent which modifies the structure of the bulk water. The extent to which the water in seawater is modified by the salt is related to a measurable physical property.

In this paper I am considering the question of water structure in seawater by determining experimentally (1) the compressibility of water in single salt solutions and (2) how the major sea salts of seawater affect the compressibility of water in seawater. The compressibility approach to the problem is analogous to that taken for determining hydration numbers, with the distinction that the results will be discussed in terms of structure breaking and structure making.

EXPERIMENTAL METHODS

A specially designed high pressure differential densimeter was constructed to measure differences in compressibilities. By comparing differences in compressibilities, denoted by $\Delta\beta$, it should be possible to determine the extent to which ion types influence water structure. The theory, design and fabrication, and operation of the apparatus is described in detail in another paper (Duedall and Paulowich, in press). The densimeter is schematically illustrated in Fig. 1 and consists of three main parts: 1) a pair of closely matched stainless steel bellows units which are mounted independently of each other but arranged back-to-back inside the same pressure vessel; 2) two closely matched electromechanical transducer assemblies; and, 3) a specially constructed detector assembly. One bellows is filled with a reference solution and the other one is filled with a test solution. As hydrostatic pressure is applied to the bellows assembly, the resulting changes in compressed volumes of the solutions in the bellows are determined by sensing the differential axial compression of the bellows using linear variable differential transformers. The core of each transformer unit is mechanically linked to each bellows while the windings of the unit are situated outside the pressure vessel. The output voltages from transformers are fed into an operational amplifier which functions to rectify the voltage signals from the two transformers and to subtract the two signals; this difference voltage is directly related to the difference between the compressibilities of the two solutions. The relationship between $\Delta\beta$ and the experimental measurements can be found in the paper by Duedall and Paulowich.

The compressibility experiments were conducted in the following manner. An initial series of experiments were carried out to determine $\Delta\beta$ for systems in which the reference water was ordinary water (i.e. distilled water). The salts tested were for the most part the sea salts NaCl, Na_2SO_4, $MgSO_4$, and $MgCl_2$; some additional measurements were made for LiCl, KCl, CsCl, NaF, NaI, $CaCl_2$ and $BaCl_2$. The second series of experiments were runs in which the reference solution was seawater and the salts tested were the sea salts. The $\Delta\beta$ measurements for the ordinary water runs provide the basis for a comparison of the water in seawater with ordinary water. The $\Delta\beta$ values in these two series of runs were determined at nominal concentrations of 0.13 m (m = molal) and 0.26 m at temperatures of 2°C and 15°C.

EXPERIMENTAL RESULTS

Figs. 2 and 3 show $\Delta\beta$, at 15°C, as a function of hydrostatic pressure. In Fig. 2, the reference solution is ordinary water, and the test salt concentration is 0.26 m. In Fig. 3, the reference solution is seawater (35 °/oo), and the test salt concentration is 0.26 m. An evident feature of the data shown in Figs. 2 and 3 is that the results for certain salt types tend to group together; that is, particular salt types produce about the same effect on the compressibility of the water in the test solution. This behavior is especially true for the alkali-metal chlorides where it can be seen that LiCl produces about the same decrease in compressibility as does CsCl in spite of the fact that the crystallographic radii of the two cations differ by almost a factor of three. The $\Delta\beta$ for $MgCl_2$ and $CaCl_2$ are about twice the values for the alkali-metal chlorides. If these data are considered on a equivalent concentration basis instead of a mole basis (to normalize the chloride concentration effect), then $1/2\,\Delta\beta$ for $MgCl_2$, $CaCl_2$, and $BaCl_2$ is about the same as the $\Delta\beta$ value for each of the alkali-metal chlorides. Figs. 2 and 3 also show that the sulfate salts affect compressibility more than the chlorides and other halide salts. The interpretation of the sulfate results is, however, not straightforward for several reasons. First, SO_4^{2-} by virtue of its tetra-oxy structure and its minus two charge would not be expected to interact with water in the same way as Cl^- does. Second, SO_4^{2-} forms ion-pairs with cations and it is usually assumed that Cl^- does not form ion-pairs in solutions whose composition is similar to that studied in this work. The structure of the water near or within sulfate metal ion-pairs may be very different from that near free SO_4^{2-}. That sulfate salts produced the greatest effect on compressibility must be due in some way to a relatively large interaction of the sulfate ion with neighboring water molecules. Nearby water may hydrogen bond with oxygen atoms of the sulfate ion, because hydrogen bonds are present in the crystalline hydrates of sulfate salts (Falk and Knop, 1972). Samoilov (1972), however, implies that the sulfate ion does not bond appreciably with water. He bases his reasoning on the reported shift in the IR absorption spectra of the hydrated salts of sodium perchlorate and sodium sulfate. Apparently when these hydrates are dissolved in water, the spectra indicating hydrogen bonding undergoes a shift in the direction of decreased hydrogen bonding. Subramanian and Fisher (1972) also concluded (based partly on work reported by Samoilov) that perchlorate does not hydrogen bond with water in solution.

There is plenty of experimental evidence (summarized by Kavanau, 1964) to show that anions do not interact with water in the same way as do cations. Anions probably break down water structure, while cations enforce water structure. Actually, cations also break down water, but the reorientation of the water by the cation is thought of as structure making. The mechanisms of structure breaking and structure making will be discussed in the next section. $\Delta\beta$ can be related to entropy of transfer, excess hydrogen bond breaking, and to the effective radii of ions in solution.

STRUCTURE BREAKING BY IONS

Greyson (1967) determined entropies of transfer, ΔS from D_2O to H_2O, for most of the alkali-metal halides. His results showed that the halides have a greater structure altering effect on water than do the alkali-metal ions. Greyson reasoned that the entropy of transfer for the salt is due entirely to the difference between the structure of D_2O and H_2O. A large negative entropy of transfer may be interpreted as due to structure breaking of D_2O by the salt. Fig. 4 shows a comparison Greyson's entropy of transfer with the $\Delta \beta$ for corresponding salts. The fact that both lines (Fig. 4) are sloping in the same direction suggests that the variation of $\Delta \beta$ among different salts is due to the variation of the amount of broken water structure. It is obvious from the slopes of the dashed lines (Fig. 4) that Greyson's EMF method is more sensitive to structure breaking by anions than is the compressibility method.

Hindman (1962) used a structural parameter, which he called n_i, in a chemical shift formula to account for the number of additional hydrogen bonds either formed or broken due to the action of a ion. n_i may be considered an adjustable parameter which permits the assignment of zero values for the hydration of anions. The use of such a parameter provided Hindman with a convenient way to overcome a negative hydration number, however, the initial outcome of a computed negative hydration number demonstrates rather clearly that (1) anions are not hydrated, and (2) anions are structure breakers to a much greater extent than are cations. Fig. 5 shows a plot of $\Delta \beta$ versus n_i. As previously mentioned, n_i is the number of hydrogen bonds formed in excess of those involved in the reorientation of water near an ion; n_i is zero for cations (Li^+ is an exception; see Fig. 5), but n_i is negative for anions (F^- is an exception) because "excess" hydrogen bonds are broken by anions. $\Delta \beta$ for a series of halides having a common cation should correlate with the n_i for anions because the compressibility of a solution containing a strong structure breaker, such as I^-, will not be as great as for a solution containing a weak structure breaker (e.g., Cl^- or F^-).

That size is an important factor in the propensity of ions to cause structure breaking is also shown by the work of Padova (1963) who calculated the intrinsic "effective" radii (r_e) of ions in aqueous solutions. Fig. 6 shows $\Delta \beta$ plotted as a function of r_e for the series NaF, NaCl, and NaI. The $\Delta \beta$ values correlate with r_e. This was not unexpected because it was already shown that the $\Delta \beta$ values for the sodium halides correlated with the degree to which

sodium halides break hydrogen bonds (Fig. 5). According to
Padova, Cl⁻ and I⁻ are structure breakers because the
r_e of ions in solution is greater than the minimum radius
necessary to cause complete dielectric saturation of the
local water structure. Fig. 7 shows a similar plot for
the cation series LiCl, NaCl, KCl, and CsCl. Here there is
no obvious correlation between the $\Delta\beta$ for the solutions
and the r_e for the cations. Fig. 8 shows another plot
of $\Delta\beta$ versus r_e, but for the series of alkaline-earth
chlorides $MgCl_2$, $CaCl_2$, and $BaCl_2$. Here there exists a
good correlation between $\Delta\beta$ and r_e. It is particularly
interesting that for the alkaline-earth salts, the slope
of $\Delta\beta$ versus r_e is positive, but for the halide salts the
slope of $\Delta\beta$ versus r_e is negative. Both F⁻ and Mg^{2+} are
usually considered strong struture makers, yet they appear
to affect $\Delta\beta$ in relatively opposite ways. The same analogy
can, of course, be applied to I⁻ and Ba^{2+} which are generally
classed as structure breakers. (However Ba^{2+}, according
to Padova, is a structure maker.) Such inverse effects, as
shown in Figs. 6 and 8, give added support to the belief
that anions alter water structure in a very different way
than do cations.

SUMMARY AND CONCLUSIONS

The previous discussion dealt mainly with the correlation of the compressibility data for single salt solutions with properties of solutions which have been used to determine structure breaking and structure making. Because the $\Delta\beta$ values for seawater as a reference solution differ little from the $\Delta\beta$ values for water as a reference solution, the ideas of structure breaking and making apply to seawater as well as to ordinary water. The correlations presented here have not led to any "predictive" model for the structure of water near an ion. The present approach has been essentially a "rationalization" of the $\Delta\beta$ results. In this regard, Holtzer and Emerson (1969) demonstrated, in a very dialectic style that "rationalization" of experimental results can lead to contradictory conclusions. As Holtzer and Emerson put it "indeed, in many cases even a single given argument is sufficiently indeterminate as to be capable of producing several contradictory conclusions."

ACKNOWLEDGMENTS

The results and discussion presented in this paper were abstracted from a draft of Ph.D. thesis to be submitted to Dalhousie University, Halifax, Nova Scotia. I am thankful to Drs. Wangersky, Cooke, and Kwak, all of Dalhousie, to Dr. Falk of the National Research Council of Canada (Halifax), and to Dr. Marshall of the Oak Ridge National Laboratory, for their helpful discussions. The experimental work reported here was done while the author was on the staff of the Fisheries Research Board of Canada, Bedford Institute of Oceanography, Dartmouth, Nova Scotia.

REFERENCES

Horne, R. A. (1969). Marine Chemistry. Wiley Interscience, New York.

Duedall, I. W. and S. Paulowich (In press). A Bellows-type Differential Compressimeter for Determining the Difference Between the Compressibilities of Two Seawater Solutions to 900 Bars. Rev. Sci. Instr.

Falk, M. and O. Knop (In press). Water in Stoichiometric Hydrates. In: Water: A Comprehensive Treatise (Ed. F. Franks), Vol. II. Plenum Press, New York.

Samoilov, O. Ya. (1972). Residence Times of Ionic Hydration. In: Water and Aqueous Solutions (Ed. R. A. Horne), Wiley Interscience, New York.

Subramanian, S. and H. F. Fisher (1972). Near-Infrared Spectral Studies on the Effects of Perchlorate and Tetrafluoroborate Ions on Water Structure. J. Phys. Chem. 76 84-89.

Greyson, J. (1967). The Influence of the Alkali Halides on the Structure of Water. J. Phys. Chem. 71 2210-2213.

Hindman, J. C. (1962). Nuclear Magnetic Resonance Effects in Aqueous Solutions of 1-1 Electrolytes. J. Chem. Phys. 43 1000-1015.

Padova, J. (1963). Ion-Solvent Interactions. II. Partial Molar Volume and Electrostriction: A Thermodynamic Approach. J. Chem. Phys. 39 1552-1557.

Holtzer, A. and M. F. Emerson (1969). On the Utility of the Concept of Water Structure in the Rationalization of the Properties of Aqueous Solutions of Proteins and small Molecules. J. Phys. Chem. 73 26-33.

Kavanau, J. L. (1964). Water and Solute-Water Interactions. Holden-Day, Inc. San Francisco.

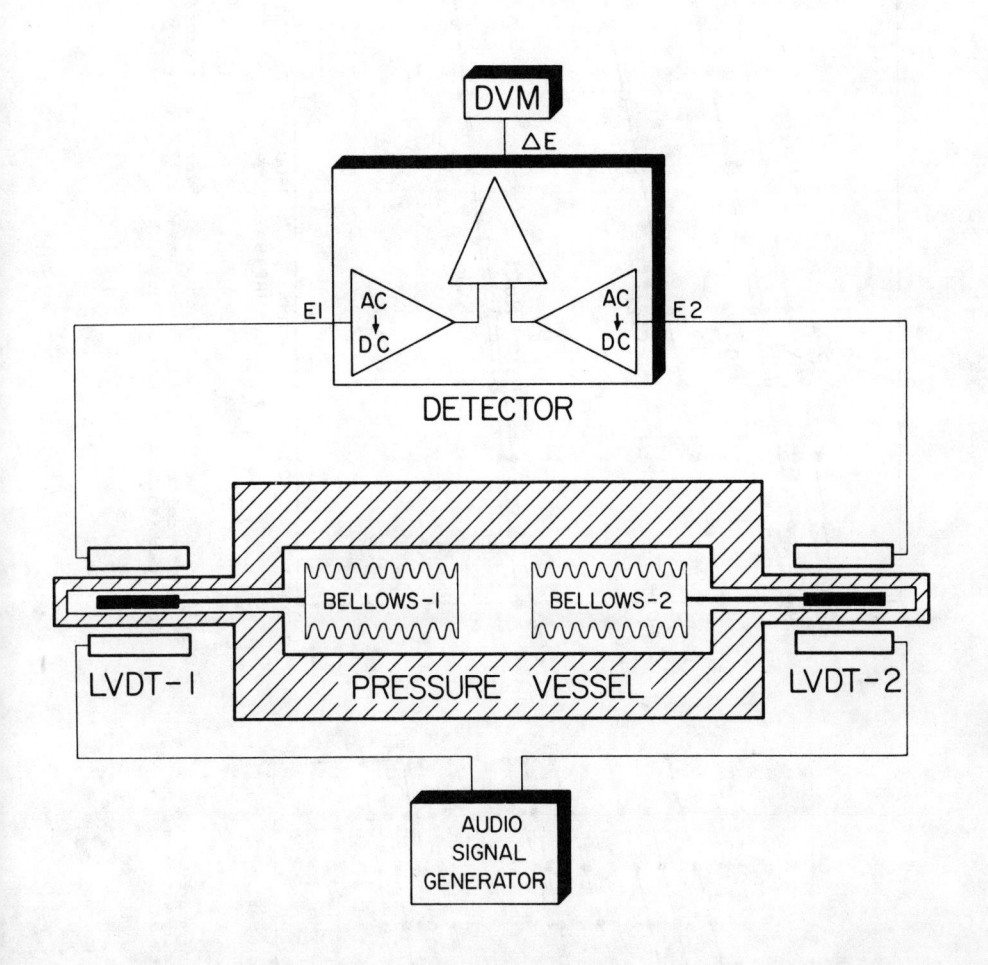

Fig. 1. The differential densimeter (Duedall and Paulowich, In press).

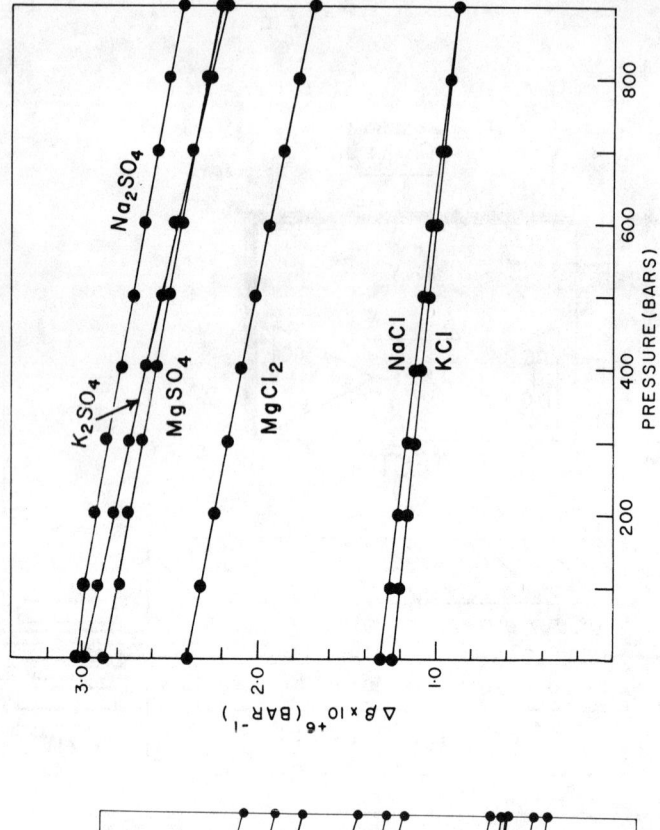

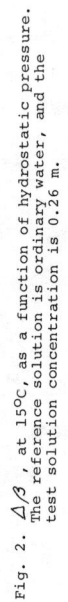

Fig. 3. $\Delta\beta$, at 15°C, as a function of hydrostatic pressure. The reference solution is seawater (35°/oo), and the test salt concentration is 0.26 m.

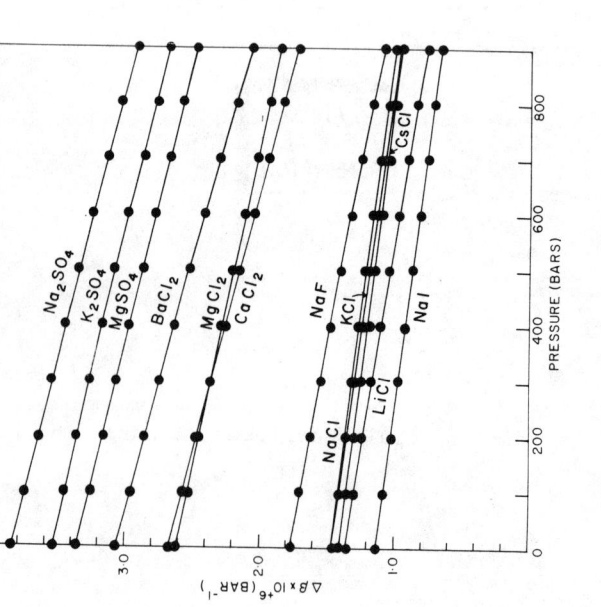

Fig. 2. $\Delta\beta$, at 15°C, as a function of hydrostatic pressure. The reference solution is ordinary water, and the test solution concentration is 0.26 m.

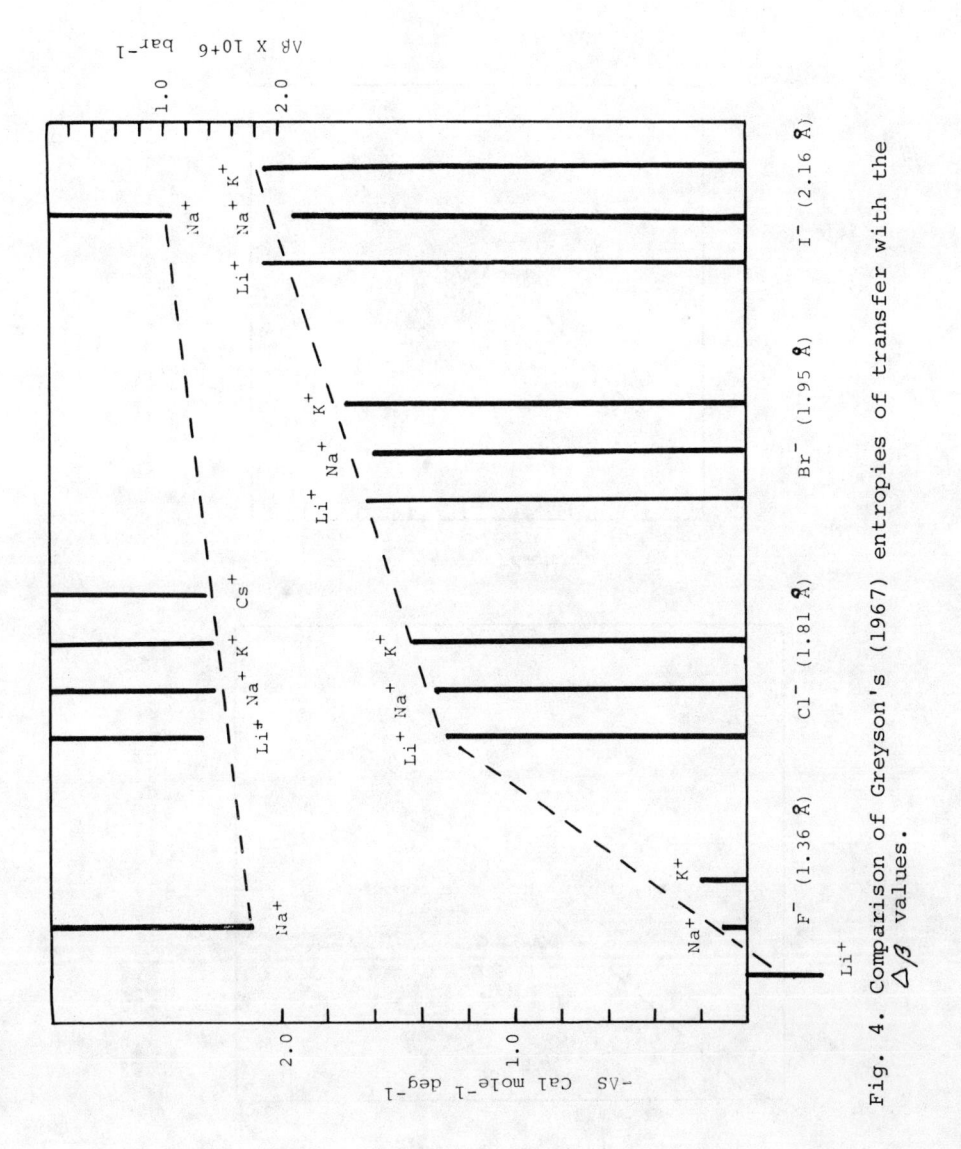

Fig. 4. Comparison of Greyson's (1967) entropies of transfer with the $\Delta\beta$ values.

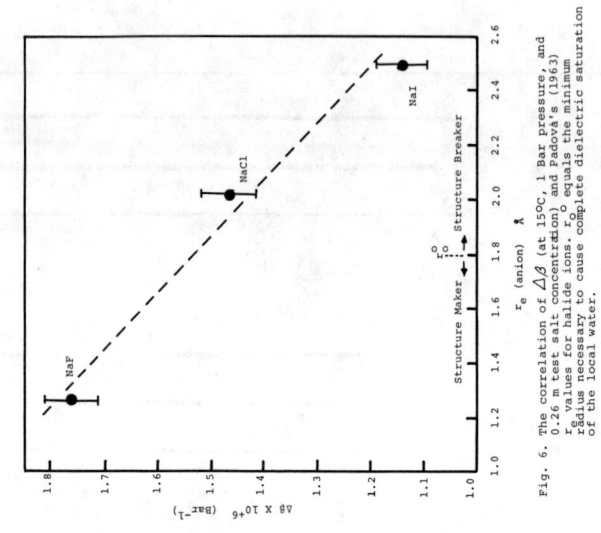

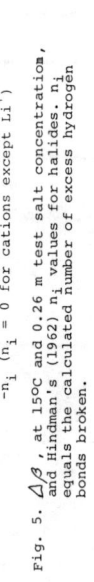

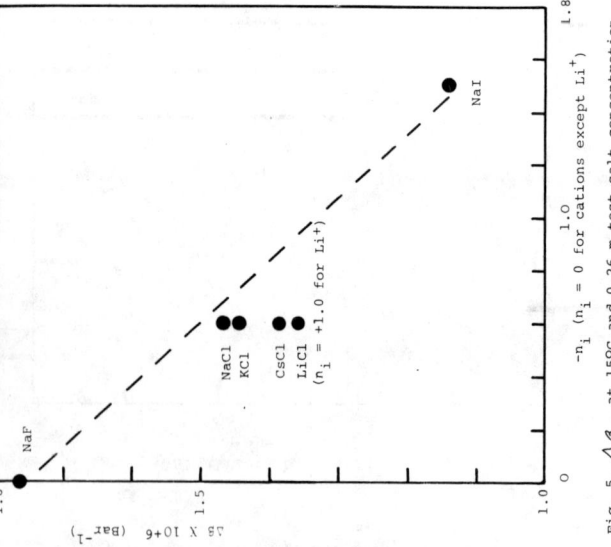

Fig. 5. $\Delta\beta$, at 15°C and 0.26 m test salt concentration, and Hindman's (1962) n_i values for halides. n_i equals the calculated number of excess hydrogen bonds broken.

Fig. 6. The correlation of $\Delta\beta$ (at 15°C, 1 Bar pressure, and 0.26 m test salt concentration) and Padova's (1963) r_e values for halide ions. r_o equals the minimum radius necessary to cause complete dielectric saturation of the local water.

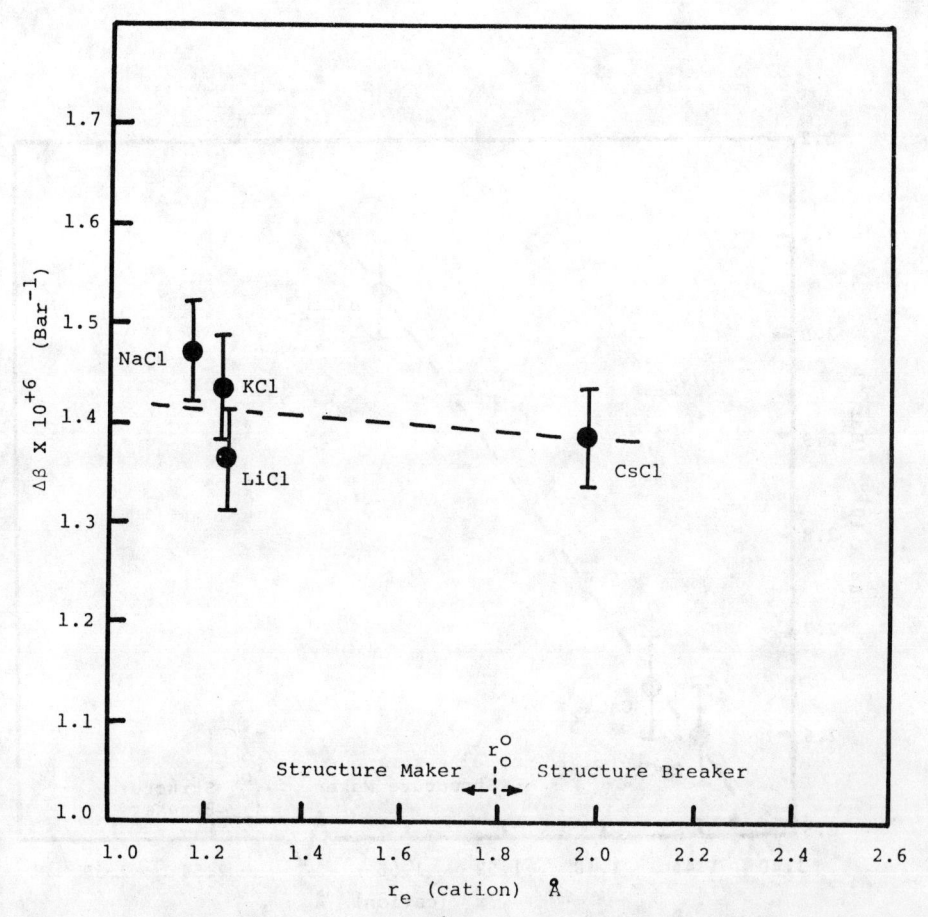

Fig. 7. The correlation of $\Delta\beta$ (at 15°C, 1 Bar pressure, and 0.26 m test salt concentration) and Padova's (1963) r_e values for alkali metal ions. See Fig. 6 for definition of r_o^o.

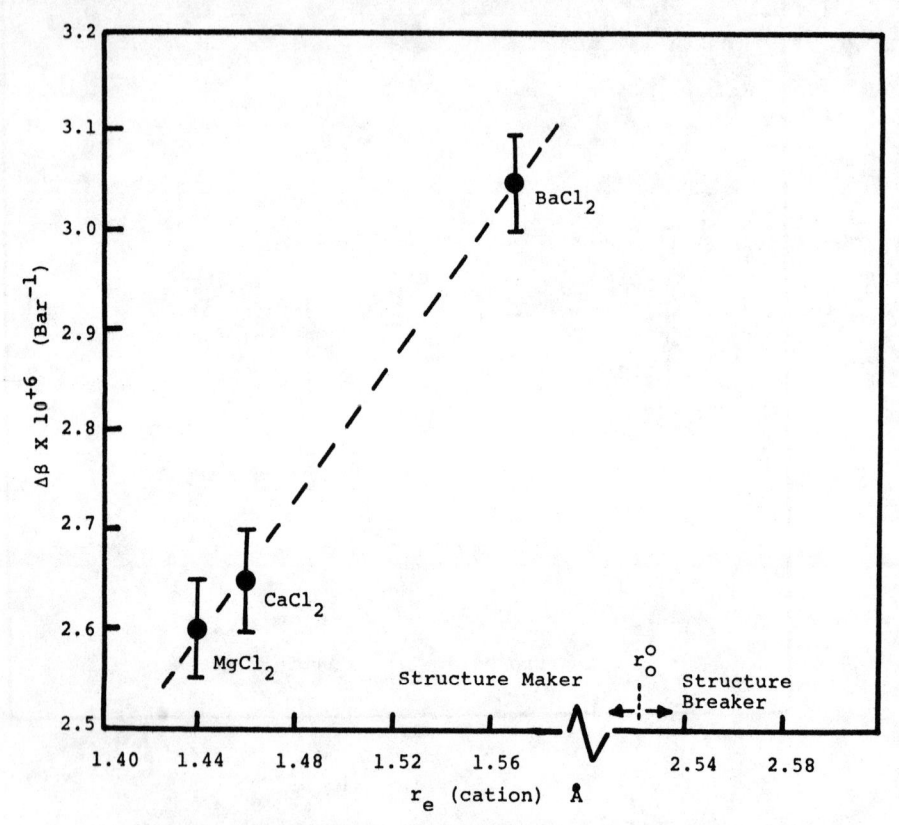

Fig. 8. The correlation of $\Delta\beta$ (at 15°C, 1 Bar pressure, and 0.26 m test salt concentration) and Padova's (1963) r_e values for the alkaline-earth ions. See Fig. 6 for definition of r_o^o.

STRESS CORROSION CRACKING OF Ti-8Al-1Mo-1V ALLOY IN NATURAL SEA WATER

W. R. Cares and M. H. Peterson

NRL Marine Corrosion Research Laboratory
P.O. Box 1739
Key West, FL 33040

Naval Research Laboratory
Washington, D. C. 20390

ABSTRACT

The stress corrosion cracking (SCC) of a Ti-8Al-1Mo-1V alloy was studied in flowing, natural sea water. Sea water presents both a corrosive and practical environment for titanium alloys. The stress intensity factor, K_{Iscc}, which measures the susceptibility of a metal to SCC, was evaluated by the step-loaded cantilever beam technique. Long holding-times between step-loadings were employed to insure that slow SCC initiation or propagation steps were not influencing the K_{Iscc} values observed. No adverse effect was found for electrolytically generated hydrogen on either normal or iron-contaminated titanium specimens; hydrogen embrittlement was not a problem for this alloy. Large degrees of cathodic polarization increased the alloy's K_{Iscc} value to 40 ksi$\sqrt{in}$. from the value of 24-25 ksi$\sqrt{in}$. observed at the alloy's natural potential; small degrees of cathodic polarization or anodic polarization decreased the K_{Iscc} value to 20-21 ksi$\sqrt{in}$. The potential of the alloy when stressed is -130 to -430 mv vs the Ag/AgCl reference electrode in sea water.

INTRODUCTION

The use of titanium alloys for corrosion resistant equipment has become commonplace in recent years. Catastrophic failure of titanium may, however, result from stress corrosion cracking, even though little or no corrosion weight loss can be measured. Natural sea water presents both a practical and a corrosive environment for titanium.

A measure of the resistance of a metal to stress corrosion cracking (SCC) is given by the stress intensity factor, K_{Iscc}. K_{Iscc} correlates to the lowest experimental stress intensity level where SCC <u>does</u> occur under given experimental conditions; it does not necessarily follow that SCC will never occur at lower stress intensity values. A common technique for evaluating K_{Iscc} involves the step loading of a cantilever beam attached to a notched, precracked metal specimen (1). Typically, for titanium, the cantilever beam is loaded over a relatively short time period. A critical question concerning this technique deals with the kinetics of the SCC initiation and growth. If either of these steps is slow, one or more step loadings past the critical loading may have been applied to the beam in the typical SCC experiment before crack growth becomes visible to the experimenter. This "over-loading" would lead to values of K_{Iscc} much larger than should have been observed. Extending the time duration between incremental loadings of the beam results in both more accurate K_{Iscc} determinations and a better insight into the nature of the kinetic processes involved in SCC.

Additional insight into the SCC process can be provided by studying the effects of polarization of the titanium specimens. Although the application of cathodic protection is known to retard SCC of titanium in some environments (2), its application has been little used in sea water (3). One mechanism proposed for the SCC of titanium alloys is that hydrogen dissolves into the titanium to form a brittle titanium hydride which rapidly cracks under the applied stress. The application of cathodic protection to a specimen can result in the generation of hydrogen which could embrittle the specimen and lead to a higher incidence of SCC failures. Previous studies have shown that the hydrogen content of titanium markedly effects the fracture of the metal specimens in air (4,5). The incorporation of iron contamination onto the titanium surface is thought to present active surface sites where hydrogen might be easily transmitted into the metal structure. Studies of electrolytically generated hydrogen on both normal and iron-contaminated titanium surfaces as a function of the applied potential have been constructed to determine the effect of electrolytically generated hydrogen on the SCC process.

EXPERIMENTAL TECHNIQUES

Titanium 8Al-1Mo-1V specimens (NRL inventory designation R7), 135 ksi yield strength, were used for all experiments. The step-loaded cantilever beam technique was used to evaluate K_{Iscc} values which were calculated by the methods of Freed and Krafft (6). Care was taken to insure that each specimen was surrounded by flowing natural sea water before the cantilever arm was attached [previous studies (7) have shown that the loading of a specimen before it is surrounded by the corrosive environment may cause a dulling of the sharp crack tip by plastic flow, thus, leading to the determination of artificially high K_{Iscc} values]. All potentials were measured and controlled against Ag/AgCl reference electrodes. In sea water these electrodes lead to the same numerical potential values as the saturated calomel electrode.

Non-Contaminated Specimens. Specimens of the alloy with WR orientation were prepared in 1-in. x 7-in. x 1/4-in. size. Each specimen was notched at the center of the 1/4-in. edge, side grooved, and fatigue precracked. The specimens were sealed into two-compartment cells so that water entered each cell in the titanium compartment (containing reference electrodes), flowed over (and through) a porous polystyrene weir into the second compartment and from there to the drain. The potentials of the specimens were controlled to within 5 mv using potentiostats (with zinc or magnesium anodes) or an Anotrol Model 4100 Potential Controller (using iron or platinum electrodes). The anodes or working electrodes were placed in the second compartment of the cells so that the water flow prevented any contamination of the water surrounding the titanium by the water surrounding the counter electrodes; this insured an uncontaminated sea water environment around each titanium specimen. Most specimens were step loaded at > 200 hour intervals; each step loading was equivalent to a 2 to 3 ksi$\sqrt{in}$. increase in the stress intensity level of the specimen. Specimens controlled at -525 and at -625 mv were loaded at $\geq$ 50 hour intervals in steps of 0.8 ksi$\sqrt{in}$.

In order to establish a natural potential for the alloy in sea water, one specimen was placed in a SCC cell in an unstressed condition and its potential monitored with time.

A second study was made by immersing the broken end of a fractured specimen into a larger corrosion cell. Water flowed through each cell at ca. 200 ml/min until a relatively stable potential was established; the water flow was then stopped for 18 days, after which flow was resumed at the original rate.

Iron-Contaminated Specimens. Iron-contaminated specimens of the alloy with RW orientation were prepared by filling two widely separated Vee notches in the 1/4-in. face of 1/4-in. x 1-in. x 10 3/4-in. specimens with an iron weld bead. A precrack was mechanically "popped-in" to each specimen so that it passed through the weld bead and into the titanium. The specimens were then side grooved.

The specimens were installed in SCC cells as previously described. Step loading was at 1/2 to 1 hour intervals. Several control specimens were broken at the "natural potential" of the alloy. A test specimen was then potentiostated at -1550 mv for six days (three days with no applied load and three days with only the cantilever arm attached) before step loading was begun. An additional control specimen at natural potential followed the same schedule as the -1550 mv specimen.

RESULTS AND DISCUSSION

The data obtained in the attempt to define the corrosion potential of the R7 alloy are shown in Fig. 1. Even after 213 days the potentials have not reached a true steady state value for either specimen, and a difference in potential of over 400 mv exists between the two specimens. This alloy behaves similarly to some other metals in sea water in that there is no single observed electrochemical potential which is characteristic of the alloy; wide variations in potential are observed between individual specimens. Similar potential results were observed for both the non-contaminated and the iron-contaminated alloy in the stressed condition. Stopping the water flow to the two control specimens did not significantly affect the potential of either specimen. The value chosen to represent the freely corroding potential of an individual specimen is that potential exhibited immediately before SCC began occurring (at which time the potential of each specimen rapidly becomes more negative). The freely corroding potentials, defined in this manner, ranged between -130 and -430 mv. It appears that the potential exhibited by this alloy is a complex function of stress on the specimen, the condition of passivating surface coatings, and/or other factors beyond the scope of this study. No direct correlation to water temperature was observed.

Figure 2 shows the observed K_{Iscc} values plotted vs the controlled potentials for the step-loaded, non-contaminated titanium specimens. The K_{Iscc} values of 24-25 ksi$\sqrt{in}$. for the specimens are in reasonable agreement with those values previously reported for unpolarized specimens of this alloy in 3.5% NaCl solution (8). The increase of the K_{Iscc} value as the potential of the titanium becomes more negative (increasing cathodic polarization) becomes linear at potentials more negative than -1000 mv. Correspondingly, under anodic polarization the K_{Iscc} values tend to reach a limiting value of about 20-21 ksi$\sqrt{in}$.

The decrease in K_{Iscc} value observed in the -425 to -875 mv region indicates a small adverse effect at low levels of cathodic polarization, whereas the K_{Iscc} value is dramatically increased at higher levels of cathodic polarization.

An apparent kinetic effect was observed at higher levels of cathodic polarization. The specimen controlled at -1550 mv failed 8.1 hours after the stress intensity was increased, and the specimen at -1350 mv failed 6.1 hours after the load increase; whereas all other specimens failed within 0.1 hour of the load increase. Evidence that this kinetic effect is due primarily to the applied potential and not to the total length of time that the sample was subjected to stress is given by a long term holding-time experiment for the sample at -325 mv (natural potential). This specimen was held for 2157 hours with no evidence of SCC occurring before its stress intensity was increased 0.8 ksi$\sqrt{in}$. units; failure occurred within five minutes of the stress intensity increase, giving a K_{Iscc} value of 24.2 ksi$\sqrt{in}$.

The specimens potentiostated at -1350 and at -1550 mv showed hydrogen evolution and calcareous deposit accumulation, both being greater for the latter specimen. There is no indication of any adverse effect upon the K_{Iscc} values for these specimens, providing additional evidence that hydrogen electrolytically generated on the surface does not further embrittle the alloy. A crack propagation mechanism involving hydrogen or hydride formation can not be disproved by the present data, however, because the residual hydrogen in the alloy may have been sufficient to produce the maximum degradation in alloy properties. The decrease in the crack growth rate shown by the increase in time-to-failure indicates that

the kinetics of crack propagation in this alloy is affected
by cathodic polarization, but the data are insufficient to
determine the mechanism by which this is effected; i.e.,
the reaction could be slowed by an electrochemically
induced change in the character of the oxide on the
titanium surface, by changes in the chemistry of the
corrodent in the crack due to secondary reactions, by the
electrochemical polarization of the titanium, or by changes
in the microstructure of the alloy due to absorbed
hydrogen.

The data for time-to-failure for the twelve specimens
show that 200 hours between step loadings is a sufficiently
long period to insure that SCC is not slowly occurring.

The data obtained for the iron-contaminated specimens
reinforce the conclusion, from the non-contaminated
specimens, that there is no additional adverse effect of
electrolytically generated hydrogen on K_{Iscc} values. The
RW orientation of these specimens led to SCC cracks that
at times only approximately followed the side grooves of
the specimen. Additionally, the "popped-in" mechanical
crack was found to lead in the center - usually about 3x
the depth at the outer edge - this fact requires the
calculation of both minimum and maximum K_{Iscc} values. Because the mechanical crack depth could be more accurately
measured at the outer edges, the minimum K_{Iscc} values
permit a more meaningful comparison between the various
specimens. Due to a larger variation in K_{Iscc} values for
the freely corroding samples of the iron-contaminated
alloy, all K_{Iscc} values for these specimens were normalized
to the mean K_{Iscc} value observed (27.3 ksi$\sqrt{in.}$) The results
are presented in Table 1.

Table 1--Electrochemical Potential and
Normalized K_{Iscc} Values for the Iron-Contaminated
Ti-8Al-1Mo-1V Alloy

Specimen	Potential mv vs Ag/AgCl	Normalized K_{Iscc} Min	Max
R7-2(A)	Natural	1.03	2.36
R7-2(B)	Natural	0.92	3.10
R7-3(A)	Natural	1.03	2.84
R7-3(B)	Natural	1.02	1.76
R7-4(B)	-1550	1.36	3.11
R7-5(B)	Natural	1.02	2.84

It is evident from the K_{Iscc} values of Table 1 that
potentiostating the titanium alloy at -1550 mv increases

the K_{Iscc} value for the specimen relative to that for the alloy at its natural potential. This is demonstrated qualitatively from the added weight (above that of the beam) required to break the specimens: 60 pounds for the specimen at -1550 mv versus 39 to 46 pounds for the specimens at natural potential.

Before loading, specimen R7-4(B) was held at -1550 mv; there was copious hydrogen evolution from the wetted areas of the specimen and a calcareous deposit build-up to ca. 1/4-in.

The readily available supply of hydrogen did not affect the iron-contaminated R7 alloy in an adverse manner. Cathodic polarization at -1550 mv increased the K_{Iscc} value above that observed at natural potential.

CONCLUSIONS

Several important conclusions result from this study:

1. The freely corroding potential of the Ti-8Al-1Mo-1V, R7 alloy is highly variable. The typical electrochemical potential range for this alloy when stressed in sea water is between -130 and -430 mv versus a Ag/AgCl reference electrode.

2. Cathodic polarization to highly negative potentials substantially increases the resistance of this alloy to SCC; however, slight cathodic polarization may actually decrease the resistance to SCC.

3. The stress corrosion resistance of the alloy is not adversely affected by electrolytically generated hydrogen.

4. A 200 hour interval between loadings is sufficient to insure that slow SCC is not occurring between the incremental loads.

5. Iron-contamination does not result in any significant change in the hydrogen-SCC susceptibility characteristics of the alloy.

ACKNOWLEDGMENT

The authors wish to thank R. L. Newbegin for specimen preparations and R. E. Groover and C. W. Billow for helpful discussions on experimental techniques.

REFERENCES

1. B. F. Brown, "The Application of Fracture Mechanics to Stress-Corrosion Cracking," Lecture presented to ASM Educational Conference on Fracture Control for Metal Structures, Philadelphia, Pa., 26-28 Jan 1970; Chicago, Ill., 21 May 1970.

2. T. R. Beck, in "The Theory of Stress Corrosion Cracking of Alloys," Ed. by J. C. Scully, NATO Scientific Affairs Division, Brussels, 1971, pp. 64-83.

3. B. F. Brown, "Titanium Alloys in the Marine Environment," presented to the Symposium on Titanium for the Chemical Engineer, A.I.Ch.E. National Meeting, Atlanta, Ga., 15-18 Feb 1970.

4. D. A. Meyn, Report of NRL Progress, Naval Research Laboratory, Washington, D.C., Dec 1971, p. 11.

5. G. Sandoz, Report of NRL Progress, Naval Research Laboratory, Washington, D.C., May 1968, p. 31.

6. C. N. Freed, and J. W. Krafft, J. Materials $\underline{1}$, 770 (1966).

7. B. F. Brown, Materials Res. & Stds., $\underline{6}$:3, March 1966, pp. 129-133.

8. G. Sandoz, and R. L. Newbegin, Report of NRL Progress, Naval Research Laboratory, Washington, D.C., Nov 1967, pp. 35-36.

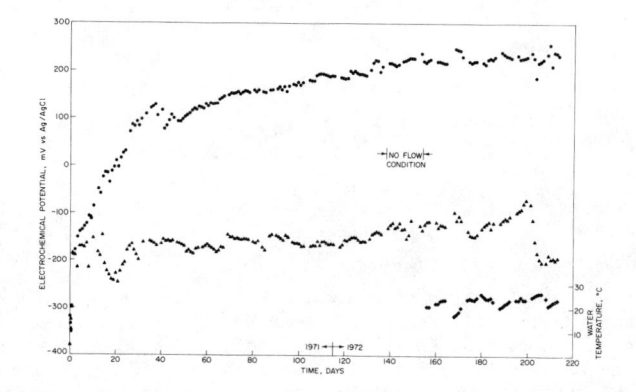

Fig. 1--Electrochemical potential (vs. Ag/AgCl) and sea water temperature versus time for two specimens of R7 alloy. ● fractured specimen, ▲ unbroken specimen, ◆ water temperature.

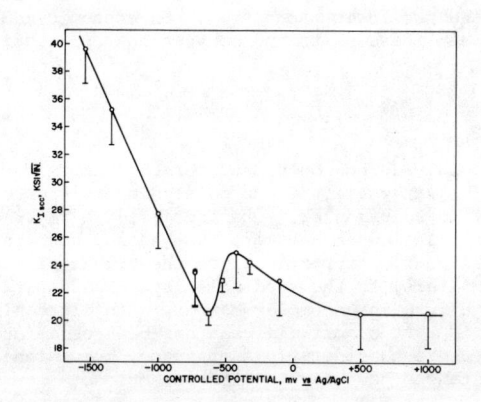

Fig. 2--Stress intensity factor, K_{Iscc}, versus controlled potential (vs. Ag/AgCl) for Ti-8Al-1Mo-1V alloy specimens cracked in the WR orientation. ○ Specimens step-loaded at ≤ 200 hr intervals. □ Specimens step-loaded at ≤ 50 hr intervals. The lower limit of the error bar represents the stress intensity value at which the specimen did not fail.

PRODUCTION OF FRESH WATER FROM THE SEA BY ELECTRODIALYSIS

Frank B. Leitz and Mauro A. Accomazzo
Ionics, Inc.
65 Grove Street
Watertown, Massachusetts 02172

ABSTRACT

Desalination of sea water by electrodialysis has long been technically feasible. Because of the dependence of costs on feed salinity, electrodialysis has been used on sea water only in special situations. Recent developments (high temperature operation, thin membranes and improved spacers) have reduced the costs of electrodialysis, making sea water desalting by electrodialysis economically competitive. The technical and economic aspects and the present status of testing of these developments are discussed. Design equations and empirical relationships for the new components are presented. Electrodialysis unit designs are based on a modular stage containing 130 cell pairs. Performance and economics are predicted for units of 3780 m^3/day (1 MGD) and two lower capacities. An economic comparison is made between these units and conventional plants in sea water service.

BACKGROUND

Electrodialysis has been used for many years for the desalination of brackish water (1). Electrodialysis operates by removing the salt from a saline solution or, more precisely, by moving the salt from one saline solution to another. The energy cost in electrodialysis is essentially linearly proportional to the amount of salt removed, or, for very salty waters, to the feed salinity. The capital costs increase, although in a more complex manner, with increasing feed salinity. The net result is that electrodialysis has been used for desalting of sea water only where considerations of space, convenience or other special circumstances have favored its selection.

Several sea water electrodialysis plants have been constructed. A 2.7 m^3/day plant was built in 1958 for operation on the particularly saline waters (43,000 mg/ℓ) of the Persian Gulf (2). For non-technical reasons, this plant was never commissioned. A 7.5 m^3/day plant was built for the U.S. Coast Guard Station in Ocracoke, North Carolina. There does not appear to be any published information on this plant. A

5 m^3 per day plant is reported in Russia (3). Research with smaller scale units has been carried on in France (4), Japan (5), and Russia (6).

The use of electrodialysis for production of potable water on small seagoing vessels and fishing boats has been investigated (7-11). Several Japanese companies (8-10) have constructed units with capacities of 0.5 to 2.0 m^3 per day for this purpose. A French company (11) has built a tiny unit with a capacity of 2 liters per day for inclusion in a survival kit. This sort of unit exploits the relatively small amount of space required by an electrodialysis unit compared to either fresh water storage or a small still.

Electrodialysis at high temperatures has been studied extensively by the authors' company (12-13), and by the Negev Institute in Beersheva, Israel (14-16).

Patents on electrodialysis frequently include a statement about the applicability to sea water desalination. This is probably more an indication of the normal hyperbole of patents than of reasonable economic evaluation or investigation.

A considerable amount of work has been done on electrodialysis of sea water for production of brines, usually concentrated twofold over sea water, which are used in salt production in Japan. This is now a fully developed industrial process in that country (17,18).

PROCESS DESCRIPTION

The basic apparatus for electrodialysis is a stack of membranes terminated on either end by an electrode. The fabrication and properties of ion-exchange membranes have been described extensively in the literature. For the present discussion a membrane is a thin sheet of material having a very low water permeability and a high ionic conductivity with a near unity transport number for ions of one type of charge or the other. Flow of the process streams is contained and directed by spacers which alternate with the membranes. The assembly of membranes spacers and electrodes is held in compression by a pair of end plates.

For water desalting the membranes are arranged alternately cation and anion. The compartments which have an anion membrane on the anode side are demineralizing compartments while the alternate compartments are concentrating compartments. The unit composed of a cation membrane, a desalting spacer, an anion membrane and a concentrating spacer is a repeating unit termed a "cell pair". Because of the thinness of the components, the stack is usually internally manifolded. The manifolds are formed by lining up holes cut in the membranes and spacers. A diagram of the process appears in Figure 1.

For design purposes it is convenient to talk about an electrical "stage", i.e. that length of flow path along which the applied voltage per cell pair is constant. In large plants one stack frequently is one stage. However, more than one stage can be put in a stack, by inclusion of more than one pair of electrodes, and two stacks run hydraulically in series with the same applied voltage can be considered as one stage.

It is generally not feasible to operate an electrodialysis unit in such a manner that the entire change in salt concentration desired is obtained in a single stage. In a continuous unit, the output from one stage is fed directly to the inlet of a second which is usually run at a lower cell pair voltage than the first. This process is continued until the desired concentration is obtained. Since the hydraulic capacity of most stacks is high, this mode is useful for applications in which large quantities of water are to be treated. For treatment of small quantities of water, several modes of operation involving recycle of product are used.

Equations are given below which relate the desalination duty to the physical and electrical properties of a stage. These provide a useful quantitative description of the process.

It is convenient to express various parameters in terms of two variables: the degree of desalting, f, and the inlet feed ratio, g.

$$f \equiv \frac{C_{di} - C_d}{C_{di}}$$

$$g \equiv C_{di}/C_{ci}$$

where
C_{di} = inlet diluate concentration - meq/ml
C_d = diluate concentration at a given point - meq/ml
C_{ci} = inlet concentrate concentration - meq/ml

A material balance around a differential length of channel gives the degree of desalting per unit length as:

$$\frac{df}{dL} = \frac{wei}{F_d C_{di} \mathscr{F}}$$

where
w = channel width - cm
e = current efficiency

i = current density - mA/cm^2
F_d = flow rate in diluting compartment - ml/sec
$\mathcal{F}$ = 96,500 ampere seconds/equiv
L = channel length - cm

The current efficiency for the process in the absence of concentration polarization is:

$$e = \bar{t}_+^c + \bar{t}_-^a - 1$$

where
$\bar{t}_+^c$ = transport number of cation in the cation membrane
$\bar{t}_-^a$ = transport number of anion in the anion membrane

Typically a plant is run with concentrate and diluate streams in parallel flow at the same velocities through spacers of the same thickness. A more general derivation which does not include some of these simplifications is given by Mason and Kirkham (19).

Under steady state conditions, with solutions and current flowing, the concentration in the diluate compartment decreases from its initial value and the concentration in the concentrate compartment increases from its initial value:

$$C_d = C_{di}(1 - f)$$

$$C_c = C_{ci}(1 + fg)$$

An average solution concentration can be defined by:

$$\frac{1}{C_a} \equiv \frac{1}{2}\left(\frac{1}{C_c} + \frac{1}{C_d}\right)$$

$$C_a = 2 C_{di} \frac{(1-f)(1+fg)}{(1+g)}$$

An effective resistance per cell pair, R_p, defined as the ratio of voltage per cell pair to current density can be expressed as function of C_a

$$R_p = \frac{a'}{C_a} + b' - c'C_a$$

This results from expressing the product, $R_p C_a$, as a power function of C_a and truncating after the third term. With values of C_a below 0.2 meq/ml, data can be satisfactorily correlated with only the first two terms.

The voltage per cell pair is essentially constant throughout a stage. The local current density is then:

$$i = \frac{1000 V_p}{R_p}$$

With i expressed as a function of f and g, the differential material balance can be integrated down the flow path. For a given demineralization, the area required per unit capacity is:

$$\frac{A_p}{F_d} = \frac{C_{di} \mathcal{F}}{1000 V_p e} \int_0^{f_o} R_p \, df$$

$$= \frac{C_{di} \mathcal{F}}{1000 V_p e} \left[\frac{a'}{2 C_{di}} \ln\left(\frac{1+gf_o}{1-f_o}\right) + b' f_o \right.$$

$$\left. - \frac{2c' C_{di}}{1+g} \left(f_o + \frac{(g-1)}{2} f_o^2 \frac{g f_o^3}{3} \right) \right]$$

Under conditions where R_p can be considered constant, a much simplified approximation can be used:

$$\frac{A_p}{F_d} = \frac{C_{di} \mathcal{F} R_{po} f_o}{1000 V_p e}$$

However this becomes inaccurate at values of f above 50% or at low values of g.

The power required for demineralization per unit capacity is:

$$\frac{P}{F_d} = \frac{V_p C_{di} \mathcal{F} f_o}{e}$$

High current density results in low area proportional equipment cost while low current density produces low power costs. Since current density varies across the stage, "current density" in the discussion on optimization refers to an average value. A reasonable estimate of the economic optimum current density can be made by assuming that the major costs can be divided into three categories: those directly proportioned to current density, those inversely proportioned to current density and those independent of current density. While most costs are not linear over the whole range of possible current densities, a reasonable linear approximation can usually be made over a limited range. This leads to an equation for total cost, d, of the form:

$$d = ai + b/i + c$$

where a, b and c are the relative proportionality constants. The economic optimum current density, i_{opt}, is:

$$i_{opt} = (b/a)^{1/2}$$

Recognizing that the major directly proportional cost is power consumption and that the major inversely proportional cost is cell pair area, this can be approximated by

$$i_{opt} = 31,600 \, (\overline{C}_s/R_p C_e)^{1/2}$$

where

$\overline{C}_s$ = amortized cost of cell pair area - $/cm^2 hr

C_e = cost of energy - $/KWH

Since the lifetime of the membranes is shorter than that of the membranes, it is more accurate to amortize the area proportional cost of equipment (less membrane cost) and the membranes separately and then sum the two factors.

Current density in an electrodialysis unit is frequently limited by polarization. The fraction of current carried through the anion membranes by anions, t_-^a, is usually 90 to 95%. The fraction of current carried in the solution by anions, t_-, is about 60% for most common electrolytes. At the anion membrane boundary of the diluting channel, there is a deficiency of anions which must be supplied by diffusion and convection. Assuming that there is a diffusion layer of thickness δ, the limiting current density which occurs when the concentration at the membrane surface goes to zero is given by:

$$i_{lim} = \frac{n c_d \mathcal{F} \mathcal{D}}{(t_-^a - t_-) \delta}$$

where
and
$\mathcal{D}$ is the diffusion constant - cm^2/sec

n is the number of charges per ion

In dilute solutions particularly, the optimum current density is generally above the limiting current density. Above the limiting current density the apparent resistance increases, the diluate stream pH falls, the concentrate stream pH rises and the probability of fouling and scaling of the membranes increases considerably. To avoid these conditions, spacers are designed to have an even velocity distribution and to give enhanced mass transfer to the membrane surfaces. This is accomplished by inserting turbulence promoting straps or loosely woven screens in the flow portion of the spacers.

Under conditions where polarization limits the current density, the effective value of δ and, hence, the limiting value of the ratio i/C_d tends to be dependent only on linear solution velocity. A typical relationship has the form:

$$(i/C_d)_{lim} = kv^{0.6-0.7}$$

For a given spacer, if R_p can be considered constant, the degree of desalting, f_o, is then uniquely a function of solution velocity.

$$f_o = \frac{(i/C_d)_o A_p e}{\mathcal{F} F_d + (i/C_d)_o A_p e}$$

For the more exact case an expression for f_o can be obtained by iteration. This value depends to a small degree on g which varies from stage to stage. If f_o is constant, or essentially so, from stage to stage, the overall demineralization, f_T, is related to f_o by:

$$(1 - f_T) = (1 - f_o)^N$$

where
        N is the number of stages.

Since the flow channels are thin, i.e. of low equivalent diameter, and long, a significant pressure drop can be experienced even at moderate solution velocities. In design of a spacer a trade-off is made between large, closely spaced turbulence promoters which give enhanced mass transfer at the cost of high pressure drop and small, sparse promoters which give lower rates of mass transfer and lower pressure drop. Because of the method in which the stack is constructed there is a maximum inlet pressure, usually taken to be 4 atm, above which the stack leaks excessively. Where a number of stacks are operated in series, present practice is to run them without interstage pumping. Consequently there is a maximum usable value of velocity.

SIGNIFICANCE OF PROCESS IMPROVEMENTS

High temperature operation has several beneficial effects on the electrodialysis process. Current densities can be increased throughout the unit. In the concentrated portion of the apparatus, current density at or near economic optimum can be used. Since the effective cell pair resistance decreases with increasing temperature, the optimum current density increases. In the dilute portions of the apparatus, where current density is limited by polarization, the rate of mass transfer increases rapidly with increasing temperature. At constant velocity the limiting value of the ratio of current density to normality, i/C_d, doubles between 25°C and 80° C. Since solution viscosity decreases with increasing temperature, a further increase in mass transfer and hence in current density can be obtained by increasing the velocity in the dilute portion of the unit while operating at the same overall pressure drop.

The purely resistive portion of the effective cell pair resistance decreases by about 2% per °C. The portion due to concentration potentials, partial polarization and other factors is only slightly affected. The net result is a decrease in R_p of about 50% between 25°C and 80°C. Since the high temperature unit is to be run at higher current densities than a source temperature unit, the power consumption is not reduced in proportion but is reduced to about 75% of that at source temperature.

The principal costs associated with high temperature operation are the capital cost of exchanging heat and the energy cost of replacing non-recoverable heat. These offset some of the benefits of high temperature operation. In brackish water desalting a rather complex tradeoff has been made between area proportional electrodialysis apparatus costs, heat exchanger costs and cost of heat (20). This study indicated that high temperature operation is less expensive than ambient operation when transferring heat is cheap relative to electrodialysis equipment. In sea water service where the direct current energy input and electrodialysis apparatus costs are both higher per unit of product treated, high temperature operation is beneficial under all reasonable circumstances. Further, the joule heating in the stack is generally sufficient to offset the non-recoverable heat. Thus the unit requires no external heat source.

The desirability of the use of thinner components is suggested by the fact that the membrane-solution interface is where the desalting occurs. Most of the energy input in the system is expended in transporting the ions through the bulk of the solutions and through the bulk of the membranes, and is merely converted into heat. Only a small fraction of the energy input into the system does the useful work of separation.

Since the ancillary apparatus cost is significant, increasing the capacity of the stack can lower unit costs. However, there is a physical limit to the size of a stack. A spacer or membrane one-half square meter in size can be manipulated (cleaned, turned over, stacked) by two men easily or one man if necessary. If the size is increased much, this is no longer possible. A stack of 300 cell pairs of standard components is about one meter high which is close to a practical limit above which the stack becomes unstable during assembly. Within the same physical limitations, however, a stack of thin components will contain twice the cell pair area. This produces a considerable savings in capital cost per square cm of area in units of reasonable size because of savings in auxiliary equipment.

NEW COMPONENTS

A thin spacer (0.05 cm) has been developed for high temperature electrodialysis. A spacer is required to provide as nearly as possible a consistent flow velocity over the membrane surface. A local velocity above average is not bad except as it indicates a velocity below average somewhere else. The manifold system must provide even feed to each spacer. Finally it must provide a satisfactory balance between the conflicting requirements for high mass transfer and low pressure drop. The spacer designed provides satisfactory support for thin membranes (0.02 to 0.03 cm) and can be used at operating temperatures up to 80° C. The coefficient of expansion of the composite structure is sufficiently close to that of the membrane so that no difficulties are experienced when the temperature of the stack is changed. The spacer (47 cm by 103 cm) is a hybrid between a screen-sheet flow spacer and a multiple parallel flow path spacer. It utilizes two commercially available plastic screens.

The Conwed netting is used in the active area due to its low pressure drop characteristics at the desired flow velocities (20 to 50 cm/sec). Essentially this netting consists of 1.5 strands/cm in the direction of flow with one strand every 2.5 cm perpendicular to flow. The strands are approximately 0.025 cm thick and their intersections are 0.05 cm thick. The polypropylene Vexar screen (10 PDS 169) is used at each end of the spacer and essentially consists of 0.025 cm strands evenly spaced about 0.16 cm apart. This material provides a greater degree of membrane support in the manifold area to prevent cross-leaks and distributes the flow evenly into the active area.

Fabrication of the spacer consists of welding two strips of Vexar screen (50 cm by 10 cm) to each of the short ends of a Conwed screen (50 cm by 85 cm), with a heated bar welder. Then the welded screen (50 cm by 105 cm) is placed in a mold and gasket and manifold areas are formed by filling the appropriate areas with silicone rubber. After the rubber is cured, the manifold holes and outside borders (47 cm by 103 cm) are die cut. The resulting spacer is shown in Figure 2.

The principal difficulty in development of membranes for high temperature has been the selection of suitable fabric backing materials. The resins for both cationic and anionic membranes are usable to 80°C for longer periods of time than membranes customarily last, provided the anion resin is not exposed to pH much above 7.

At present, three types of thin membranes have been under consideration: three-ply glass-dacron-glass backed membranes, single ply dacron backed membranes, and single-ply teflon backed membranes. The two types of single ply membrane are currently running slightly thicker than the three ply membranes. All types of membranes are currently being tested both in stacks at 80° C and in a static accelerated life test at 121° C. Assuming an activation energy of the backing decomposition reaction of 25 K cals/mole, one day in the accelerated test is equivalent to 30 days at 80° C. At present the teflon-backed membranes appear to have much the superior properties although the cost of the teflon cloth is higher than that of other materials.

PROJECTED ECONOMICS

To design an electrodialysis plant with the equations given above, three empirical relationships are needed: pressure drop versus velocity, limiting polarization parameter, i/C_d, versus velocity and resistance per cell pair versus average concentration. The constants in these relationships are functions of temperature so it is necessary either to make measurements at the proposed operating temperature or to take enough data to determine the functional relationship with temperature.

At 65° C the following relationships were determined for the new thin spacers and membranes.

$$\Delta P \text{ (in atm)} = 0.00364 \, V + 0.000224 \, V^2 \text{ (V in cm/sec)}$$

$$(i/C_d)_{lim} = 130 \, V^{0.6}$$

$$R_p = 0.334/C_a + 12.5 - 8.28 \, C_a$$

In the design of these units, the feedstock was assumed to be 35,000 mg/l salt with approximately 10% by weight of the ions being divalent. In many coastal areas the water salinity is considerably lower than 35,000 mg/l. Consequently the estimates presented below are felt to be conservative even taking into account the probable increased organic loading and relatively increased divalent ion content which may be experienced in these waters.

The process of desalting sea water leaves a brine stream concentrated twofold in salts, filtered, and about 5°C warmer than the original source. Several authors (21,22) have proposed this as a starting material for the recovery of minerals from the sea which is somewhat more useful than raw sea water. In the economic assessment of the water recovery process, no credit was taken for any value which may have been added to this stream, nor was any penalty given for its disposal.

Electrodialysis apparatus costs above roughly 4000 m^3/day or 1 MGD can be reasonably estimated as proportional to the seven-tenths power of capacity. Below 4000 m^3/day apparatus costs become an irregular function of capacity. Estimates of installed apparatus costs for standard components are given in References 23 and 24.

Calculations were made to demonstrate the effect of high temperature operation and thin components on the costs of electrodialysis. A 3780 m^3/day (1 million gallons per day) plant with sea water feed and 50% water recovery was estimated. Three cases were studied: standard components at 25°C, thin components at 25°C and thin components at 65°C. Operating conditions were given a rough optimization. Further refinement might yield small savings over these estimates. Fluid velocity was not found to have a strong effect on cell pair area required, largely because much of the unit is not limited by polarization. The velocity chosen is 24 cm/sec which requires 20 stages. The highest demineralization per stage is 31.5%. The highest current is 247 amps and voltage per cell pair ranges from 0.17 to 0.78. Electrodialysis equipment and membrane costs are present commercial costs with no technological advances required and with conservative economies of scale applied. Other economic factors were taken from a study by the Stanford Research Institute (25). These included amortization of equipment except membranes at 4% for 30 years, amortization of membranes at 4% for 5 years. Since the cost of power used in that study (0.45¢/KWH) seemed too low, a value of 0.8¢/KWH was used. The costs are given in Table 1. Briefly, significant savings are attributable to both high temperature operation and to use of thin components. The cost of 26¢/m^3 with thin components at 65° C is encouragingly close to predictions made for distillation plants of this capacity.

Two plants of smaller capacities were also estimated. These are a four-stack plant, each stage consisting of the maximum design number (130) cell pairs per stage and a single stack plant, which is close to the minimum size plant in which high temperature operation would be practical. These plants have capacities of 580 m^3/day and 145 m^3/day. These results are shown in Table 2. The cost per m^3 is significantly higher in the lower capacities. However, the cost is still considered competitive considering the capacity of the units.

PROJECT STATUS

Contract 14-30-3084 with the Office of Saline Water, U.S. Department of the Interior was recently begun. This present program calls for construction and testing of two 130 cell pair stages. These

Table 1

Cost Comparison of
3780 m^3/day (1 MGD) Sea Water Electrodialysis Plants

APPARATUS	Standard	Thin Spacers & Membranes	Thin Spacers & Membranes
Temperature	25°C	25°C	65°C
Number of Stacks	114	39	26
Stack Power, kw	5450	3500	1910
Cell Pair Area, cm^2	10.0x10^7	9.12x10^7	6.12x10^7
CAPITAL COSTS ($)			
Stacks, Pumps, etc.	3.43x10^6	1.64x10^6	1.10x10^6
Rectifier	0.197x10^6	0.126x10^6	0.069x10^6
Heat Exchanger	-	-	0.096x10^6
FIXED COSTS ($/m^3)			
Membrane Amort.	0.182	0.136	0.091
Other Stack Amort.	0.111	0.040	0.027
Rectifier Amort.	0.009	0.006	0.003
Exchanger Amort.	-	-	0.004
O & M COSTS ($/m^3)			
Energy	0.277	0.178	0.097
Chemicals	0.009	0.009	0.009
Supplies & Spares	0.006	0.006	0.006
O & M Labor	0.021	0.018	0.024
TOTAL COSTS	0.615	0.393	0.261

Table 2

Cost Comparison of
Sea Water Electrodialysis Plants of Various Capacities

CAPACITY, m^3/day	3780	580	145
Number of Stacks	26	4	1
Stack Power, kw	1910	293	73
Cell Pair Area, cm^2	6.12×10^7	0.94×10^7	0.235×10^7
CAPITAL COSTS ($)			
Stacks, pumps, etc.	1.10×10^6	0.21×10^6	0.075×10^6
Rectifier	0.069×10^6	0.0212×10^6	0.0073×10^6
Heat Exchanger	0.096×10^6	0.0295×10^6	0.0074×10^6
FIXED COSTS ($/m^3)			
Membrane Amort.	0.091	0.091	0.091
Other Stack Amort.	0.027	0.040	0.066
Rectifier Amort.	0.003	0.006	0.009
Exchanger Amort.	0.004	0.009	0.009
O & M COSTS ($/m^3)			
Energy	0.097	0.107	0.118
Chemicals	0.009	0.010	0.036
Supplies & Spares	0.006	0.006	0.006
O & M Labor	0.024	0.039	0.103
TOTAL COSTS ($/m^3)	0.261	0.308	0.438

will be tested with synthetic solutions for performance and reproducibility over a range of operating temperatures and feed salinities, and subsequently tested in the field on sea water.

ACKNOWLEDGMENTS

This work has been made possible by the continued sponsorship of the Office of Saline Water, U.S. Department of the Interior under Contracts 14-01-0001-2322 and 14-30-2963.

REFERENCES

1. Leitz, F. B. and W. A. McRae, "The Current Status of Electrodialysis", presented at the OSW Symposium on Membrane Processes for Desalination, Roswell, New Mexico, November 1970.

2. Powell, J. H. and E. M. Guild, Proc. of the Symp. on Salinity Problems in the Arid Zones, Teheran, Persia, p. 363, UNESCO Publication (1961).

3. Gaidaymov, V. B., et al., "Ionoobmen. Membrany Elektrodialize" K.M. Saldadze, ed. "Khimiye" press, Leningrad, 171-6 (1970).

4. Carriere, F., Chim. Mod. 15 (92), 97,100,106 (1970).

5. Azumi, T. et al., Nippon Kaisui Gakkai-Shi, 22 (6), 383 (1969).

6. Orzherovskii, M., O. Zakharchuk and V. Zagoruiko, Morskoi Flot 20 (9), 24 (1960).

7. Kapinski, J. and N. Chlubek, Chemik (Gliwice), 21 (5), 174 (1968); CA: 69, 99218m.

8. Itoi, S. Desalination, 2, 378 (1967).

9. Tsunoda, Y. and M. Kato, Desalination, 3, 66 (1967).

10. Matsuda, T., S. Ogawa and Y. Onoue, Desalination, 3, 295 (1967).

11. Lemaigen, J. Desalination, 3, 203 (1967).

12. Leitz, F. B. and H. I. Viklund, Summary Reports on High Temperature Electrodialysis - Phases I and II, OSW Contract 14-01-0001-1158, U.S. Dept. of Interior (1968) and (1969). (Unpublished)

13. Leitz, F. B., M. A. Accomazzo and H. I. Viklund, Summary Report on High Temperature Electrodialysis, Phase III, OSW Contract 14-01-0001-2322, U.S. Department of the Interior (1970). (Unpubl.)

14. Forgacs, Ch., Intern. Symp. Water Desalination, 1st, Washington, D.C., Vol. III, p. 155 (1965).

15. Bejarano, T., Ch. Forgacs and J. Rabinowitz, Desalination 3, 129 (1967).

16. Forgacs, Ch., L. Koslowsky and J. Rabinowitz, Desalination, 5, 349 (1968).

17. Tsunoda, Y., Intern. Symp. Water Desalination, 1st, Washington, D.C., Vol. III, p.325 (1965).

18. Kaho, M. and T. Wanatabe, Ind. Water Eng., 6(11), 30 (1969).

19. Mason, E. A. and T. A. Kirkham, Chem. Eng. Progress, Symp. Series, No. 24, 55, 173 (1959).

20. Leitz, F. B. and W. A. McRae, "Evaluation of High Temperature Electrodialysis", paper presented at the 157th Annual Meeting ACS, New York, September 1969.

21. Gilliland, E. R., Intern. Symp. Water Desalination, 1st Washington, D.C., Vol. III, p. 309 (1965).

22. Katz, W. E., Industrial Water Engineering, 8(6), 29 (1971).

23. Tallmadge, J. A., J. A. Butt and M. J. Soloman, Ind. Eng. Chem., 56, 45 (1964).

24. Weinberger, A. J. and D. F. Delapp, Chem. Eng. Progress, 60, 56 (1964).

25. Clark, C. F., O.S.W. Research and Development Report, No. 495, U.S. Department of the Interior, Washington, D.C. (1969).

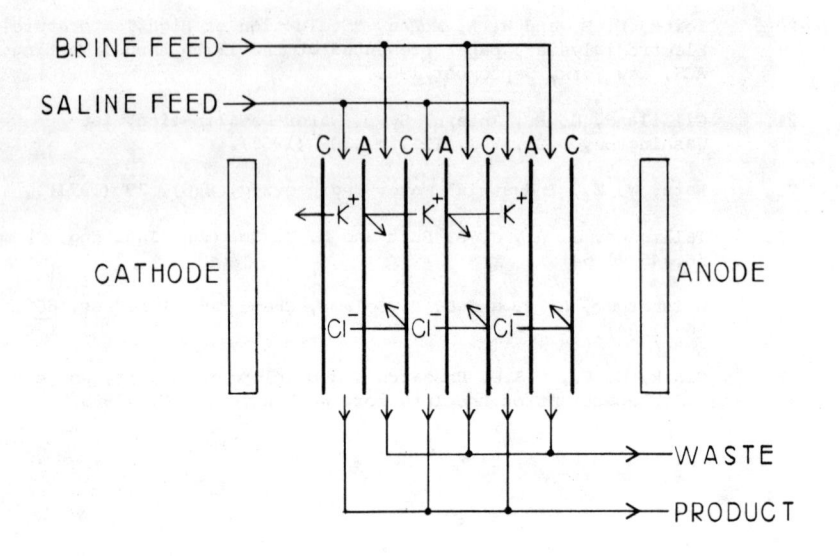

Figure 1. Diagram of Electrodialysis Process

Figure 2. High Temperature Electrodialysis Spacer

PURITY OF DISTILLATE IN MULTI-STAGE FLASH DESALINATION PLANT

Kenkichi Izumi

Hitachi Research Laboratory, Hitachi, Ltd.

3-1-1 Saiwai-cho, Hitachi, Ibaraki, Japan

The purity requirements of the distillate, one of the performance factors of the multi-stage flash desalination plant, are very different. The high purity water is obtained by removal of entrainment from vapor streams.
The entrained particles are separated from the vapor in the flash box and the demister. The separation of entrainment from the vapor in the flash box depends on its structure. A high weir type flash box was developed, and the empirical equation between the desalting factor and the vapor expansion rate in the flash box is given.
A wire-mesh type demister was used in the experimental work. And it was confirmed that the separation of entrainment in the demister was related to the vapor velocity through the demister.
Determinations of purity were estimated by measuring the electric conductivity of the solutions. The relation between the total dissolved solids and the electric conductivity of the solutions was examined.

1. Introduction

Today, many kinds of desalination plants are developed in many countries of the world. A multi-stage flash evaporator is the most popular desalination plant among them.

The purity of the distillate is one of the performance factors of the multi-stage flash desalination plant. The purity requirements of the distillate are very different according to its uses. For example, the requirements of the purity are 500 ppm TDS (Total Dissolved Solids) in conventional water, 50 ppm TDS in drinking water and 0.05 ppm TDS in boiler feed water.

The purity of the distillate in the flash evaporator is affected by entrainment of the small liquid particles in the vapor evaporated from the brine. At first, entrainment is separated from the vapor in the flash box. This separation of entrainment depends on the structure of the flash box. Second, the remaining entrainment is removed by the demister. It is said that the removal efficiency of the demister is related to the vapor velocity through the demister (8).

2. Theory of carry-over in the flash evaporator

2.1 Theory of carry-over in usual evaporators

The purity of the distillate is lowered by carry-over of the evaporating brine. Usually carry-over occurs in three stages (1).

(1) Entrainment: The small liquid particles generated on the surface of the evaporating liquid are carried to the condenser by the vapor.

(2) Splashing: When the surface of the liquid is violently agitated by boiling, the liquid splashes directly into the condenser.

(3) Foaming: Many bubbles form on the surface of the evaporating liquid, and pile up on each other. Consequently bubbles are carried over into the condenser.

Entrainment is the cause of carry-over only at a low boil-up rate (defined as evaporation rate per unit evaporating surface). As the boil-up rate rises, splashing of the evaporating liquid occurs in addition to entrainment. Foaming of the liquid depends on the nature of the solution and evaporating conditions.

There are many studies of entrainment in the evaporator. The mechanism of the generation of entrained particles from single bubbles has been made clear (2, 3, 4). When a single bubble bursts on the surface of the evaporating liquid, two kinds of particles are generated. One is a small particle ($\approx 50\mu$), the other is a large particle (≈ 1 mm). The small particles are fragments of the hemisphere thin film of the bursting bubble and the large particles are made by the jet action of the liquid surface after the bursting.

Because of gravity, large particles fall on the evaporating surface if the height of the vapor space is higher than 0.6 meters. On the other hand, small particles are carried over in the vapor, so an effective demister is necessary (4).

In practical evaporators, the mechanism of the generation of liquid particles is very complicated (5, 6), because the bubbles interfere with each other and the evaporating surface is disturbed by boiling and splashing. Therefore, there is no complete theory of carry-over in the evaporator.

In the flash evaporators, the evaporating liquid flows through the flash chamber; and so the phenomenon is more complicated than on the usual evaporator.

2.2 The structure of the flash evaporator and theory of carry-over in the flash evaporator

A typical inside structure of the flash evaporator is shown in Fig. 1. The tube arrangement of this flash evaporator is a longitudinal type.

The evaporator vessel is generally divided into a plurality of flash evaporating stages by partition walls. Each stage is composed of a flash chamber and a condenser. In the flash chamber, there is an inlet and an outlet for the brine. The flash box, which consists of orifice, weir and splashing plate, is fitted in the inlet for the brine. In the condenser, many heat transfer tubes are arranged, and a distillate chamber is installed under the tubes. A demister is set in the vapor passage which leads from the flash chamber to the condenser.

Examples of evaporating conditions in the flash evaporator are shown in Fig. 2. In the flash evaporator, the main cause of carry-over is entrainment at a low boil-up rate, too (Fig. 2-1). As the boil-up rate rises, splashing of the evaporating liquid occurs (Fig. 2-2). When foaming occurs (Fig. 2-3), we use an antifoam agent to prevent foaming.

In the flash evaporator, the entrained particles are generated in the flash box as the brine evaporates. These particles are separated in the flash box at first. The entrained particles in the vapor streams are removed by the demister.

2.3 Desalting factors in the flash evaporator

In order to examine the purity of the distillate in the flash evaporator, an overall desalting factor, $(DF)_o$, is defined as follows.

$$(DF)_o = C_1/C_3 \quad \dots\dots\dots\dots\dots\dots\dots\dots\dots\dots \quad (1)$$

where C_1 = concentration of brine (ppm TDS)
C_3 = purity of distillate (ppm TDS)

An overall desalting factor, $(DF)_o$, is given by the product of a desalting factor in the flash box, $(DF)_f$ and a desalting factor in the demister, $(DF)_d$.

$$(DF)_o = (DF)_f \times (DF)_d \quad \dots\dots\dots\dots\dots\dots\dots \quad (2)$$

where $(DF)_f = C_1/C_2 \quad \dots\dots\dots\dots\dots\dots\dots\dots\dots \quad (3)$

$(DF)_d = C_2'/C_3' \doteqdot C_2/C_3 \quad \dots\dots\dots\dots\dots\dots \quad (4)$

C_2 = purity of flashing vapor (ppm TDS)
C_2' = purity of vapor at the inlet of demister (ppm TDS)
C_3' = purity of vapor at the outlet of demister (ppm TDS)

Concentration, C_1 and purities, C_2, C_2', C_3', C_3 are shown in Fig. 1.

3. Experimental equipment and experimental method

3.1 Experimental equipment (a 10-stage flash evaporator)

The experimental equipment is a 10-stage flash evaporator which has a capacity of 50 cubic meters per day. The flow diagram of a recycle type 10-stage flash evaporator is shown in Fig. 3. The evaporator consists of a heat recovery section and a heat rejection section. The heat recovery section extends from the first stage to the eighth stage and the heat rejection section is the ninth and tenth stages.

A 10-stage flash evaporator is shown in Fig. 4. The heat recovery section consists of two modules and both of them have four stages. The flash chamber of each stage in the heat recovery section is 0.1 m wide, 0.7 m long and 1.7 m high. In each condenser of these stages, the heat transfer tubes are vertically located. The flash chamber of the ninth stage is 1.2 m wide, 0.75 m long and 1.35 m high. And the flash chamber of the tenth stage is 1.2 m wide, 1.05 m long and 1.35 m high. In the condenser of the ninth and tenth stages, the heat transfer tubes are horizontally located.

The desalting factors in five kinds of flash boxes were examined in the flash chamber of the heat rejection section. The flash chambers in the heat rejection section are much wider than in the heat recovery section, and the boil-up rate in the flash chamber of the heat rejection section is widely changeable by altering the width of the flash chamber.

A wire-mesh demister was used in this flash evaporator. The desalting factors in the demister were examined in the flash chamber of all stages.

3.2 The method of measuring the purity of solutions

The purity of solutions is expressed by parts per million of total dissolved solids, ppm TDS, in the solutions. There is a relationship between the total dissolved solids in the solutions and the electric conductivity of the solutions. Therefore the electric conductivity of the solutions was measured and the purity of the solutions was estimated by the value of the electric conductivity.

Sampling points in vapor and liquid lines are shown as symbol S in Fig. 3. The vapor was extracted from the flash chamber and was condensed by using a vapor sampling apparatus shown in Fig. 5. The vapor sampling apparatus is made of glass and is divided into two sections. The upper section of the vapor sampling apparatus is a condenser of vapor and its lower section is a cooler of condensate. Feed sea water, blow-down brine and distillate were extracted through the cooler which was used to keep their temperature constant.

Fifteen minutes were required to get one sample of the condensate by using the vapor sampling apparatus.

3.3 The relationship between the total dissolved solids and the electric conductivity of the saline water.

The relationship between the total dissolved solids, C in the solutions and the electric conductivity, σ of the solutions is shown in Fig. 6.

The electric conductivities of the solutions were measured by a Hitachi electric conductivity meter. And the total dissolved solids in the solutions were estimated by chemical analysis, sodium flame photometry and chloride ion analysis.

In Fig. 6 the measured values are plotted on a straight line. The empirical equation between the total dissolved solids, C (ppm TDS) in the solutions and the electric conductivity, σ ($\mu\mho$/cm) of the solutions is given as follows:

$$C = 0.83 \, \sigma^{0.82} \quad \ldots \quad (\sigma > 50 \, \mu\mho/\text{cm}) \quad \ldots \ldots \ldots \quad (5)$$

At the high purity water of which the electric conductivity is below 50 $\mu\mho$/cm, the electric conductivity does not decrease so much as the total dissolved solids decreases. The electric conductivity of this water is affected by dissolved gases.

In general the electric conductivity of the solution increases as its temperature rises. In the electric conductivity meter, the value of the electric conductivity of the solution which has an arbitrary temperature is corrected to the value at 25°C. The relationships between the electric conductivity, σ and the temperature, T of the solutions are shown in Fig. 7. As shown in Fig. 7 the corrections on the temperature of the solutions are quite good.

4. Experimental results and discussions

4.1 Desalting performance of flash boxes

4.1.1 Structure of a basic flash box and its functions

As shown in Fig. 1, a basic flash box consists of orifice, weir and splashing plate. The flash box has essentially three functions as follows:

(1) Perfect flash evaporation to equilibration

(2) Perfect inter-stage sealing

(3) Effective brine flow control

When high-purity water is required, a fourth function is added to the above three functions.

(4) Effective removal of entrainment from vapor streams

4.1.2 Desalting factors in the butterfly-valve type flash box

The desalting factors in five kinds of flash boxes were experimentally examined. The shapes of the flash boxes, that is, the butterfly-valve type, the weir type, the shower type, the high weir type and the double weir type, are shown in Table 1.

At first desalting factors in the butterfly-valve type flash box were examined. The feature of this flash box is to be able to change the brine flow and flashing conditions by opening the butterfly valve which is installed at the outlet of the box.

Desalting performances of the butterfly-valve type flash box are shown in Fig. 8. The desalting factors, $(DF)_f$ are plotted against the boil-up rate, B for three different rotating angles of the valve plate, 0°, 90°, 120° in Fig. 8. The rotating angles of the valve plate are measured from a horizontal line. When the valve plate is set horizontally (0°), vertically (90°) and inclinedly (120°), the valve is closed, half opened and full opened respectively.

In both cases of the valve angle 90° and 120°, the desalting factor does not decrease so much as the boil-up rate rises. On the other hand, in case of the valve angle 0° the desalting factor decreases very rapidly from 1,000 to 10 as the boil-up rate rises. It is considered that since the valve closed at the valve angle 0° and the brine is incompletely evaporated in the flash box, the brine is suddenly flashed at the outlet of the flash box and many particles are generated.

Experiments showed that the desalting factors in the butterfly-valve type flash box depend on the valve position of the butterfly valve plate. Therefore it becomes clear that the desalting factors in the flash box are affected by the structure of the flash box.

4.1.3 Desalting performance of various flash boxes

The desalting factors in the weir type, the shower type, the high weir type and the double weir type flash box were experimentally examined after the butterfly-valve type flash box. The shapes of all kinds of flash boxes and typical experimental data are shown in Table 1.

In Fig. 9 the desalting factors, $(DF)_f$ the butterfly-valve type (120°), the weir type, the high weir type and the double weir type flash box are plotted against the boil-up rate, B.

The weir type flash box has a simple structure and good control of the brine flow. The values of the desalting factors in this weir type flash box are as high as those in the butterfly-valve type flash box at the valve angle 120°.

The shower type flash box is designed as the extreme condition of the butterfly-valve type flash box at the valve angle 0°. The values of the desalting factors in the shower type flash box are very low as shown in Table 1, because many small particles are generated in the violently showering nozzle.

The high weir type flash box is developed from the weir type flash box. The values of desalting factors in the high weir type flash box are higher than those in the weir type flash box. As shown in Fig. 9, the highest value of the desalting factor, 50,000 was obtained.

The double weir type flash box consists of a low weir, a high weir and a splashing plate. This flash box has **the same functions as the high** weir type flash box at steady state. At starting of the plant, the brine flows over the low weir and through the passage under the high weir, so it is not necessary that the brine flows beyond the high weir. Therefore the brine flows at higher rate in the double weir type flash box than in the high weir type at starting. The values of the desalting factors in the double weir type flash box are a little lower than those in the high weir type.

From experiments with five kinds of flash boxes, the design principles of the flash box for high purity water are given as follows:

(1) Mix sufficiently the flashing vapor with the brine in the flash box and catch the entrained particles in the brine.

(2) Make the kinetic energy of the brine flowing off the flash box as low as possible.

(3) Change the direction of vapor streams suddenly and remove entrainment from the vapor.

4.1.4 Empirical equation of the desalting factor in the flash box

It is very difficult to relate exactly the desalting factors in the flash box to the structure of the flash box and the conditions of flash evaporation, because this phenomenon is very complicated and desalting factors are affected by many parameters, such as, boil-up rate, vapor expansion rate, brine flow rate, pressure drop, temperature and concentration.

According to investigations on the relation between desalting factors in the flash box and evaporating conditions, it becomes clear that the desalting factor decreases as vapor expansion rate increases.

The desalting factors, $(DF)_f$ in all kinds of flash boxes are plotted against the vapor expansion rate, G in each flash box in Fig. 10. Though the experimental values are very scattered, the empirical equation between desalting factors, $(DF)_f$ and vapor expansion rate, G (1/s) is shown as follows:

$$(DF)_f = 6.0 \times 10^5 \; G^{-1.7} \quad \ldots\ldots\ldots\ldots\ldots\ldots \quad (6)$$

where $G = F \cdot \gamma / V$ (1/s)
  F : evaporating rate in the flash box (kg/s)
  γ : specific volume of vapor (m³/kg)
  V : volume occupied by vapor in the flash box (m³)

Since the highest value of the desalting factors in the flash box is 50,000 at the high weir type flash box, the demister is necessary to obtain a high purity water.

4.2 Desalting performance of the demister

There are many kinds of demisters which remove entrainment particles from vapor streams, and there is a suitable type of the demister according to the diameter of the entrained particles (7).

A wire-mesh type demister which was made by Japan Mesh Industrial Co. was used in this flash evaporator, because it is considered that the entrained particles in the flash evaporator range from 1μ to 1 mm in diameter. The structure of this demister is shown in Fig. 11 and its specifications are given as follows:

Material   : Stainless steel, Wire diameter: 0.254 mm
Density   : 193 kg/m³   , Free volume : 97.5 per cent
Surface area: 375 m²/m³   , Thickness   : 100 mm

The wire-mesh type demister consists of multi-layer knitted wire fabric, with each layer crimped and matted to provide interlacings of wires in **successive layers**. **The entrained particles are removed from vapor streams by impaction on the wires** (7, 8).

Desalting factors, $(DF)_d$ in the wire-mesh type demister are plotted against linear velocity, v of the vapor through the demister in Fig. 12. In this case, it is considered that the re-entrainment velocity is 6.0 m/s as shown in Fig. 12. The empirical equation between the desalting factor, $(DF)_d$ and the linear velocity, v of the vapor through the demister is obtained as follows:

$$(DF)_d = 15.0 \ v^{0.23} \quad \quad \quad \quad \quad \quad (7)$$

When the vapor velocity is beyond the re-entrainment velocity, the desalting factors in the demister fall rapidly. Therefore the vapor velocity must be fixed below the re-entrainment velocity, 6.0 m/s.

Demisters were installed at a height of 0.6 m above the brine level. This must be done because the big particles begin to fall at this height.

The pressure drop is one of the performance factors of the demister. The pressure drop caused by the wire-mesh type demister is small, for example, 60 mm of water per 100 mm of demister thickness at vapor velocity 6.0 m/s.

4.3 The method for producing high purity water

From experiments the highest value of purity, 0.2 ppm TDS was obtained by using the high weir type flash box and the wire-mesh type demister. However it is difficult to produce the high purity water which contains dissolved solids below 1 ppm TDS at high boil-up rate. Hence, a method by which high-purity water, ranging from 1 ppm TDS to 0.05 ppm TDS, can be produced must be contrived.

The high purity water can be obtained by re-distilling the distillate. In each stage of the multi-stage flash evaporator, the distillate evaporates as the brine does. But the vapor produced from the distillate is mixed with the vapor produced from the brine and then flows into the condenser. Therefore if the condenser which is exclusively used for the vapor produced from the distillate is installed, the high purity water is obtained.

It is possible to apply this idea to all flash evaporators. As an example, the flow diagram of Hitachi multi-effect multi-stage flash evaporation plant with a re-distiller is shown in Fig. 13.

5. Conclusions

From experiments on the purity of the distillate in a 10-stage flash evaporator, the following results were obtained:

(1) Entrainment is initially separated from the vapor in the flash box and then the remaining entrainment is removed by the demister.

(2) It becomes clear that the desalting factor in the flash box depends on the structure of the flash box.

(3) The high weir type flash box in which the highest desalting factor, 50,000 was obtained was developed.

(4) **An empirical equation between the desalting factors and the vapor expansion rate in the flash box was obtained.**

(5) The desalting performance of the wire-mesh type demister is given.

(6) The flash evaporator with a re-distiller was designed in order to produce high purity water.

Acknowledgement

The author wish to thank Dr. T. Tejima and Dr. S. Takahashi for their helpful advice.

References

(1) O'connell, H.E. & Pettyjohn, E.S. : Trans. Am. Inst. Chem. Engrs, 42, 795-814 (1946)
(2) Garner, F.H., Ellis, S.R.M. & Lacey, J.A. : Trans. Instn Chem. Engrs, 32, 222-235 (1954)
(3) Newitt, D.M., Dombrowski, N. & Knelman, F.H. : Trans. Instn Chem. Engrs, 32, 244-261 (1954)
(4) Mitsuishi, N., Matsuda, Y. Yamamoto, Y. & Oyama, Y. : Chem. Engng in Japan, 22, 680-687 (1958)
(5) Manowitz, B., Brett, R.H. & Horrigan, R.V. : Chem. Engng Progr., 51, 313-319 (1955)
(6) Aiba, S. & Yamada, T. : A.I.Ch.E. Journal, 5, 506-509 (1959)

(7) Reed, M. : Proc. Instn Mech. Engrs, *178*, Part 1, No.4, 91-105 (1963-64)
(8) Carpenter, C.L. & Othmer, D.F. : A.I.Ch.E. Journal, *1*, 549-557 (1955)

(1) Entrainment (2) Splashing (3) Foaming

Fig. 2. Three stages of carry-over in the flash evaporator

Fig. 4. The outside view of a 10-stage flash evaporator

(1) An outside view (2) The detail

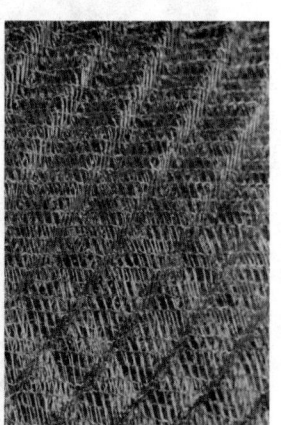

Fig. 11. A wire-mesh type demister

Table 1. The shape of five kinds of flash boxes and the purity of the distillate in each flash box

No.		1			2	3	4	5
Type		Butterfly-valve			Weir	Shower	High weir	Double weir
		120°	90°	0°				
Shape								
Q	t/h	10.0	10.0	10.0	10.0	7.5	10.0	10.0
T	°C	82.0	80.0	82.0	76.5	51.0	57.0	65.0
D	kg/h	225	260	225	148	273	345	214
B	kg/m²·h	251	290	251	173	226	285	177
C_1	ppmTDS	60,000	61,000	60,000	63,000	60,000	61,000	61,000
C_2	ppmTDS	13	50	1360	6.6	1250	3.0	3.5
DF	-	4,600	1,220	43	9,550	48	20,300	17,400

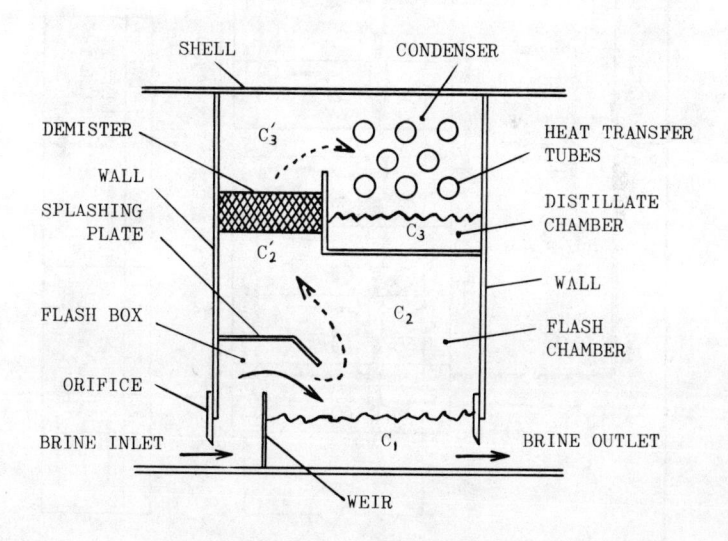

Fig. 1. The inside structure of the flash evaporator

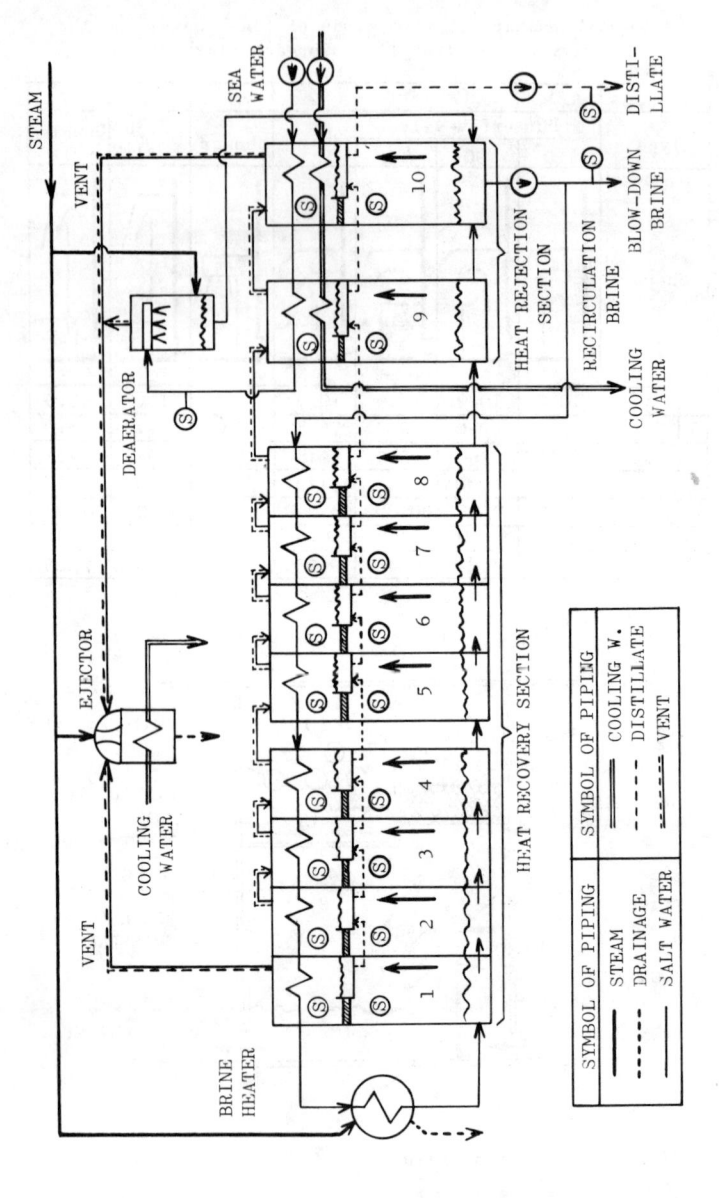

Fig. 3. The flow diagram of a 10-stage flash evaporator

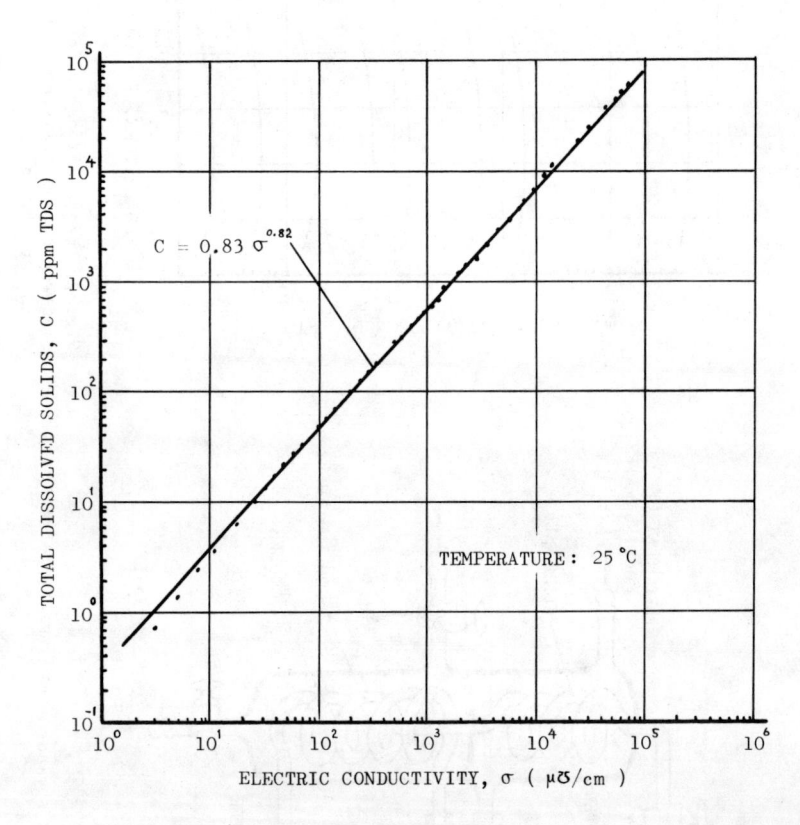

Fig. 6. The relationship between total dissolved solids and the electric conductivity of the solutions

$C = 0.83\, \sigma^{0.82}$

TEMPERATURE : 25 °C

Fig. 7. The relationship between the electric conductivity and the temperature of the solutions

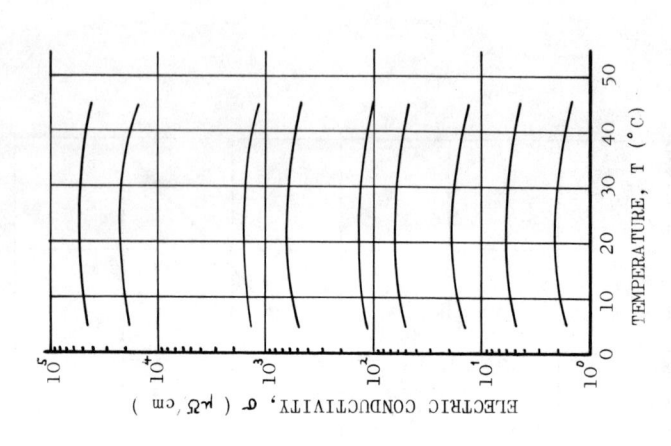

Fig. 5. A vapor sampling apparatus

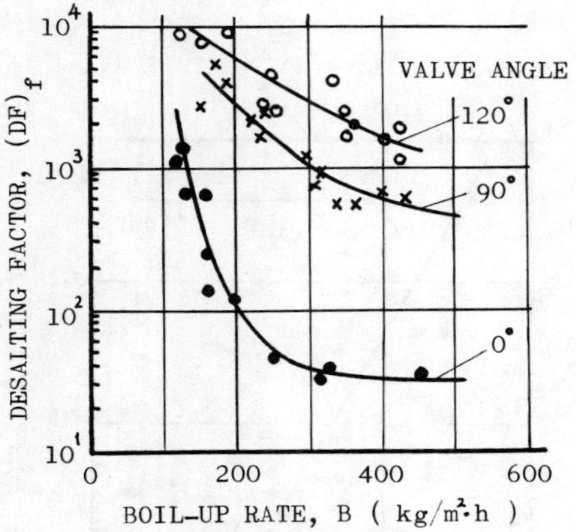

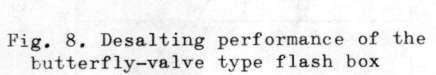

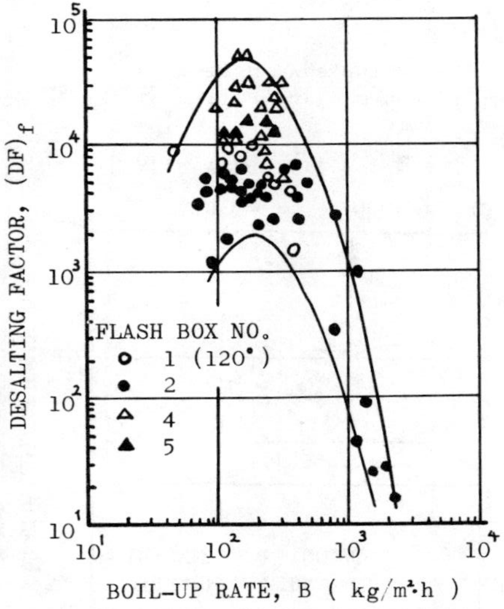

Fig. 9. Desalting performance of the various flash boxes

Fig. 8. Desalting performance of the butterfly-valve type flash box

313

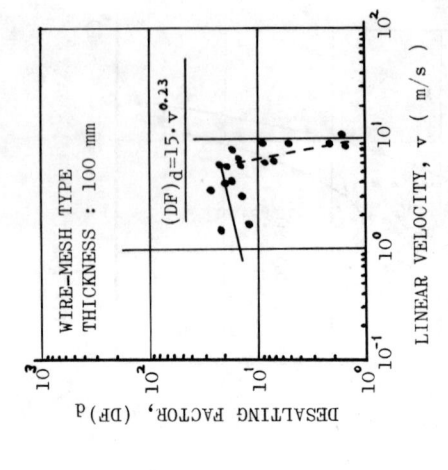

Fig. 12. Desalting performance of the wire-mesh type demister

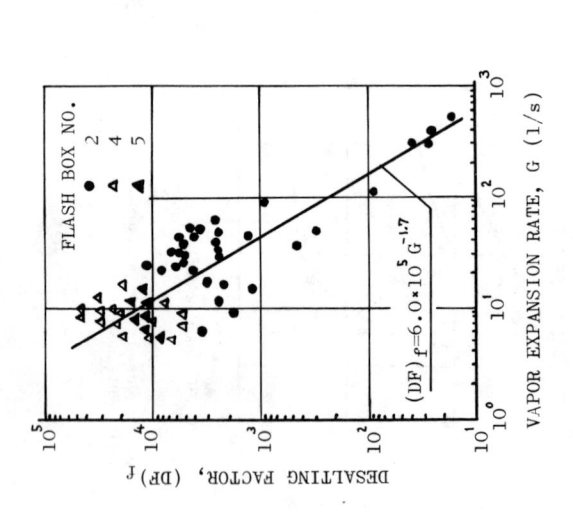

Fig. 10. The relationship between the desalting factor and the vapor expansion rate in the flash boxes

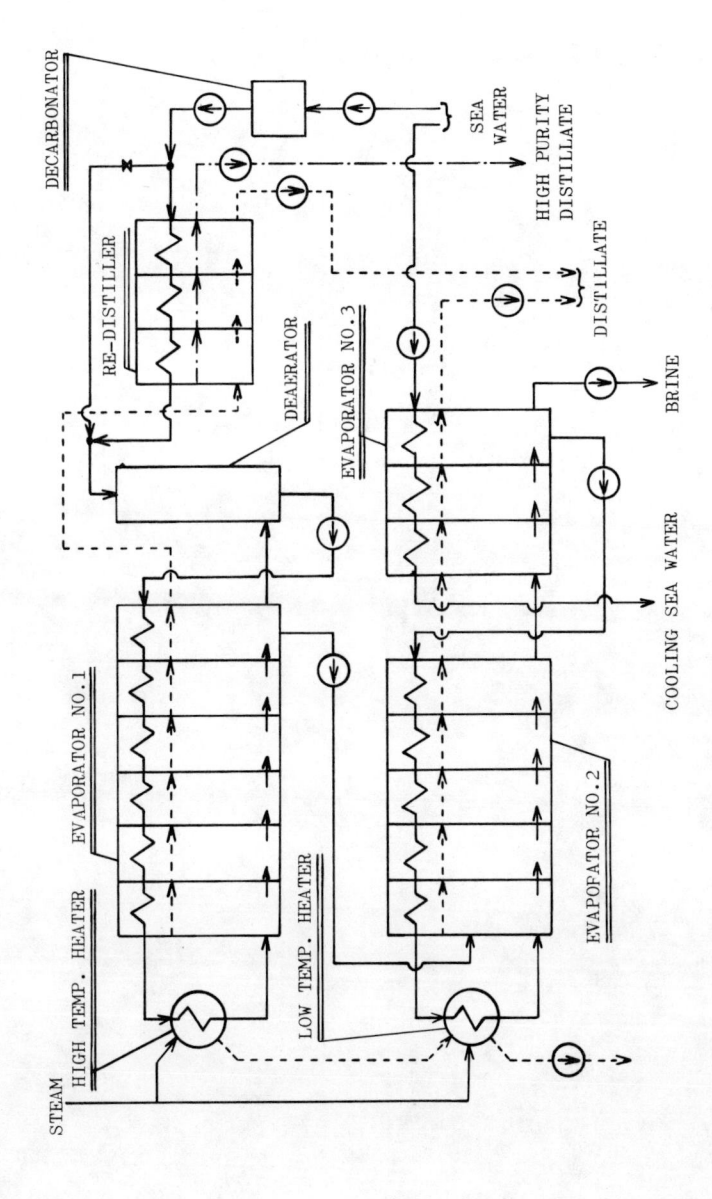

Fig. 13. The flow diagram of Hitachi ME type multi-stage flash evaporation plant with re-distiller

THE USE OF ELECTROCHEMICAL DEVICES

IN CHEMICAL OCEANOGRAPHY

Chairmen:
G.C. Whitnack
H.V. Weiss

ANODIC STRIPPING VOLTAMMETRY OF TRACE METALS IN SEAWATER

Alberto Zirino[1], Stephen H. Lieberman[1]
and
Michael L. Healy[2]

[1] Chemical Oceanography Branch
Naval Undersea Center, San Diego, CA 92132

[2] Department of Oceanography
University of Washington, Seattle, WA 98105

ABSTRACT

A review of past and current applications of anodic stripping voltammetry to the study of trace metals in the marine environment is presented. The uses of thin film and hanging mercury drop electrodes are discussed and new voltammetric techniques such as AC, continuous flow and differential pulse anodic stripping are presented. Current studies on the speciation of Cd, Cu, Pb and Zn in seawater by anodic stripping voltammetry are discussed in terms of their significance to marine chemistry.

INTRODUCTION

Recently, considerable attention has been given to the application of anodic stripping voltammetry (ASV) to the study of certain trace transition elements in seawater and other natural water systems. Because this method is extremely sensitive and requires few or no reagents for the analysis, it is useful for the study of Cu, Zn, Cd, Pb and other metals which amalgamate reversibly with Hg.

Early papers approached seawater as a solution of inorganic electrolytes whose trace metal content could be determined simply by standard additions. Newer work has considered seawater as a heterogeneous mixture containing trace metals in liquid and solid (particulate) phases and attempts have been made to distinguish between organic and inorganic trace metal species. The latter work has illustrated the potential of ASV as an investigative tool for the study of trace metal speciation in the marine environment.

BACKGROUND

Whitnack[1] first demonstrated that polarographic analysis could be applied to seawater. Later Whitnack[2] showed the feasibility of detecting and estimating trace amounts of Cd, Co, Cr (as $CrO_4^=$), Cu, IO_4^-, Mn, Ni, Pb and Zn in seawater using differential cathode ray polarography with derivative readout. Following Whitnack, several investigators independently explored the possibilities of extending the limit of detection in saline waters by applying the technique of anodic stripping. Ariel and Eisner[3] determined the concentration of Cd and Zn in a sample of Dead Sea brine using the hanging mercury drop electrode (HMDE) of Gerisher[4] described by Shain and Lewison[5]. In a subsequent work, Ariel et al.[6] determined Cu in Dead Sea brine by electroplating Cu from the brine and stripping into 0.5 N NH_4OH - 0.5 N NH_4Cl. In the new medium, the Cu signal was clearly resolved from the mercury dissolution wave. Macchi[7] was able to measure Zn in Mediterranean seawater with high precision with ordinary capillary electrodes and dropping times of approximately one minute. His fast sweep technique with oscillographic readout yielded measurements that were indicative of the "ionic" component of Zn (i.e., that fraction reducible at the electrode) in seawater. Whitnack and Sasselli[8] extended the range of seawater determinations by measuring Cd, Cu, Pb and Zn in Pacific Ocean samples using the HMDE. Their results, obtained by ASV, were comparable to those obtained by single sweep differential oscillo-polarography.

The work conducted in the past decade established ASV as an important new method for the investigation of certain trace metals in seawater. Recent work has been directed towards the development of more sensitive stripping techniques and towards using ASV in conjunction with other methods to study the speciation of trace metals in marine waters.

RECENT METHODS

At least five new techniques have been developed which are promising investigative tools for the study of trace metals in the marine environment:

(1) ASV with a dropping mercury electrode with drop times as long as one day. The electrode was described by Kemula and Kublik[9] and has been modified and extensively used to study Zn in seawater by

Bernhard et al.[10] The advantages of this electrode are high hydrogen overvoltage and good reproducibility, permitting the observation of Zn in seawater at very low pH values. The investigators have used it to determine spatial and temporal distributions of Zn in the Gulf of La Spezia and Gulf of Taranto and to study the physico-chemical states of Zn in seawater. Smith and Redmond[11] with the Kemula electrode studied the behavior of 17 ions added to seawater and found that Cd, Co, Cr, Cu, Ni, Sb, Sn, Pb, V and Zn gave current peaks. Ni interfered with the determination of Zn while V contributed to the Cu peak current. However, the extent of the interferences was slight because Ni had a current sensitivity only 7% that of Zn, and V had a current sensitivity 13% that of Cu. The determinations of Cd and Pb appeared to be free from interference.

(2) ASV with matched electrodes used in a differential manner[12]. Two similar electrodes are immersed in the same cell but are connected to separate inputs of a differential amplifier. During electrolysis, metals are plated on one electrode while the other is temporarily switched out of the cell circuit. After the electrolysis is completed, the second electrode is reinserted into the circuit and the potential sweep is applied to both electrodes. The residual current signal of the second electrode is similar to that of the working electrode and is subtracted from it by the differential amplifier. Because nearly all residual currents are cancelled, greater amplification of the feradaic component is possible, and plating times can be shortened. Differential anodic stripping voltammetry (DASV) has been used at sea to measure the zinc content of open ocean waters[12,13].

(3) Voltammetry with the mercury composite graphite electrode (MCGE). Although stable film electrodes are somewhat difficult to produce, they are highly sensitive, display sharp, well defined peaks for several metals in seawater and have the added advantage that Cu peaks are well resolved, even in a saline medium. This electrode was developed by Matson[14] and has been used to investigate trace metals in seawater by Matson[15], Fitzgerald[16], Seitz[17], Gilbert[18] and Barsdate and Nebert[19]. Relatively lower hydrogen overvoltage precludes its use to determine Zn under acid conditions, but Cd, Pb and Cu can be easily observed at pH>3. Because the electrode is solid, it is ideal for field use and has been extensively used at sea by Fitzgerald[16] to study the concentration and speciation of Pb and Cu. Seitz[17] made a thorough study of the application of the MCGE to the determination of metals in seawater by ASV. He observed that the measurement of Cd and Pb in seawater were virtually interference-free while Zn determinations were unreliable because of intermetallic interferences from Ni, Cu and Co. Interferences in the analysis of Cu could be reduced by selecting a plating potential which would not reduce Ni and Zn. Seitz[17] also suggested that under proper conditions Tl and Bi could be determined in seawater with the MCGE. Zirino[20], using matched MCGE's and fast sweep oscillographic DASV, measured Pb concentrations

as low as 0.03 ug/l in coastal waters of the Gulf of California. In an effort to extend the application of ASV in saline waters, Gilbert[18] developed methods for the determination of Ag, Bi, Sb and Tl in seawater using wax impregnated graphite and mercury coated electrodes.

(4) Phase selective Alternating Current and Differential Pulse ASV. These methods illustrate the increased possibilities of studying trace metals in seawater which arise when the stripping cycle is carried out by methods other than a linear scan. Seitz[17] increased the sensitivity of Pb determinations by a factor of two and of Cd determinations by a factor of five with phase selective AC stripping. This process was particularly effective in resolving Zn peaks in raw seawater from the irreversible hydrogen current. In our laboratory (NUC), the measurement of Zn, Cd and Pb in raw seawater was carried out by differential pulse polarography ASV[21] on a tubular MCGE. This technique was found to be considerably more sensitive than linear scan ASV (Fig. 1).

(5) ASV with continuous flow systems. Lieberman and Zirino (in preparation) have combined the high sensitivity of the MCGE with the increase in convective transport and the ease of sample handling inherent with tubular electrodes to form a tubular mercury graphite (TMGE) flow system. Because electrode instability increased in this dynamic situation, a means of maintaining a consistently active mercury coat was developed. A cyclic approach was devised which permitted alternate plating from a mercury solution or a sample solution. With a one-inch TMGE and a flow of 160 ml/min., this system was evaluated to be approximately eight times more sensitive than stationary techniques.

Clearly, technical advances have made the measurement and study of many trace metals in seawater possible. Now an effort must be made to interpret peak currents and peak potentials in terms of trace metal speciation.

STUDIES OF TRACE METALS SPECIATION IN SEAWATER BY ASV

(1) Zinc. This element is the most abundant of the metals that are present in seawater and amalgamate reversibly with Hg, and consequently has been most extensively studied. Voltammetric investigations of the nature of Zn in seawater began with the observation that the Zn peak current increased when the pH of the seawater sample was lowered[12,22,23]. When seawater from the Gulf of La Spezia and the Gulf of Taranto was analyzed for Zn at pH values ranging from 8.1 to 1.4, a characteristic titration curve resulted which was interpreted by Piro et al.[24], Piro[25] and Branica et al.[26] as indicating the release of Zn to the electrode by ionic, particulate and organically complexed forms naturally present in seawater (Fig. 2). These

investigators suggested that the initial increase in peak current occurring when the pH is lowered from approximately 8.1 to 6 is caused by the solution of particulate Zn (as, for instance, from the solution of hydroxy colloids or desorbtion of Zn from other particulate matter) while the increase in peak current from pH 6 to 1.4 is caused by the release of Zn from organic ligands inert at pH 8. The plateau in the curve observable between pH 4 and 6 was ascribed to organic ligands which compete for the Zn ion as it released from the particulate matter. If the organic matter is first destroyed with a strongly oxidizing UV light[16], the plateau is eliminated.

At pH 8.1, and at a potential of -1.25 to -1.30 versus the saturated calomel electrode (SCE), only 10 to 15% of the total Zn is available to the HMDE. This fraction is presumably composed of Zn ion, Zn bound to labile organic and inorganic complexes and Zn bound to directly reducible complexes. Measurements made over a two-year period produced dozens of nearly identical titration curves which were also remarkably constant with geography and depth. Such constancy suggests that the three forms of Zn exist in equilibrium. Recent Zn - pH titrations of samples from the same locations have produced curves without the familiar plateau. Apparently the organic ligands which competed for Zn from pH 4 to pH 6 are no longer present in the water.

Piro et al.[27] have conducted several experiments in order to substantiate the three-component model. Using potentiostatic electrolysis combined with radiometric determinations they were able to observe the partition of stable Zn and ^{65}Zn into the three postulated physico-chemical states. The polarographically available Zn was determined by ASV as the pH was varied. Kinetic studies were conducted by removing one or two of the Zn fractions in a mercury pool electrode and observing the redistribution among the various Zn species. They concluded that "by lowering the pH of seawater all three forms can be transferred into ionic Zn in approximately one minute". However, "when returning from pH 2 to high pH values the equilibrium between the different physico-chemical states is not so quickly re-established." There is a marked influence of pH on the rate of distribution of Zn in the various forms. These investigators observed that at pH 8, the organically complexed fraction (not polarographically observable unless the pH is lowered below 4) was remarkably inert to ^{65}Zn exchange, even after one year of equilibration. The presence of an inert organic Zn fraction in seawater was also noted independently by Fukai[28] using radiometric and spectrophotometric techniques.

In a field experiment, Zirino and Healy[29] observed that Zn (and Pb) peak potentials shifted approximately +30 mV vs. SCE when the pH of seawater was lowered from approximately 8 to 5.6. They suggested that the peak potential shift indicated Zn association with OH⁻ or possibly $CO_3^=$ at pH 8.1. Association with these anions and subsequent agglomeration into colloidal particles would produce the observed shift in peak potential and decrease in peak currents. This suggestion was strengthened by the thermodynamic calculations of Zirino and Yamamoto[30]. Their inorganic model indicated that at pH 8, Zn in seawater was primarily present as $Zn(OH)_2^0$.

In summary, the anodic stripping analysis of Zn in seawater has suggested this model (Fig. 3): at pH 8.1 only a small fraction of the total Zn is directly reducible at the electrode. This fraction is composed of ionic and inorganic forms as well as reducible organic Zn complexes. The bulk of the Zn is present in inert forms which may be either particulate (including colloidal) or organic or both and which release Zn into solution when the seawater is acidified.

The competition for Zn at pH 8.1 by organic and inorganic ligands is demonstrated in Figure 4. Curve A is the trace observed when a filtered seawater sample from a fjord in British Columbia is analyzed by ASV. Curve B is the trace produced by the same sample after the organic matter has been destroyed with an ultraviolet lamp. After the irradiation the Zn peak has diminished and an adsorption peak (possibly $Zn(OH)_2^0$) can be observed at 0.9 V vs. SCE. The experiment suggests that much of the Zn originally observed was sequestered by labile organic ligands which were then destroyed by the irradiation. Without organic matter Zn was free to agglomerate into inorganic colloids and particles. The results for Pb also present in the sample suggest that prior to the irradiation Pb was sequestered by inert organic ligands which, when destroyed, released ionic Pb to the solution.

(2) <u>Copper</u>. Fitzgerald[16] has conducted the most extensive voltammetric investigation of Cu in seawater. He noted that, similarly to Zn, the Cu peak current increased with decreasing pH, and interpreted this to be caused by Cu ions released when hydrogen ions competed for the ligand sites. Natural organic sequestering agents were also characterized as "weak" or "strong" according to the following: weakly chelated Cu was that portion observed by the electrode when the pH was lowered to 3, while strongly chelated Cu required UV irradiation of the sample before it could be reduced at the electrode. Once the irradiation had been carried out, lowering the pH to 3 did not make more Cu available to the electrode. The presence of organically sequestered Cu was also suggested by other independent tests. Fitzgerald observed that an average of 63% of the Cu in near-shore waters around Woods Hole, Mass., was organically complexed and that Sargasso Sea samples analyzed aboard ship gave similar results.

Barsdate and Nebert[19] report the results of several investigations of Cu in high latitudes. Using the MCGE, short plating periods and rapidly responding instrumentation, these investigators analyzed many samples from the Bering Sea, Alaskan coastal areas and high altitude lakes. Their results indicated that most of the Cu present was acid-exchangeable and could be detected by lowering the pH. Large complexes, verified by dialysis and radiochemical experiments, rarely exceeded 10 to 20% of the total exchangeable metals. Higher concentrations were found in polluted areas and humic-rich fresh waters.

Fitzgerald's observations concerning the speciation of Cu in seawater differ from those made for Zn by Piro et al.[27] and Branica et al.[26] in three respects: 1) Fitzgerald states that the contribution of particulate Cu to the total observed Cu is a minimum, 2) the UV irradiation makes more Cu observable at the electrode than acid titration, and 3) unlike Zn, addition of acid to organic-free seawater does not produce an increase in the Cu peak current. These observations point out what may be a fundamental difference in the speciation of Cu and Zn in seawater: Zn has a tendency to form inorganic colloids, even in the presence of organic matter, while Cu does not. The thermodynamic model of Zirino and Yamamoto[30] indicates that both Cu and Zn form hydroxide complexes at pH 8.1 and that their seawater chemistry should be similar. The differences noted above indicate that either the model is incomplete, or that in seawater, there is a substantial difference in the reaction rates of these two metals with the available inorganic ligands.

(3) <u>Lead and Cadmium</u>. Few systematic studies of these two metals in seawater have been made, but some information concerning their concentration and speciation in seawater is available. Zirino and Healy[12] report that the Pb peak current increases when the pH of the sample is lowered, while the peak current of Cd does not. The Pb peak potential also shifts approximately +30 mV vs. SCE when the pH of seawater is lowered to 5.6 while that of Cd, present in the same sample, does not[29]. The model of Zirino and Yamamoto[30] predicts that Pb in seawater is associated with $CO_3^=$ while Cd is largely present as $CdCl^+$ and $CdCl_2^0$. Thus the model appears to support the experimental observations.

Fitzgerald[16] measured Pb and Cd in the upper 1000 meters at three locations in the western Sargasso Sea and found that the average "free" (that which is directly observable at the electrode at pH 8) Pb concentration at each of the three stations was, respectively, 0.7, 0.4 and 0.2 ug/l, while the average "free" Cd concentrations were respectively, 0.04, 0.04 and 0.03 ug/l. A significant portion of the Pb in the samples was judged to be sequestered by weak organic ligands.

Baier[31] studied Pb in several seawaters ranging from fjord water from British Columbia and plume water from San Francisco Bay to coastal eastern tropical Pacific Ocean water and Caribbean Sea waters. He applied the technique of successive standard additions of metals to the field samples, and attributed the decrease in signal response with successive Pb additions to the effects of organic ligands. He also suggested that in the sulfide-bearing waters of Saanich Inlet, British Columbia, organic ligands maintained Pb at concentrations higher than would be predicted from the solubility of PbS.

ASV DIAGNOSTIC CRITERIA FOR ORGANIC CHELATORS

Most of the recent investigators have described various ASV procedures for the characterization of trace metal binding in marine waters. Matson[15] and Allen et al.[32] outlined some basic approaches for the use of ASV data as diagnostic criteria. These approaches included the techniques of successive standard additions, acid titrations and metal exchange. The method of successive standard additions resembles a complexometric titration and for a simple metal ligand system, i.e.,

$$\text{METAL} + \text{LIGAND} \underset{k_b}{\overset{k_f}{\rightleftarrows}} \text{COMPLEX}$$

information about the stability constant can be obtained from the slopes and inflection points of the plots of peak current vs. metal concentration. However, marine samples tend to contain complicated mixtures of ligands and the resulting titration curves are difficult to interpret. Nevertheless, applications of this technique tend to demonstrate the "chelation capacity" of the sample. This capacity may be an important oceanographic variable and show a functional relationship with marine productivity[33].

One diagnostic technique seldom applied is that of ligand addition. Two experiments of this nature were performed recently at the University of Washington during some initial investigations of the effects of sewage effluents on the oceans. Samples of primary-treated, unchlorinated effluent were obtained from the Hyperion plant in Los Angeles in March, 1972, and frozen. One of these samples was thawed and added to samples of a stock of Bering Sea water which was initially filtered and had been aging for five years. The metal concentrations in this water were low, and the seawater samples spiked to give the initial concentrations shown in Table I. In both experiments the samples contained 10% v/v effluent after effluent addition. Since different electrodes were used for each experiment, the sensitivities are not comparable. The metals were plated for 10 minutes on a MCGE.

The results of experiment I at the natural pH of seawater show decreases in the signals from all three metals after effluent addition. Binding of the metals by constituents in the effluent is evident. The results of experiment II, carried out at pH 5, show a marked decrease in the Cd and Cu peak heights which is typical of strong complexation by organic ligands. Further binding occurs at a slow rate as seen by the continued decrease after the analysis was repeated 24 hours later. The increase in the magnitude of the lead peak after 24 hours may be due either to release of lead by some constituent in the effluent or to some experimental artifact, such as contamination or desorption.

One could not imply on the basis of these few data that organics in sewage effluent bind the trace metals in seawater; however, the experiment points out that the effects of effluents can be witnessed by the response of the electrode. Such effects are of immediate concern to the environmental scientist.

TABLE I.

Exp. I. pH = 8.1	Cd	Pb	Cu
Initial concentration in micrograms per liter	1.68	2.80	1.61
Stripping peak current in microamperes			
Before effluent addition	1.20	0.30	0.06
After effluent addition	0.64	0.12	0.02
Exp. II. pH = 5	Cd	Pb	Cu
Initial concentration in micrograms per liter	0.56	1.80	0.97
Stripping peak current in microamperes			
Before effluent addition	0.35	2.20	2.56
After effluent addition	0.18	2.00	0.98
24 hours after effluent addition	0.09	2.90	0.62

CONCLUSION

The application of ASV to marine chemistry continues to increase our knowledge of the trace metal chemistry in the marine environment. At last, oceanographers can observe "directly" some transition metal ions in seawater without prior chemical concentration steps and are able to interpret experimental results in terms of metal and ligand activities as well as concentrations.

As with all newly applied techniques, ASV has pointed out new research problems in oceanography. Because the various voltammetrically observed trace metal fractions have biological and geochemical significance, they must be identified and their distribution in the oceans determined. The importance of these species as regulators of trace metal concentrations in the oceans[34] must be studied and their role in the control of oceanic primary production[33] must be determined. The implications of the results of the many diverse but related experiments by the cited investigators are important in other areas of study. Both toxicity levels and growth enhancements determined by the addition of simple inorganic salts of trace metals may not reflect the operation of natural processes at all. The biological system is forced to respond to the introduction of species completely out of the natural equilibrium[35]. Also, geochemical experiments based on the introduction of radionuclides may be falsely interpreted. The pathways and cycling of the isotopes involve phase distributions of the chemical species. The simple inorganic fraction may be the least important. Indeed, not only the amount but also the chemical form must be taken into account. ASV is becoming an increasingly effective method to gain knowledge about the chemical form.

REFERENCES

1. G. C. Whitnack, J. Electroanal. Chem., 2 (1961) 110-115.

2. G. C. Whitnack, in Graham Hills, Polarography 1964, Vol. 1, McMillan, London, 1966, pp. 641-651.

3. M. Ariel and U. Eisner, J. Electroanal. Chem., 5 (1963) 362-374.

4. H. Gerisher, Z. Physik. Chem., 202 (1953) 302.

5. I. Shain and J. Lewinson, Anal. Chem., 33 (1961) 187.

6. M. Ariel, U. Eisner and S. Gottesfeld, J. Electroanal. Chem., 7 (1964) 307-314.

7. G. Macchi, J. Electroanal. Chem., 9 (1964) 290-298.

8. G. C. Whitnack and R. Sasselli, Anal. Chim. Acta, 47 (1969) 267-274.

9. W. Kemula and Z. Kublik, Advan. Anal. Chem. Instr., 2 (1963) 123-177.

10. M. Bernhard, M. Branica and A. Piro, "Zinc in Seawater I: Distribution of Zn in the Ligurian Sea and Gulf of Taranto Under Special Consideration of the Sampling Procedure." (In Press).

11. J. D. Smith and J. D. Redmond, J. Electroanal. Chem., 33 (1971) 169-175.

12. A. Zirino and M. L. Healy, Environ. Sci. & Tech., 6 (1972) 243-249.

13. A. Zirino and M. L. Healy, Limnol. Oceanog., 16 (1971) 773-778.

14. W. R. Matson, D. K. Roe and D. Carrit, Anal. Chem., 37 (1965) 1954-1955.

15. W. R. Matson, Ph.D. Thesis, Mass. Inst. of Technol., Cambridge, Mass., 1968.

16. W. F. Fitzgerald, Ph.D. Thesis, Mass. Inst. of Technol., Cambridge, Mass., 1970.

17. W. R. Seitz, Ph.D. Thesis, Mass. Inst. of Technol., Cambridge, Mass., 1970.

18. T. R. Gilbert, Ph.D. Thesis, Mass. Inst. of Technol., Cambridge, Mass., 1971.

19. R. J. Barsdate and M. Nebert, in D. W. Hood, Proceedings of the Symposium for Bering Sea Study., Univ. of Alaska, Institute of Marine Science, Occ. Publ. No. 2 (1972). (In Press).

20. A. Zirino, Alpha Helix Research Program (1972).

21. H. Siegerman and G. O'Dom, Amer. Lab., June, 1972, pp. 62-69.

22. A. Zirino and M. L. Healy, Transactions, AGU, 49 (1968) 697, (Abstract).

23. A. Piro, Rev. Oceanogr. Med., Tome XX, (1970) 133-149.

24. A. Piro, M. Verzi and C. Papucci, Pubbl. Staz. Zool. Napoli, 37 (1969) 298-310.

25. A. Piro, in Proc. Second ENEA Seminar, Hamburg, 1971, pp. 77-84.

26. M. Branica, M. Bernhard and A. Piro, "Zinc in Seawater II: Determination of the Physico-chemical States of Zinc in Seawater." 1972, (In Press).

27. A. Piro, M. Bernhard, M. Branica and M. Verzi, in Proc. Symposium on the Interaction of Radioactive Contaminants with the Constituents of the Marine Environment. Seattle, 1972, IAEA. (In Press).

28. R. Fukai, in Proc. International Symposium on Hydrogeochemistry and Biogeochemistry, IAGC, Sept. 1970, Tokyo.

29. A. Zirino and M. L. Healy, Limnol. Oceanog., 15 (1970) 956-958.

30. A. Zirino and S. Yamamoto, Limnol. Oceanog., 17 (1972), (In Press).

31. R. Baier, Ph.D. Thesis, Univ. of Washington, Seattle, 1972.

32. H. E. Allen, W. R. Matson and K. H. Mancy, J. Water Poll. Cont. Fed., 42 (1970) 573-581.

33. R. T. Barber and J. H. Ryther, J. Exp. Mar. Biol. Ecol. 3 (1969) 101-109.

34. K. B. Krauskopf, Geochim. Cosmochim. Acta, 9 (1956) 1-32.

35. M. Berhard and A. Zattera, in D. J. Nelson and F. C. Evans, Symposium on Radioecology, Ann Arbor, Mich., March 1969.

Figure 1. Differential pulse and linear sweep ASV of a seawater sample on a tubular MCGE. Conditions: 5 min. electrolysis, pH 8.1, 2 ppb Zn. Current: DPASV 10 µA/in., LSASV 1 µA/in.

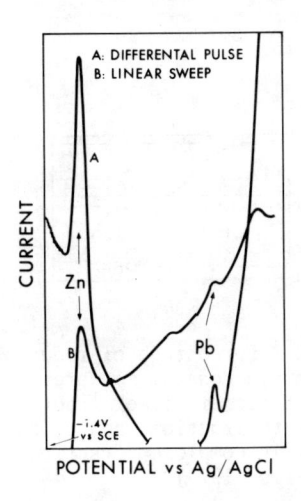

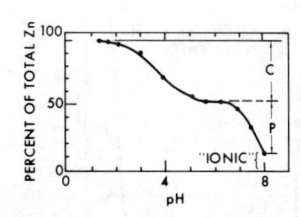

Figure 2. Percentage of reducible Zn vs. pH in coastal seawater. I indicates ionic or "free" fraction, P particulate fraction, and C, the organically complexed fraction. (After Piro, 1970).

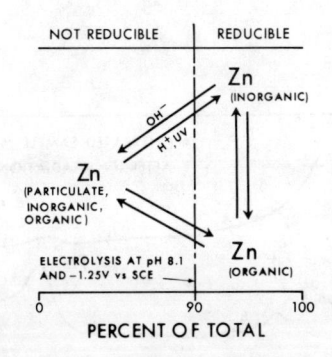

Figure 3. Speciation of Zn in seawater as suggested by ASV.

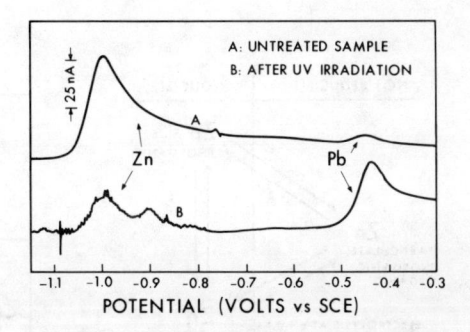

Figure 4. Differential ASV of seawater from a fjord: effects of UV irradiation.

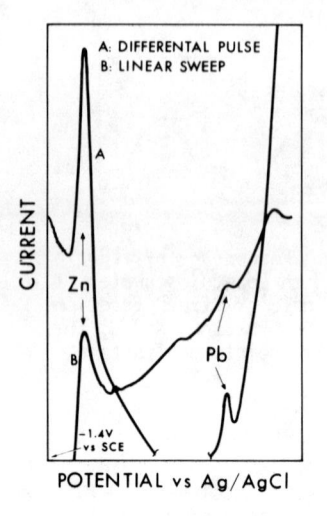

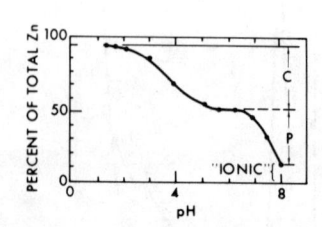

Figure 2. Percentage of reducible Zn vs. pH in coastal seawater. I indicates ionic or "free" fraction, P particulate fraction, and C, the organically complexed fraction. (After Piro, 1970).

Figure 1. Differential pulse and linear sweep ASV of a seawater sample on a tubular MCGE. Conditions: 5 min. electrolysis, pH 8.1, 2 ppb Zn. Current: DPASV 10 µA/in., LSASV 1 µA/in.

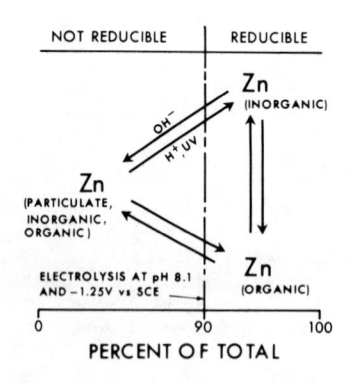

Figure 3. Speciation of Zn in seawater as suggested by ASV.

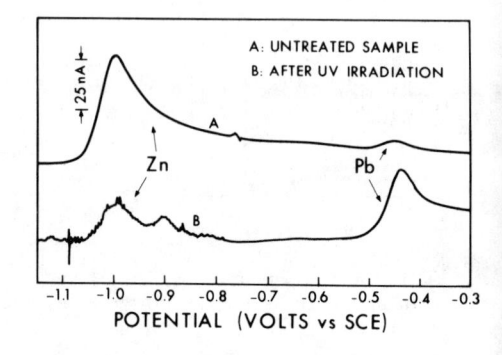

Figure 4. Differential ASV of seawater from a fjord: effects of UV irradiation.

A TECHNIQUE FOR MEASURING CARBON DIOXIDE
HYDRATION KINETICS IN SEAWATER

James H. Mathewson
Department of Chemistry
California State University, San Diego, California 92115

ABSTRACT

A procedure for the measurement of the rate of carbon dioxide-bicarbonate interconversion, which can be readily performed at sea, is described. The change in pH after injection of a CO_2 solution is followed with a glass electrode.

INTRODUCTION

Chemical Oceanography has traditionally been a service field concerned primarily with the analysis of seawater for constituents of interest to physical oceanographers, biologists and geologists. Analytical procedures have usually provided only total concentrations or concentrations remaining after vaguely defined filtrations through bacteriological filters. Limitations on ship operational time and hydrowire sampling techniques preclude the determination of detailed variations of concentrations in either space or time. Large scale profiles or contours using steady state levels of oceanic components together with isotope data are adequate for the construction of models for processes with long time constants and global dimensions. However, for detailed descriptions of the rapid dynamics of nutrients, trace metals, atmospheric gases and pollutants, continuous, real-time, in situ concentrations of seawater constituents are required. Determination of the chemical speciation of substances including more sophisticated size fractionations are also necessary. Finally, rate constants for the processes of transfer and transformation of components are needed for modeling systems in rapid flux.

The rates of transfer of carbon IV species between components of the ecosphere has been an active area of research for many years (1). Photosynthetic uptake of carbon dioxide and its respiratory release constitute the fundamental biogeochemical cycle which mediates energy flow in ecosystems. Calcium carbonate deposition and dissolution and the transfer of carbon dioxide across the air-sea interface are important aspects of the carbon cycle in the oceans.

The process for the transfer of atmospheric carbon dioxide to oceanic or cellular bicarbonate consists of several steps:

$$CO_2(gas) \underset{k_{-1}}{\overset{k_1}{\rightleftharpoons}} CO_2(absorbed) \qquad (1)$$

Absorption in process 1 may be adsorption at an interface or membrane, or solution. Hydration may take two paths, 2a and 2b or 3, depending on pH:

$$CO_2 \rightleftharpoons HCO_3^- + H^+ \overset{K_a^{-1}}{\rightleftharpoons} H_2CO_3 \qquad (2a)$$

$$CO_2 \rightleftharpoons H_2CO_3 \overset{K_a}{\rightleftharpoons} HCO_3^- + H^+ \qquad (2b)$$

$$CO_2 + OH^- \underset{k_{-OH^-}}{\overset{k_{OH^-}}{\rightleftharpoons}} HCO_3^- \qquad (3)$$

The undissociated carbonic acid is not an obligatory intermediate. Direct hydration to bicarbonate or carbonic acid are kinetically equivalent pathways and the observed rate constants are composites (2). The kinetics of reactions 2 and 3 have been thoroughly investigated by a wide variety of techniques (3), but direct measurements of the rate constants of the individual steps have only recently been reported for artificial seawater (4).

Biochemists have been prominent in the study of carbon IV transformations because of the importance of these reactions in photosynthesis and respiration and also in cation exchange and the deposition of calcareous tissues. A class of enzymes, the carbonic anhydrases, which catalyze reaction 2 are found in all higher life forms and some bacteria (5). In a recent report of variations in the rate of equilibration of atmospheric carbon dioxide with two different seawater samples, the suggestion was made that one of the samples contained catalytic activity attributable to carbonic anhydrase (6). To test this hypothesis, a carbonic anhydrase assay was adapted for use as a carbon dioxide hydration rate measurement which could be readily performed at sea on fresh samples.

METHODS

A method involving pH measurements on samples in which nitrogen is used to flush out carbon dioxide was tried initially. The technique suffers from experimental and data handling complications primarily due to the dependence of process 1 on surface effects (4). A method readily adaptable to shipboard use is that of Davis in which the pH drop of a phosphate buffer is followed with a recording pH meter after injection of a carbon dioxide solution (7). Introducing carbon dioxide in solution has the advantage of avoidance of process 1 and also of reaction 3 since the pH is lowered below the competing range for the hydroxide reaction (3). Since seawater is already a bicarbonate solution, phosphate is not required in the adapted procedure.

The method was used aboard the USNS De Steiguer (T-AGOR-12) during the MINOX I cruise from the Naval Undersea Research and Development Center, July 9-August 4, 1970. Kinetic measurements were made on 150 different samples obtained at eight stations off the west coast of Mexico ranging in depth from the surface to 5000 meters.

Samples were obtained from Nisskin bottles immediately after the completion of casts through septums placed over the sampling valves by means of 50 ml syringes which were filled to hold 60 ml. A reaction vessel consisting of a magnetically stirred round bottom flask was filled through a three-way stainless steel syringe petcock and emptied by inserting a flexible polyethylene tube fitted to a syringe through a vent opening. The chamber was flushed with sample water and then filled with 30 ml of sample that had been chilled to 5 ± 0.1°C. The reactant solution which was kept in a gas equilibrating chamber also maintained at 5°C was slowly drawn into a chilled syringe and promptly injected into the magnetically stirred sample. The total volume of the reaction vessel was just over 32 ml so that the exposure of the solution to the atmosphere was negligible.

For most samples a "standard assay" was used in which the sample was injected with 2 ml of standard "Copenhagen" seawater (chlorinity = 19.374 °/oo) saturated with carbon dioxide at atmospheric pressure and 5°C. Standard seawater reactant was used to avoid ionic strength changes at the reference liquid junction on mixing. This solution is 50.53 mM in CO_2. The pH of the reaction mixture was measured with a Corning semi-micro combination electrode with a ceramic junction and silver/silver chloride reference and recorded with a Malmstadt recording pH meter (Heathkit Model EU-20-11). The glass electrode was standardized against Beckman buffers at pH 7.41 at 25°C (7.50 at 5°C) and at pH 4.00 at 25°C. The meter was initially adjusted using a Heathkit pH meter standardizing circuit.

Response time of the system was 4 pH unit/sec change in pen deflection when 2 ml of 50 mM HCl was injected into a seawater sample. Magnitudes of rates observed during assays were of the order of 0.02 pH units/sec. After injection of the reactant, 1 sec was required for the electrode to respond (mixing plus glass surface diffusion). In a few runs a sharp negative response of 0.1 - 0.7 pH units followed by a recovery to higher pH within another 3 seconds occurred, after which the sample followed the usually observed pH curve.

The initial response of the system varies with sample pH (due to the presence of carbonate and to the effect of reaction 3). This was verified experimentally by taking surface samples, bubbling carbon dioxide through them to drive down the pH to values in deeper water (equivalent to the natural respiratory activity), then performing an assay; Rates were identical with deep samples. The converse experiment was performed in which a deep sample was flushed with nitrogen until the pH reached a surface value. The response then followed closely that of a genuine surface sample.

The method was tested by performing kinetic runs on fresh water systems and on standard seawater.

The hypothesis that deep seawater contains carbonic anhydrase was tested by introducing the potent carbonic anhydrase inhibitor "acetozolamide" into a typical deep sample; no change in rate was observed. Introduction of 1.0 µg/l of bovine carbonic anhydrase (Worthington Biochemical) increased the rate of a typical sample three-fold.

RESULTS

The response of a typical sample at the average pH of seawater (7.8) under standard assay conditions is shown in Figure 1. Figure 2 shows the response on varying initial carbon dioxide concentrations.

The system is essentially a buffer that is out of equilibrium at the time of addition of an acid, carbon dioxide, to its conjugate base, bicarbonate. The time for relaxation to the equilibrium pH depends on the rate of carbon dioxide hydration. The pH no longer changes when the rate of dehydration equals the rate of hydration. During the course of the reaction the total concentration of bicarbonate, constrained by charge balance, changes very little. The pH is determined by the bicarbonate-carbonic acid ratio.

The rate expression for reaction 2 is:

$$\frac{-dCO_2}{dt} = k_{CO_2}(CO_2) - k_{-CO_2}(a_{H^+})(HCO_3^-) \quad (4)$$

Using the mixed equilibrium constant for the first ionization of carbonic acid in seawater:

$$K_1 = \frac{a_{H^+}(HCO_3^-)}{(CO_2)}, \quad (5)$$

the rate expression can be put in the form:

$$\frac{(HCO_3^-)}{K_1}\frac{d(a_{H^+})}{dt} = k_{CO_2}(CO_2) - \frac{k_{CO_2}}{K_1}(HCO_3)(a_{H^+}) \quad (6)$$

During the initial phase of the reaction before the pH drops below 7.5, reaction of carbon dioxide with carbonate alters the initial bicarbonate concentration, but thereafter it remains essentially constant and equal to the carbonate alkalinity.

$$CO_3^{2-} + CO_2 + H_2O = 2HCO_3^- \quad (7)$$

With constant bicarbonate, the rate expression becomes:

$$\frac{d(a_{H^+})}{dt} = k_{CO_2}\left(\frac{K_1(CO_2)}{(HCO_3^-)} - a_{H^+}\right) \quad (8)$$

Fitting the data to equation 8 requires evaluation of carbon dioxide concentration changes. During the first half of the reaction (10-60 sec), carbon dioxide is nearly constant. Using values of CO_2 and bicarbonate corrected for reaction 7, a constant, $H°$, is calculated:

$$H° = \frac{K_1(CO_2)°}{(HCO_3^-)°} \quad (9)$$

The integrated rate expression then becomes:

$$\ln(H° - a_{H^+}) = -k_{CO_2}t + \ln(H° - a_{H^+})_{initial} \quad (10)$$

Plots of this integrated rate expression were used to calculate the rate constant for twenty samples from various depths and locations. A mean of $3.6 \times 10^{-3} sec^{-1}$ with a standard deviation of 0.3 was found.

Carbon dioxide concentration variation, temperature control and curve fitting probably contribute the most significant errors. Differences in proton activity are used so that the standardization against a buffer of different ionic strength and composition does not alter k_{CO_2}.

There is no discernable trend in values with depth or location. Using an Arrhenius plot the value of k_{CO_2} found for 5°C was in good agreement with other reported values at higher temperatures suggesting that k_{CO_2} is insensitive to solutes, corroborating earlier work (8). The existence of carbonic anhydrase in seawater outside of organisms was not confirmed.

ACKNOWLEDGEMENTS

I am indebted to Dr. George Pickwell, Naval Undersea Center, for the opportunity to make collections of samples and perform the experiments at sea on the MINOX I cruise. Laboratory checks with fresh water were made by Bruce Anderson at CSUSD. I wish to thank Dr. Alberto Zirino, NUC, and Professor H. E. O'Neal, CSUSD, for invaluable advice during this investigation.

REFERENCES

1. a. G. Skirrow, "The Dissolved Gases - Carbon Dioxide," Chemical Oceanography, J. P. Riley, G. Skirrow (eds.), Vol. 1, Academic Press, N.Y., 1965, pp. 227-322.

   b. P. E. Cloud, "Carbonate Precipitation and Dissolution in the Marine Environment, op. cit., Vol. 2, Chapter 17.

2. J. T. Edsall, "Carbon Dioxide, Carbonic Acid and Bicarbonate Ion: Physical Properties and Kinetics of Interconversion," in CO_2: Chemical, Biochemical and Physiological Aspects, NASA Publication SP-188, Washington, D.C., 1969, pp. 15-27.

3. D. M. Kern, "The Hydration of Carbon Dioxide," J. Chem. Educ., 37, 14-23 (1960).

4. R. F. Miller, D. C. Berkshire, J. J. Kelley, D. W. Hood, "Methods for Determination of Reaction Rates of Carbon Dioxide with Water and Hydroxyl Ion in Seawater," Env. Sci. and Technol., 5, 127-133 (1971).

5. a. S. Lindskog, L. E. L. Henderson, K. K. Kannan, A. Liljas, P. O. Nyman, B. Strandberg, "Carbonic Anhydrase," The Enzymes, P. D. Boyer, ed., Vol. V, 3rd Ed., Academic Press, N.Y., 1971.

   b. T. H. Maren, "Carbonic Anhydrase: Chemistry, Physiology and Inhibition," Physiol. Rev., 47, 595 (1967).

   c. R. E. Forster, J. T. Edsall, A. B. Otis, F. J. W. Roughton (eds.), CO_2: Chemical, Biochemical and Physiological Aspects, NASA, Washington, D.C., 1969.

6. R. Berger, W. F. Libby, "Equilibration of Atmospheric Carbon Dioxide with Seawater: Possible Enzymatic Control of the Rate," Science, 164, 1395-1397 (1969).

7. R. P. Davis, "The Kinetics of the Reaction of Human Erythrocyte Carbonic Anhydrase: Basic Mechanism and Effect of Electrolytes on Enzyme Activity," J. Am. Chem. Soc., 80, 5209-5214 (1958).

8. G. A. Mills, H. C. Urey, "The Kinetics of Isotopic Exchange Between Carbon Dioxide, Bicarbonate Ion and Carbonate Ion and Water, J. Am. Chem. Soc., 62, 1019-1026 (1940).

FIGURE 1. TYPICAL pH RESPONSE UNDER STANDARD CONDITIONS

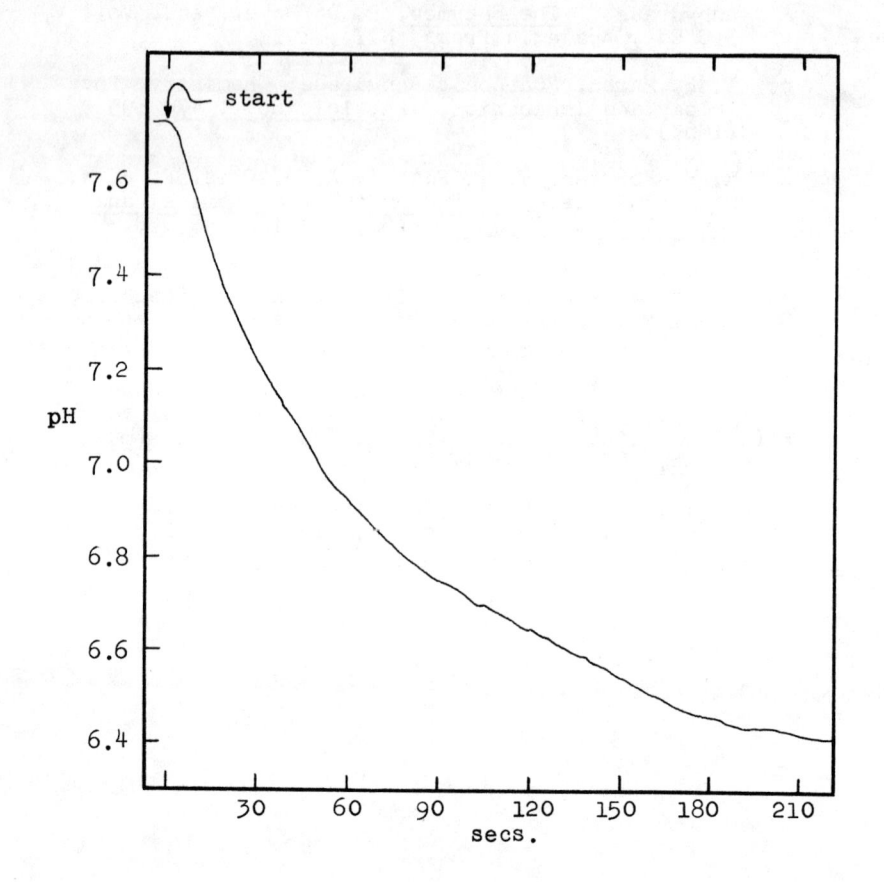

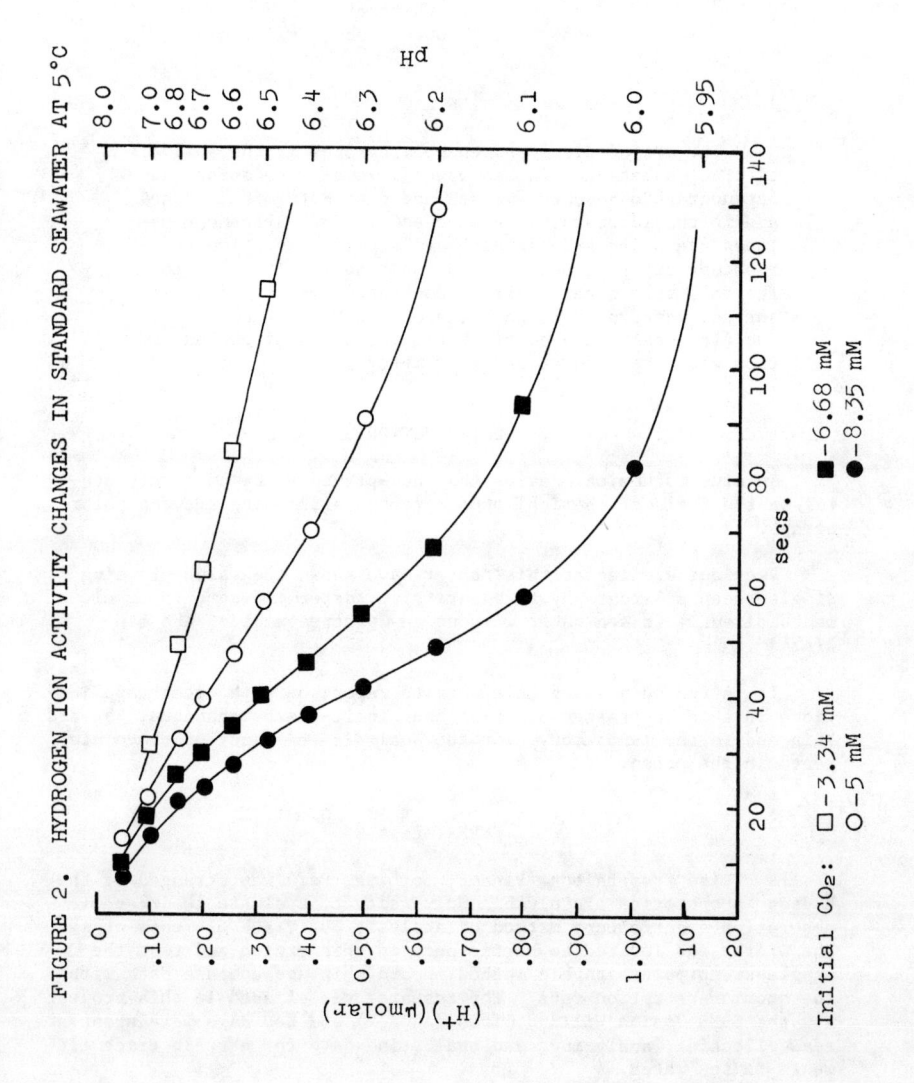

FIGURE 2. HYDROGEN ION ACTIVITY CHANGES IN STANDARD SEAWATER AT 5°C

Initial CO_2: ☐ – 3.34 mM   ■ – 6.68 mM
              ○ – 5 mM   ● – 8.35 mM

RECENT APPLICATIONS OF SINGLE-SWEEP POLAROGRAPHY, ABOARD SHIP AND IN
THE LABORATORY, TO THE ANALYSIS OF TRACE ELEMENTS IN SEA WATER

Gerald C. Whitnack
Naval Weapons Center, China Lake, California

ABSTRACT

The use of single-sweep polarography in the analysis of trace substances in sea water is evaluated aboard an oceanographic research vessel and data determined at sea and in the laboratory are discussed. Good agreement between the polarographic data and atomic absorption spectroscopy data was obtained on the same samples of water. The data show a very uniform concentration of copper in surface water over large areas of the South American Pacific Ocean but a considerable variation of concentration with depth to around 5000 meters.

INTRODUCTION

An electrochemical device that has proved to be extremely useful in the field of chemical oceanography is the single-sweep polarograph.

Previous studies at this center have shown the value of using single-sweep polarography to identify and determine some trace elements directly in sea water without any pretreatment of the sample.[1,2]

It is the purpose of this work to report on some recent applications in this laboratory of using the single-sweep technique,[3] aboard ship and in the laboratory, for the analysis and study of trace elements in the ocean.

EXPERIMENTAL

A cruise from Balboa, Panama, to Lima, Peru was arranged by the Hudson Laboratories of Columbia University to evaluate their atomic absorption spectroscopy method of analysis for trace elements at sea. The writer was invited to participate in this cruise and test the single-sweep polarographic method aboard ship and compare data with the atomic absorption data. The research vessel used in this project was the USNS Josiah Willard Gibbs (AGOR-V) and ten days were spent at sea collecting, analyzing, and evaluating data for certain trace elements in sea water.

Many samples of sea water were taken at the surface and at various depths to approximately 5000 meters. The sampling technique used was to lower teflon-coated Nansen bottles attached to a steel cable to the desired depth with a winch and then to send a weighted messenger down the steel cable to close the bottles. Surface samples were sometimes obtained by casting into the sea an acid-cleaned polyethylene bottle or bucket tied to a nylon cord. The samples were analyzed aboard ship as soon as they were taken and then refrigerated aboard ship for future analysis in the laboratory at home. The sea water samples were not filtered before the analyses were made aboard ship. Some samples were filtered for later examination in the laboratory at home.

The polarographic instrumentation used in this study consisted of a Davis Differential Cathode-Ray "Polarotrace", Model A-1660, manufactured by Southern Analytical Instruments, Ltd., of England and a Moseley Model 2D, X-Y recorder, manufactured by F. L. Moseley Company, Pasadena, California. The dropping mercury electrodes used had drop times of about 7 seconds in distilled water on open circuit and m = 4 to 6 mg. per drop. All measurements of peak currents were made at 25°C ± 0.1°C. All sweep voltages are referred to the mercury pool. Each sample is purged of dissolved oxygen with purified nitrogen before analyzing for the trace elements. The nitrogen is passed over copper turnings @ 450°C and then thru distilled water before entering the sample. All data is recorded on 8 1/2 by 11-inch paper of 10 by 10 to the one-half inch. Quartz cells of 5-ml capacity are used for the samples of sea water in analysis to prevent loss or gain of a trace element to or from the container during the time of analysis.

The atomic absorption apparatus used by the Hudson Laboratory personnel was modified Perkin-Elmer instrumentation using a hollow cathode lamp of the trace element under study as a light source and a burner to vaporize the sample. The methodology is described in detail in a published work of Fabricand et al.[4]

METHODOLOGY

The single-sweep polarographic procedure used in this study is a standard addition technique, previously developed in this laboratory.[2]

A 2-ml sample of sea water is placed in a quartz polarographic cell after addition of C.P. mercury to the cell as the anode; the dropping mercury electrode is then inserted into the solution and the sample flushed with nitrogen for 3-5 minutes to remove the dissolved oxygen from the sea water. Single-sweep polarograms are then recorded for some trace elements such as copper (Cu^{2+}), cadmium (Cd^{2+}), nickel (Ni^{2+}), zinc (Zn^{2+}), iodine (IO_3^-), ($CrO_4^=$) and manganese (Mn^{2+}) respectively by setting the start potential dial of the "polarotrace" to record increments in voltage of 450 millivolts per sweep from zero to about -1.9 volts. For example, to see if copper is present, the sweep voltage is started at -0.05 volts and the voltage range from -0.05 volts to -0.500 volts is then covered each sweep. The sweep rate is 2 seconds per sweep and a 5-second delay allows for a current-voltage curve of copper (with a copper peak about -0.25 volts) to be seen every 7 seconds on the cathode ray screen of the instrument. After a few sweeps the current-voltage curve can be recorded on graph paper with the X-Y recorder for a permanent record. The signal one gets is a simple d.c. microampere signal and can be digitally recorded or, with modern electronics, put on tape, etc., for computer use. Each sample of sea water is studied first in this fashion and a series of current voltage curves are recorded over the entire voltage range. Typical single-sweep polarograms obtained with this technique have been reported in earlier work by Whitnack.[2] To obtain a quantitative result for the recorded wave height, a second current-voltage curve is obtained after a standard micro-addition of the respective trace element suspected of being present is made. The concentration of the trace element in the sample is then calculated from a ratio of the two respective current-voltage curve wave heights.

After looking for certain simple ionic species of trace elements that are known to occur in natural water, additional information can be obtained from each sample by adding 2 or 3 drops of chemical solutions such as pure HCl, NaOH, NH$_4$OH, KCNS, and tartaric acid-HCl mixture, to the 2 ml of natural water in the polarographic cell. This provides media in which complex ion species are formed with ions such as Pb^{II}, As^{III}, Sb^{III}, Sb^{V}, Se^{IV}, Ni^{II}, As^{III}, Cr^{III}, and Cr^{VI}. These species exhibit different reduction potentials and help in identifying and measuring the concentration of a particular trace element in sea water. This takes very little more operator time and gives one a lot more valuable trace element data per sample. Some examples of complex ion species formed in this manner are shown in Figs. 1 and 2. Chromium, as either Cr^{III} or Cr^{VI} can be determined with this technique in sea water and well water and in most natural water samples. The Cr^{III} specie gives a well-defined polarogram in the presence of KCNS and acetic acid, while the Cr^{VI} specie will not be seen in this media. The Cr^{VI} specie can then be measured in a NaOH media where it gives a well-defined polarogram and the Cr^{III} specie is not seen. Fig. 3 shows typical data for chromium in well water. Sea water samples containing added chromium were found to give the same results.

Solid chemicals that are reagent grade chemically pure or before use are especially purified by electrolysis over a mercury pool can be added directly to the sea water in the polarographic cell to also form complex ions with some trace elements found in sea water. An example of this technique is to determine As^{III} with a HClO$_4$ addition to sea water and then determine As^{V} by the addition of solid pyrogallol to form the As^{V}-pyrogallol polarographically active specie.[5] A typical single-sweep polarogram for arsenic^{V} in sea water containing pyrogallol and HClO$_4$ is shown in Fig. 4.

The ratio of I⁻ to IO_3^- and the concentrations at a part-per-billion range of I⁻ and IO_3^- in sea water is difficult to determine, if not impossible to do accurately by any previously known analytical technique. With a simple standard addition technique, a single-sweep polarographic method has now been developed in this laboratory to do this. Data obtained with this technique on several samples of sea water collected during the cruise reported herein are shown in Table 1. It can be seen that the data compare favorably with those reported previously by Barkley and Thompson using an amperometric procedure.[6] The polarographic method is simple, rapid, and accurate and consists of measuring the IO_3^- wave directly on a 2-ml sample of the natural sea water and then adding a few drops of chlorine water to the sample to convert the I⁻ to IO_3^- and again measuring the IO_3^- wave. Typical data are shown in Fig. 5. The calculation of the I⁻ and IO_3^- concentrations are then made from the measurement of the two wave heights. The method is sensitive to 3-5 parts-per-billion IO_3^- and I⁻ respectively and a sample of sea water can be analyzed in about 5-10 minutes for both IO_3^- and I⁻. For conversion of the I⁻ to IO_3^- a 5-ml sample of sea water is placed in a 10-ml volumetric flask. Then 2 drops of 0.1N H_2SO_4 and 2 drops of a 10% solution of fresh chlorine water is added to the sample and the contents swirled well for a minute. Finally 1-ml of a 0.1N LiOH solution is added and the contents diluted to volume with distilled water. A 2-ml aliquot of this solution is then placed into the 5-ml quartz polarographic cell for measurement of IO_3^-. The difference between the two wave heights measured for IO_3^- is that due to the I⁻ content of the sample. The IO_3^- wave height equivalent in concentration is obtained by a standard addition of a known IO_3^- concentration to the sample and a calculation of the ratio of the two wave heights.

RESULTS AND DISCUSSION

The trace elements discussed in this work give well-defined single-sweep polarograms that are proportional to concentration over the range 5×10^{-7} to 5×10^{-9} g/ml.

The polarographic data reported here are for a specific ion concentration of a trace element present in the sea water while the atomic absorption data are for the total amount of the element present in the sea water. For example, in Table 2 the polarographic data reported for copper is for Cu^{2+} only. The atomic absorption data is for total copper. From the data shown it appears that the copper found in the South American waters sampled is predominantly Cu^{2+}.

In case of the iodine data the polarograph could determine both the iodide (I⁻) and iodate (IO_3^-) content of the sea water while atomic absorption could not be used for this element.

The determination of arsenic in sea water by single-sweep polarography allows one a method to obtain the amount of As^{III} and As^V respectively. The atomic absorption analysis will determine total arsenic only and does not have the sensitivity to obtain reliable data for arsenic at the low levels usually found in sea water.

The sensitivity (5 X 10^{-9} g/ml) of the single-sweep polarographic technique used at this center for trace element analyses can be realized aboard ship in reasonably calm seas. In rough weather the sensitivity decreased to around 5 X 10^{-8} g/ml. However, even in weather where the ship rolled and pitched violently, the dropping mercury electrode with the automatic device that knocks the drop off at the end of each sweep, performed quite satisfactorily. The speed of the ship to about 15 knots did not appear to affect the sensitivity of the analysis.

The atomic absorption instrument was unable to continue analysis of the sea water for more than a half-hour at a time before the burner became plugged with salt and needed to be cleaned. The single-sweep polarograph operated during the entire 10-day cruise without any difficulty on a continuous 24-hour basis.

The surface samples of sea water analyzed at sea by single-sweep polarography showed very uniform concentrations of copper over large areas of the Pacific Ocean that were covered. This data is in agreement with previous atomic absorption data for copper in the Atlantic Ocean reported by Fabricand, et al.[4] The samples taken at depth showed higher copper concentrations in general and a variation in copper content with depth and station (Table 3).

Most of the polarographic data reported in the tables were obtained aboard ship and the samples were introduced directly from the sampling vessel into the polarographic cell for analysis. It has been found in this laboratory that sea water cannot be stored very long in polyethylene bottles before traces of chromium, zinc, copper, etc. are leached out of the plastic or lost to the plastic. Table 4 shows the effect on copper analysis of storing sea water in plastic for a short period of time. It can be seen that the copper content of refrigerated sea water samples stored in polyethylene bottles is lower than in fresh samples analyzed direct from the sampling vessel.

CONCLUSION

Trace elements identified as copper, zinc, manganese, iodine, and arsenic were found in nearly all samples of sea water examined aboard ship. These elements were found to give results proportional to concentration over the range studied of 5 to 500 parts-per-billion.

Good agreement between single-sweep polarographic data and atomic absorption spectroscopy data for copper in sea water was obtained aboard ship during a cruise.

Single-sweep polarography appears to offer a distinct advantage over other trace element analytical methods in its ability to analyze directly for ionic species in sea water. Thus, examples of determining Cu^{2+}, Mn^{2+}, As^{III}- As^{V}, I^{-}-IO_3^{-}, and Cr^{III}- Cr^{VI} in parts-per-billion, as reported in this study, would seem to show that single sweep polarography is an extremely specific and useful technique to the chemical oceanographer and those interested in trace element studies in general in the marine environment.

It is suggested that modern polarographic equipment could be installed readily aboard oceanographic research vessels or other ships at sea to provide valuable data for the oceanographer, research scientist, or personnel aboard these ships. The cost today of a single-sweep polarograph similar to the one used in this investigation is $4,500. Recently, an instrument called the "Polarographic Analyzer" PAR Model 174, manufactured by Princeton Applied Research Corporation of Princeton, New Jersey has appeared, which will cost only $2,000 and should be as useful aboard ship as the one used in this work. It should also be possible to automate a modern polarographic method of analysis to record continuous data for trace elements, dissolved oxygen, chlorinity, and other parameters of interest aboard ships at sea.

ACKNOWLEDGMENTS

The author wishes to acknowledge the interest and support of Dr. B. P. Fabricand in arranging for a research vessel to conduct part of this work at sea and to the high standard of cooperation by the crew of the USNS Josiah Willard Gibbs.

REFERENCES

1. G. C. Whitnack, J. Electroanal. Chem., Vol. 2, 110-5 (1961).

2. G. C. Whitnack, "Polarography 1964," Edited by Graham Hills, Vol. 1, p. 641, Macmillan, London.

3. Jiri Vogel, "Progress in Polarography," Edited by P. Zuman and I. M. Kolthoff, Vol. II, p. 429 (1962), Interscience Publishers.

4. B. P. Fabricand, R. R. Sawyer, S. G. Ungar, and S. Adler, Geochim Cosmochim Acta, Vol. 26, p. 1023-27 (1962).

5. Sister Mary Cletus White and Allen J. Bard, Analyt. Chem., Vol. 38, No. 1, p. 61 (1966).

6. R. A. Barkley and T. G. Thompson, Analyt. Chem., Vol. 32, p. 154 (1960).

FIG. 1. Arsenic^{III} and Selenium^{IV} in Well Water

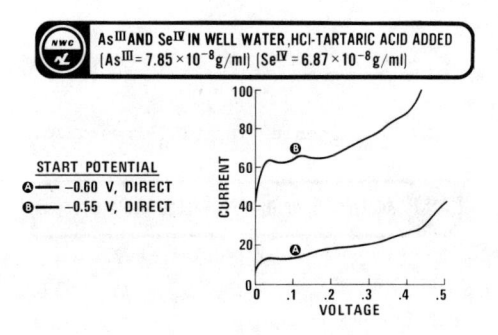

As^{III} AND Se^{IV} IN WELL WATER, HCl-TARTARIC ACID ADDED
(As^{III} = 7.85 × 10^{-8} g/ml) (Se^{IV} = 6.87 × 10^{-6} g/ml)

START POTENTIAL
Ⓐ —— -0.60 V, DIRECT
Ⓑ —— -0.55 V, DIRECT

FIG. 2. Se^{IV} in Well Water

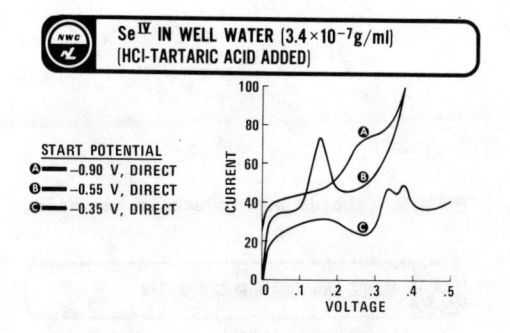

Se^{IV} IN WELL WATER (3.4 × 10^{-7} g/ml)
(HCl-TARTARIC ACID ADDED)

START POTENTIAL
Ⓐ —— -0.90 V, DIRECT
Ⓑ —— -0.55 V, DIRECT
Ⓒ —— -0.35 V, DIRECT

FIG. 3. Chromium in Well Water

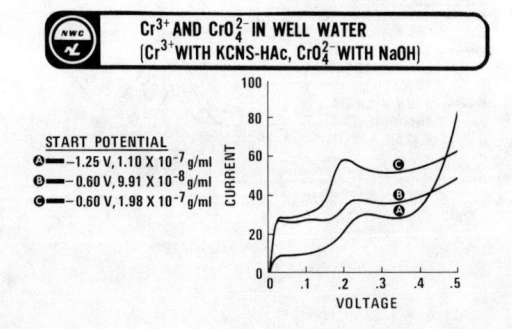

Cr^{3+} AND CrO$_4^{2-}$ IN WELL WATER
(Cr^{3+} WITH KCNS-HAc, CrO$_4^{2-}$ WITH NaOH)

START POTENTIAL
Ⓐ —— -1.25 V, 1.10 × 10^{-7} g/ml
Ⓑ —— -0.60 V, 9.91 × 10^{-8} g/ml
Ⓒ —— -0.60 V, 1.98 × 10^{-7} g/ml

FIG. 4. Arsenic^V in HClO$_4$-Pryogallol

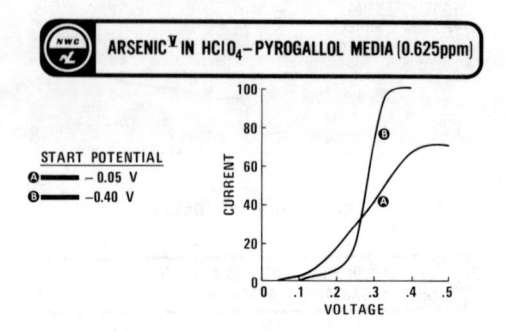

ARSENIC^V IN HClO$_4$-PYROGALLOL MEDIA (0.625ppm)

START POTENTIAL
- Ⓐ —— -0.05 V
- Ⓑ —— -0.40 V

FIG. 5. Iodide and Iodate in Sea Water

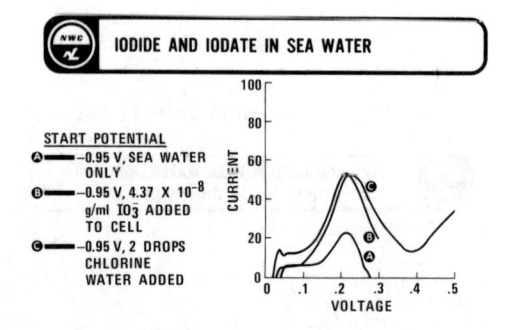

IODIDE AND IODATE IN SEA WATER

START POTENTIAL
- Ⓐ —— -0.95 V, SEA WATER ONLY
- Ⓑ —— -0.95 V, 4.37 X 10^{-8} g/ml IO$_3$ ADDED TO CELL
- Ⓒ —— -0.95 V, 2 DROPS CHLORINE WATER ADDED

Table 1. Concentration of Iodide and Iodate in Sea Water
(μg per liter)

Sample	Iodide	Iodate	Total
Surface, Pacific Open Ocean Water	12.0	2.3	14.3
Surface, Pacific Water Near San Diego	17.6	9.2	26.8
Station A, Pacific Open Ocean Water (depth, ft)			
50	20.8	10.3	31.1
100	18.2	4.6	22.8
200	13.2	35.4	48.6
400	22.2	16.0	38.2
500	25.8	6.3	32.1
1,600	7.1	34.8	41.9
6,000	< 1.0	30.8	30.8
10,090	11.2	31.6	42.8

Barkley and Thompson, 1960, reported 20-40 μg/liter $IO_3^- + I^-$ in sea water

TABLE 2. Copper in Surface Water

Sample no.	Copper, ppb	
	Polarography	Atomic absorption
1	10	8
2	7	6
3	12	8
4	8	7
5	10	8
6	5	4

TABLE 4. Effect of Polyethylene on Copper Concentration in Sea Water (Station C)

Depth, m	Copper, ppb	
	Shipboard	Storage[a]
2,544	50	33
3,325	40	25
4,032	50	27
4,259	30	25
4,486	50	37
4,713	30	34
4,940	30	<10

[a] Stored 3 weeks in polyethylene bottle (10 days in refrigerator aboard ship).

TABLE 3. Copper Concentration With Depth

Depth, m	Station	Copper, ppb
15	B	≤5
30	B	≤5
60	B	≤5
121	B	100
242	B	130
485	B	30
1,272	B	80
1,817	B	80
2,544	B	65
2,544	C	50
2,786	B	120
3,325	C	40
3,805	C	30
4,032	C	50
4,259	C	30
4,486	C	50
4,713	C	30
4,940	C	30

ELECTROCHEMICAL MONITORING AND CONTROL

OF MARINE POLLUTION

Chairmen:
H.V. Weiss
G.C. Whitnack

OIL POLLUTANTS IN THE MARINE ENVIRONMENT

Sachio Yamamoto
Naval Undersea Center
San Diego, California 92132

ABSTRACT

An overview of the study of oil pollutants in the marine environment is presented. The problem is divided into four categories: clean-up, identification, fate and effect. The principal analytical methods that are currently used are reviewed and evaluated in terms of their applicability to these studies.

INTRODUCTION

Recent oil spill incidents such as the Torrey Canyon accident and the Santa Barbara blowout have increased public awareness and concern about oil pollutants in the marine environment. However, accidental spills, which attract the greatest attention because of visibility and damage to beaches and harbors constitute only about 10% of the petroleum lost directly to the sea (1,2). The remainder comes from tanker operations, other ships, offshore oil production, refinery operations and industrial and automotive wastes. An even greater source of input of petroleum hydrocarbons may be from atmospheric transport of petroleum products emitted to the atmosphere through incomplete combustion and evaporation. Estimates indicate that this source may be as great as five times the direct input (1). Inputs from all of these sources can be expected to increase with increased production of oil in the future.

Oil pollution from direct sources are localized primarily in the coastal zone, harbors and estuaries where the bulk of the activities associated with production, processing and transport of petroleum takes place. These zones are of great importance to man and are considered to be highly susceptible to damage by pollution (1).

These factors have spurred both government and industry to increase their efforts in the study of marine oil pollution. The purpose of this paper is to review the nature of these studies and the principal analytical methods that are being utilized.

ASPECTS OF OIL POLLUTION STUDIES

Oil pollution studies can be generally categorized as follows: (i) clean-up, (ii) detection, (iii) fate of the pollutant in the

marine environment and (iv) the effect of these contaminants upon the
marine escosystem. A fifth category of equal importance but not
discussed in this report is prevention.

Clean-up includes containment, removal, sinking, dispersal and
combustion of spilled oils. Chemical agents are frequently used to
effect or aid these procedures. Among the types of chemicals used are
dispersants, sinking agents, sorbents, combustion promoters, gelling
agents, biodegradants and beach cleaners (3). Obviously clean-up by
physical containment and removal is preferable; a fact reinforced by
unfortunate experiences in the past in which dispersants used to combat
spills had apparently damaged the environment more than did the oil
(4). However, cost-effectiveness considerations often point to the
desirability of using chemical agents and, as a result, studies of the
effectiveness and toxicity of these agents have become important (3,5,6).

The primary concern of the second category, detection, is to
determine polluter responsibility and both passive methods, i.e.,
fingerprinting of oils, and active tagging techniques, i.e., the
incorporation of certain compounds into oils, are being developed.
Among oil pollution studies detection by passive means has attracted
the widest attention of analytical chemists.

Study of the fate of oil as well as the chemicals used in combating
spills concerns the chemical, physical and biological actions which
influence the distribution, concentration, persistence and form of oil
in various parts of the aquatic environment. One of the first thorough
treatments of this subject is that of ZoBell (7). Petroleum hydrocarbons
in the sea are dispersed and diluted by air and water movements. These
substances also evaporate and undergo chemical as well as microbial
oxidation. ZoBell reports that nearly all oils are susceptible to
microbial oxidation and that this process is most rapid at temperatures
ranging from 15° to 35° C and when the hydrocarbon molecule is in
intimate contact with water. Thus, biodegradation occurs more rapidly
in turbulent waters than in calm areas. This fact was demonstrated in
laboratory experiments by Button (8). Approximately two-thirds of the
hydrocarbon is oxidized to H_2O and CO_2 while the remainder is converted
to bacterial cells. Hence, oil persists in seawater only when
protected from bacterial action. Oils are also adsorbed onto clay,
silt, detritus and other particulate matter suspended in the sea and are
eventually carried to the sea floor. In the surf zone they are coated
onto beach sand. Thus, much of the oil ends up in the sediment or on
the beach where it also undergoes auto- or bacterial oxidation. Rates
of oxidation in the sediments, however, are relatively slow. Blumer,
et al. (9) found that No. 2 fuel oil persisted essentially unaltered
in the sediments for two months after a spill.

Study of the effect of oil and chemical agents used in clean-up
processes considers any changes which might occur in the ecosystem as
a result of exposure to these foreign substances. The kinds of effects

produced by hydrocarbon pollutants upon the marine environment have been discussed in numerous reports (e.g. Ref. 1,7). These include (i) poisoning of marine animals (including birds) and plants, (ii) disruption of the ecosystem to cause long-term destruction of marine life, and (iii) degradation of the marine environment for human use by reducing economic and aesthetic values. Poisoning of marine life as a result of accidental spills have been widely publicized. Spills occurring in deep waters (e.g. Santa Barbara) have caused relatively light damage to life while those occurring in closed or shallow areas such as the wreck of the Tampico Maru off of Baja California have destroyed substantial quantities of marine life (7). Long-term disruptions can result from loss of the young or from destruction of food sources.

ANALYSIS OF OIL POLLUTANTS

Types of Materials

Oil pollution studies require the detection and identification in the marine environment (i.e. water, fauna, and sediments) of oil pollutants, their degradation products, active tags, or the chemicals (and their degradation products) used in clean-up procedures. The purposes of the analyses are to establish responsibility for the pollution, assess water quality or to determine the fate and effect of oils. Hence, the choice of analytical methods to be used is to a great extent dictated by the nature and purpose of the analysis as well as by more practical considerations such as cost and time required for analysis.

In analyzing oil pollutants one is confronted by the great complexity of the samples analyzed. There are over 1000 kinds of crude oil produced, stored, transported and refined and no two crudes have yet been found to be identical. Each crude contains large numbers of natural elements (10). From these crudes over 2000 separate and distinct products and several thousand derivatives are made, many of which are transported by ships. However, most of the oil pollution is due to spills of crudes and residual-type fuels. Although the composition of crudes and refined materials differ according to their sources, a given type of fuel or product will have a degree of commonality because they generally must conform to required specifications. This commonality permits the determination of the type of pollutant material from bulk properties. The common properties could also be used in studying weathering rates or persistence of a given fuel (11).

Pollutant hydrocarbons must also be distinguished from those that occur naturally. The general composition of natural marine and petroleum hydrocarbons are compared in Table 1. Some of these properties can be used to distinguish pollutant from natural hydrocarbons.

Table 1. Comparison of Petroleum and Natural Marine Hydrocarbons

	Petroleum[a]	Natural Hydrocarbons
Molecular Weight	16– 20,000+	Single compound often predominates in a compound group

Hydrocarbon Types	Paraffins
	  Low Iso/Normal Ratio
	Olefins
	  Usually none
	Cycloparaffins
	Aromatics
	  Less than paraffins and cycloparaffins
		Straight chain paraffins[b] and Olefins predominate (Iso/Normal 0.01–0.1)
		
		Cycloparaffins and aromatics less than in petroleum

Oxygen Content	< 2% Mostly phenols and carboxylic acids	Fatty acids and phenols
Nitrogen	0.05–0.8% ca. half are basic pyridines and quinoline	Present in proteinaceous[c] material; most dissolved organics in sea are nitrogen compounds
Sulfur	Trace – 5% mercaptans, aliphatic and cyclic sulfide	Some mercaptans[d] reported in anoxic basins
Metals Ni, V	0.01–0.05% 5 – 40 ppm	-----
Stable Isotope Ratios[e] $^{13}C/^{12}C$ ($\delta\ ^{13}C$) $^{34}S/^{32}S$ ($\delta\ ^{34}S$)	 –31 to –21 –10 to +16	 –20 to –5 +17 to +22 (for seawater sulfate)

(a) Ref. 12
(b) Ref. 13
(c) Ref. 14,15
(d) Ref. 16
(e) Ref. 17

A large number of chemical agents for treating oil spills are available. A compilation of such agents published in 1970 lists well over 100 chemicals (18).

A number of active tags which can be incorporated into oils have been proposed (19). They include halogenated aromatics, organometallics, radionuclides and particulate material (including coded microspheroids). The last is believed to be preferable. Ideally, active tags should be readily identifiable, non-degradable, and have no effect upon the environment. Considerable study would be required to establish whether any potential tags meets such requirments.

Analytical Methods

As mentioned at the outset, the choice of analytical method depends upon the purpose of the analysis. To determine product type, information on bulk properties may suffice, while for studies of the fate of oil molecular differentiation may be required. In general high selectivity and the measurement of many parameters will yield the most information. The latter is particularly true for the identification of the pollutant.

Bulk Properties. Many of the important bulk properties and their typical values for gasoline, fuel oil and crude oil are listed in Table 2. These properties are useful in characterizing product type and can be measured by well established standard procedures which do not require expensive instruments. Hence, if the requirement is to determine whether a pollutant is or is not jet fuel the determination of bulk properties may be sufficient. A thorough discussion of this subject is given by Kawahara (20). However, there are some drawbacks: first, the sample material must not be too greatly dispersed; second, weathering may alter considerably some of these properties; and third, the method is not too useful in identifying specific pollutants.

Infrared Spectroscopy. Infrared (IR) methods have been developed by Kawahara (21,22) for differentiating petroleum products and by Mattson, et al. (23) for identifying crude oils. Kawahara's method involves the measurements of infrared absorbance ratios using six wavenumbers. Various product types have characteristic sets of ratios and, hence, can be differentiated. Two key ratios, 810 cm^{-1}/1375 cm^{-1} and 810 cm^{-1}/720 cm^{-1}, were used for initial classification and others were used for confirmation. Mattson, et al. instead of using a few selected peaks used the entire IR spectrum. This required greater sensitivity which was achieved by means of the internal reflectance technique. These workers were able to differentiate between natural seeps and spilled oil in Santa Barbara. In a follow-up study Mattson (24) obtained IR fingerprints of 40 crude and residual fuel oils but ran no blind samples to assess the efficacy of the method for identification purposes. While the IR method may not be as effective as neutron activation analysis (see below), it is less expensive and provides information regarding the molecular nature of the materials of interest,

Table 2. Typical Bulk Properties for Gasoline,
No. 6 Fuel Oil and Crude Oil[a]

Property	Gasoline	No. 6 Fuel Oil	Crude Oil
Solubility in			
Pentane	Very soluble	Insoluble	Partially soluble
Ether	Very soluble	Insoluble	Partially soluble
Chloroform	Very soluble	Soluble	Soluble
Gravity, API	58-62	-3 - 23	13.5-33.5
Distillation Range, °C	35-210	250-680	5- > 450
Molecular Weight	72-170	---	16- > 20,000
Sulfur	< 0.25%	Trace - 5%	Trace - 5%
Viscosity, Saybolt	---	300 @ 50°C	50 @ 38°C
Comments	high lead		

(a) Most of the values were obtained from Ref. 20

e.g., carbon type and functional groups, and hence is applicable to studies of fate and effect of oil components in the environment.

Fluorescence Spectroscopy. Fluorescence methods have been used by Thruston and Knight (26) and Parker and his co-workers (27) for characterizing spilled oils. In Thruston and Knight's method the oil sample is diluted in cyclohexane and the fluorescence emission spectrum is measured. They found that all crude oils and residual fuels have a maximum fluorescence at 386 nm and shoulders at 405 and 440 nm when excited by 386 nm radiation. However, the relative intensities of the maxima and shoulders differ with the oil and this property was used to fingerprint oils. The fluorescent components were found to be among the heavy components of oil and, hence, evaporation losses had little effect on the method. Parker, et al. excited their samples over a range of wavelengths and scanned the emission spectra at 20 excitation wavelengths. The excitation spectrum and the corresponding fluorescence spectrum were combined in a single plot to obtain a contour map fingerprint. These workers used the method to study the fate of oil in the marine environment by following changes in gross characteristics.

Visible and Ultraviolet Spectroscopy. Both are widely used analytical tools which, subsequent to separation, can provide information, regarding individual compounds, compound type, aromatics, and unsaturates. For example, ultraviolet (UV) spectroscopy has been used to identify metalloporphyrins in crude oils and to characterize aromatics in certain distillate cuts. Thus, these methods could be used in studies which require knowledge of the average molecular composition of the pollutant. However, the methods lack specificity and must be used in conjunction with highly selective separation procedures. To date no visible or UV techniques have been developed specifically for use in studying oil pollutants in the oceans.

Nuclear Magnetic Resonance. NMR has greater specificity than either visible or UV spectroscopy and can be used to determine structure, conformation, functional groups and compound types. A number of workers have used proton NMR to characterize whole as well as fractionated oils. Hence, the method is potentially applicable for fingerprinting oils and for differentiating petroleum products. However, its best use would be in studies such as fate and effect in which information about the general molecular nature of the pollutant material is required.

Mass Spectrometry. Mass spectrometry (MS) has been used as an analytical tool for years by oil chemists for both qualitative and quantitative analysis. Unlike absorption methods such as UV, IR, and NMR single compounds can be analyzed by MS since every organic compound exhibits a characteristic fragmentation spectrum. In addition, mixtures containing several components can also be analyzed. However, the more complex the mixture the more overlap there will be in the peaks and computer methods are required to unscramble spectra. Considerable progress has been made in recent years in improving computer techniques and in coupling computers directly to mass spectrometers (25). Group analysis can be conducted on mixtures such as petroleum which contains a very large number of compounds and the relative amounts of chemical groups such as n-paraffins, naphthenes, hetero-compounds, etc. can be estimated. Consequently, MS has been used in all aspects of oil pollution studies, particularly in combination with gas chromatography. Isotope ratios are also determined by mass spectrometry and, as mentioned earlier, $^{13}C/^{12}C$ and $^{34}S/^{32}S$ can be used to ascertain the origin of the hydrocarbon.

Gas Chromatography. No tool is more widely used in hydrocarbon analysis than gas chromatography and numerous textbooks and papers have been written on the subject. The efficacy of the technique for separations is exemplified by the work of Polgár, et al. (28) who quantified mixtures of saturates through C_8 containing 88 compounds and of alkenes through C_7 containing 63 compounds. Thermal conductivity

or flame ionization detectors are generally used for hydrocarbon
analysis; the latter is more popular in pollutant measurements because
of its greater sensitivity. Electron-capture detectors have been used
for the microdetermination of hetero-compounds such as phenols and
mercaptans (29). The gas chromatograph has been employed in most studies
of the fate and effect of oils, measurements of the levels of pollutant
and other hydrocarbons in natural waters, and characterization of
pollutant material. It is most useful in the C_1 to C_{40} range; beyond
C_{40} liquid chromatographic methods are used because of low volatility.

Liquid Chromatography. There are four categories of liquid
chromatographic methods: (i) liquid-solid (adsorption), (ii) liquid-
liquid (partition), (iii) ion-exchange, and (iv) exclusion (gel
permeation and filtration). Thin-layer chromatography (TLC) is a form
of liquid-solid chromatography, although the latter generally refers to
a column method. There has recently been renewed interest in liquid
chromatography and a number of books on the subject have appeared. Of
the four methods ion-exchange is least useful in oil pollution studies,
although it could be used in studying organic acids in petroleum and
chemical clean-up agents such as detergents. The other methods have
been used by many for separation and identification of petroleum
components. Gel permeation chromatography (GPC) separates species
according to their physical size in solution which is roughly proportional to molecular weight. Workers have also shown that separation
of individual species is possible by GPC. Liquid-liquid and liquid-
solid chromatography are highly selective and have been used to
separate into individual components saturated hydrocarbons, mono-,
di-, tri-, and tetraaromatics and polar hetero-compounds in petroleum.

Combined Methods. Combination of highly selective separative
techniques such as gas and liquid chromatography with IR, UV, NMR, or
MS systems provide powerful analytical tools. Among these the gas
chromatograph - mass spectrometer combination is the most highly
developed and widely used.

Trace Element Analysis. All crude oils contain a large number of
trace elements whose reported concentrations range from less than 10^{-7}
ppm for gold to 200 ppm for vanadium among metals (30) and up to 5% for
sulfur. The relative concentrations of these elements are different
for each oil and this fact is being exploited for fingerprinting
purposes. Methods range from measurements of Ni/V ratios to the
determination of ten or more elements. Obviously, it is preferable to
measure as many parameters as practicable. Emission spectrography and
x-ray fluorescence analysis have long been used by oil chemists for
metal analysis. The sensitivity of the former is relatively poor; its
detection limit for most metals in petroleum is about 0 - 10 ppm. X-ray
fluorescence analysis is somewhat more sensitive and since the metals
are in low atomic number matrices, matix effects would not be severe.
If it is shown that measurement of 3 or 4 of the more abundant metals
suffice or that Ni/V ratios are reliable for identification purposes,

non-dispersive x-ray fluorescence analysis (31) would provide a simple and rapid means of fingerprinting. However, of all the methods for trace analysis neutron activation analysis (NAA) is the most sensitive. Lukens and his co-workers have developed a highly reliable NAA method for fingerprinting oils. It involves the irradiation of a suspect sample and pollutant sample and the comparison of their gamma-ray spectra. They obtained spectral patterns of nearly 300 oils and demonstrated that they could "ascertain with 99.999% certainty" whether or not two samples are of the same oil. The chief disadvantages of NAA are its high cost and the special skills required of the analysts. Finally, while trace metal analysis is promising for fingerprinting it is not particularly useful in other areas of oil pollution studies.

Electrochemical Analysis. Electroanalytical methods have only seen limited use in petroleum analysis. Only about twenty examples are cited in the 1971 Analytical Reviews (34). These include polarographic determination of V, Ni, Co, and Fe and halogens in crude oils, amperometric and potentiometric titration of sulfur compounds, microcoulometric analysis of nitrogen, sulfur and chlorine compounds, and A. C. polarographic determination of dissolved oxygen. Electrochemical methods have high specificity for some of the hetero-compounds and should be applicable to the study of such materials in the environment. A number of highly sensitive polarographic techniques have been developed for trace metal analysis which can be utilized for fingerprinting purposes.

CONCLUSION

In reviewing the literature it was evident that much study is yet required in all areas of the problem of oil pollutants in the marine environment. In clean-up studies the fate and effect of chemical agents and particularly their degradation products need be more fully considered. Identification of oils has attracted much attention but there should be greater emphasis on rapid and simple methods that can be used by technicians. The fate and effect of these pollutants ultimately determine their impact upon the environment and an understanding of these factors is of vital importance.

Numerous analytical tools are available for use in these studies and the choice depends upon the nature of the problem. In general, methods such as the combination of gas chromatography and mass spectrometry, which have high specificity and selectivity, are the most useful.

REFERENCES

1. R. Revelle, E. Wenk, B. H. Ketchum and E. R. Corino, "Ocean Pollution by Petroleum Hydrocarbons," In Man's Impact on Terrestrial and Oceanic Ecosystems," Ed. W. H. Matthews, F. E. Smith and E. D. Goldberg, MIT Press, Cambridge, 1971.

2. J. E. Moss, "Petroleum - the Problem," In Impingement of Man on the Oceans, Ed. D. W. Hood, Wiley-Interscience, New York, 1971.

3. J. R. Blacklaw, J. A. Strand and P. C. Walkup, "Assessment of Oil Spill Treating Agent Test Methods," Proc. of Joint Conference on Prevention and Control of Oil Spills, June 15-17, 1971, Washington, D. C.

4. J. E. Smith, Ed., "Torrey Canyon Pollution and Marine Life," Cambridge University Press, New York, 1968.

5. G. P. Canevari, "Oil Spill Dispersants - Current Status and Future Outlook," Proc. of Joint Conference on Prevention and Control of Oil Spills, June 15-17, 1971, Washington, D. C.

6. R. T. Dewling, J. S. Dorrier and G. D. Pence, Jr., "Dispersant Use vs Water Quality," Ibid.

7. C. E. ZoBell, "The Occurrence, Effects, and Fate of Oil Polluting the Sea," Proc. of the International Conference on Water Pollution Research, Section 3, No. 48 (London) September, 1962.

8. D. K. Button, "Petroleum - Biological Effects in the Marine Environment," In Impingement of Man on the Oceans, Ed. D. W. Hood, Wiley-Interscience, New York, 1971.

9. M. Blumer, G. Souza and J. Sass, "Hydrocarbon Pollution of Edible Shellfish by an Oil Spill," Marine Biology, 5, 195, (1970).

10. H. Powell and E. V. Whitehead, "Modern Contributions to the Study of Petroleum Constitution," In Modern Chemistry in Industry, Ed. J. G. Gregory, London, Society of Chemical Industry, IUPAC Symposium 11-14 March 1968.

11. F. K. Kawahara, Federal Water Pollution Control Administration, Analytical Quality Control Laboratory, Cincinnati, Ohio, Private communications.

12. Encyclopedia of Science and Technology, Volume 10, McGraw-Hill.

13. M. Blumer, "Dissolved Organic Compounds in Seawater: Saturated and Olefinic Hydrocarbons and Singly Branched Fatty Acids," In

Proc. Symposium Organic Matter in Natural Waters, Ed. D. W. Hood, Inst. Mar. Sci. Occasional Publ. No. 1, University of Alaska (1970).

14. E. K. Duursma, "The Dissolved Organic Constituents of Seawater," In Chemical Oceanography, Volume 1, Academic Press, London (1965).

15. E. T. Degens, "Molecular Nature of Nitrogenous Compounds in Seawater and Recent Marine Sediments," In Proc. Symposium Organic Matter in Natural Waters, Ed. D. W. Hood, Inst. Mar. Sci. Occasional Publ. No. 1, University of Alaska (1970).

16. D. D. Adams and F. A. Richards, "Dissolved Organic Matter in an Anoxic Fjord with Special Reference to the Presence of Mercaptans," Deep-Sea Research, 15, 471 (1968).

17. P. L. Parker, "Petroleum - Stable Isotope Ratio Variations," In Impingement of Man on the Oceans, Ed. D. W. Hood, Wiley-Interscience, New York, 1971.

18. "Oil Spill Treating Agents - A Compendium," Battelle Northwest Laboratories, 1 May 1970.

19. J. Horowitz, et al., "Identification of Oil Spills: Comparison of Several Methods" In Proc. Joint Conference on Prevention and Control of Oil Spills, API-FWPCA, New York, 15-17 December 1969.

20. F. K. Kawahara, "Laboratory Guide for the Identification of Petroleum Products," Federal Water Pollution Control Adminstration, Analytical Quality Control Laboratory, Cincinnati, Ohio, January 1969.

21. _____, "Identification and Differentiation of Heavy Residual Oil and Asphalt Pollutants in Surface Waters by Comparative Ratios of Infrared Absorbances," Environ. Sci. Tech., 3, 150, (1969).

22. _____ and D. G. Ballinger, "Characterization of Oil Slicks on Surface Waters," Ind. Eng. Chem. Prod. Res. Develop. 9, 553 (1970).

23. J. S. Mattson, et al., "A Rapid, Nondestructive Technique for Infrared Identification of Crude Oils by Internal Reflectance Spectrometry," Anal. Chem., 42, 234 (1970).

24. J. S. Mattson, "'Fingerprinting' of Oil by Infrared Spectrometry," Anal. Chem., 43, 1872 (1971).

25. A. L. Burlingame and G. A. Johanson, "Mass Spectrometry," Anal. Chem., 44, 337R (1972).

26. A. D. Thruston, Jr. and R. W. Knight, "Characterization of Crude and Residual-Type Oils by Fluorescence Spectroscopy," Environ. Sci. Tech., 5, 64 (1971).

27. M. Freegarde, et al., "Oil Spilt at Sea: Its Identification, Determination and Ultimate Fate," Laboratory Practice, 20, 35 (1971).

28. A. G. Polgár, J. J. Holst, and S. Groennings, "Determination of Alkanes and Cycloalkanes through C_8 and Alkenes through C_7 by Capillary Gas Chromatography," Anal. Chem., 34, 1226 (1962).

29. F. K. Kawahara, "Microdetermination of Phenols and Mercaptans by Electron Capture Gas Chromatography," Anal. Chem., 40, 1009 (1968).

30. R. H. Filby, "An Investigation of Trace Elements in Petroleum Using Neutron Activation Analysis," Ph. D. Thesis, Washington State University, 1971.

31. S. Yamamoto, "A Radioisotope X-Ray Fluorescence Spectrometer with a High-Resolution Semiconductor Detector," Anal. Chem., 41, 337 (1969).

32. H. R. Lukens, et al., "Development of Nuclear Analytical Techniques for Oil Slick Identification (Phase I)," Gulf General Atomics Inc., GA 9889, January 1970.

33. _____, "Development of Nuclear Analytical Techniques for Oil Slick Identification (Phase IIA, Final Report), Gulf Radiation Technology, Gulf-RT-10684, June 1971.

34. R. W. King, "Petroleum," Anal. Chem., 43, 162R (1971).

DISCUSSION

Raymond Jasinski, Texas Inst. Inc., Dallas, Texas: There have been reports at the SAS meeting of metals contaminations by sea water confusing neutron activation analysis of oil spill samples, particularly in view of the statistical nature of the total analytical scheme. What is the status of this problem?

Sachio Yamamoto: In the neutron activation method referred to in this report, the oil samples were scrubbed three times with de-ionized water. Measurement of Na and Cl showed that contamination from sea salts was virtually eliminated by this procedure.

BIOELECTRIC POTENTIAL MEASUREMENTS OF LIVING
BIOLOGICAL MEMBRANES AS POLLUTION INDICATORS

Anitra Thorhaug and Marcella Fernandez
University of Miami
School of Medicine
P.O. Box 875, Biscayne Annex
Miami, Florida 33156

ABSTRACT

The temperature dependency of the bioelectric potential of the giant marine algal cell Valonia was found to abruptly change near 15 and 31°C (Thorhaug, 1971). Thousands of Valonia cells from 8 parts of the tropics were subjected to a series of temperatures in a polythermostat; lethal limits were 15°C and 31.5°C (Thorhaug, 1970). Thus, heated industrial effluents in the tropics were predicted to be potentially deleterious to benthic macroalgae. After three years of field study this prediction appears valid (Thorhaug, et al., 1972). We propose herein the use of electrical measurements of the living giant marine algal membrane as a pollution index of injury, giving numerous examples.

INTRODUCTION

The electrical and chemical properties of cells have been studied for more than one hundred years as indicators of cellular metabolism. Encouraged by the work of early physiologists, physical chemists such as Arrhenius, Boltzman and Nernst applied their ideas to living tissues and cells, working with Loeb and Bernstein. In 1922 an amazingly modern volume summarizing much of this work was written by Osterhout(1). He explored the processes of injury, recovery and death in relation to electrical conductivity and permeability of various chemicals into cells. He concluded that changes in the permeability properties of the cell can be followed by bioelectrical measurements. He adds, "These alternations are evidently important since they may affect all the fundamental life-processes. It enables us to predict the behavior of tissues, especially in respect to injury and recovery."

Subsequent work in the 1920's and 1930's by Osterhout and his colleagues, Blinks, Irwin, Darsie, and Damon (reviewed in Blinks(2)) further explored the effects of physical and chemical factors and electrical properties of giant plant cells. These provide excellent experimental material due to their exceptionally large size (up to centimeters for many species). Ease of manipulation, ease of fractionation of storage vacuole and protoplasmic material as well as the ability to impale the cell with large tipped glass pippettes drew investigators to this material. During this time, workers on animal cells also found suitable preparations, including the giant squid and lobster axons, red blood cells, and the frog skin.

Attention began to be focused on electrical measurements of animal cells from 1940 onward and a good many fundamental electrical properties of living cells were delineated using animal cells. Membrane capacitance and conductance, impulse stimulation and application of new techniques such as radioisotope tracers, voltage clamp and perfusion were all major developments. Much of this is historically reviewed by Cole in 1968 (3). Results were readily applied to animal physiology and simultaneously biophysics was introduced into medical physiology courses. Technology using these fundamental methods and materials grew rapidly. A good many practical applications of membrane biophysics were successfully advanced, for example the screening of drugs on the giant squid axon for testing effects on certain neurological diseases.

Renewed interest in the plant cells occurred in the late 1950's. This work (reviewed in Dainty (4) and MacRobbie (5)) has centered chiefly about bioelectric measurements and water transport and concentrations of electrolytes. Investigations led to descriptions of water and ion movement based on thermodynamic consideration. However, since far fewer workers are in the field of plant membrane biophysics than their animals counterpart, basic information is still in the first stage of accumulation for most of the plant cells, i.e. measuring concentrations, potentials and two way fluxes. The next two essential questions are not yet clear for most plant cells: the energy-yielding sequences and the means of transferring energy to the ion pump. Thus, lack of application of plant membrane technology to more practical problems is not surprising at this point.

On the other hand, there are areas with pressing problems which might benefit from use of plant membranes. We would like to apply techniques of electrical measurements of plant membranes to pollution control. The rationale is fairly simple. One group of the giant algal cells are marine. The nearshore areas in which man's activities and his pollutants have great environmental impact (Wolfe, et al.(6)) often contained on a food web based on microalgae (Blinks (7), Ryther (8), Mann (9), Thorhaug and Stearns (10)). Valonia, one of these giant cells, has long been used as a valuable laboratory tool to explore the physiology of marine macroalgae (Blinks (7)). In addition, we have recently found the macroalgae to be one of the most sensitive groups of organisms in Biscayne Bay, Florida to thermal pollution and siltation (Thorhaug and Stearns (10)). This would argue to examine macroalgae both as the basis of the food web and as one of the most sensitive living in it.

It is suggested that the bioelectric properties of the Valonia membrane system can provide a fast and relatively cheap index to pollutants (both type and safe levels). In particular we would like to use the case of thermal pollution as an example applying the old idea of Osterhout (1) that measurements of bioelectric properties allow prediction of behavior of the cell, especially with respect to injury.

THEORETICAL

This brief introduction to the application of electrochemical theory to plant cells may cause electrochemists to rejoice in their own less complex systems. (Much of this theory has been previously discussed in detail by Dainty (11) and (4)). First the giant plant cells are rather complex structurally. The cell wall which surrounds it acts as a weak acid, ion-exchange resin. Secondly, the protoplasm is bounded by two membranes, the plasmalemma and the tonoplast unlike animal cells, which have one major membrane. In preliminary approximations, the cell wall, protoplast and two membranes are often treated as a black box called "the membrane."

The electrochemical potential as a driving force for ions can be expressed

$$\bar{\mu} = \bar{\mu}^* + P\bar{V} + RT \ln C + \tau + zF\psi \qquad (1)$$

In this expression $\bar{\mu}$ is the electrochemical potential of an ion, $\bar{\mu}^*$ is the chemical potential of the ion in its standard state, P is the pressure, $\bar{V}$ the partial molar volume of the ion, C the molar concentration, τ a contribution from ions which are in or near solid - or gas-liquid interfaces - and in which z is the valence of the ion, F the Faraday constant and ψ the electrical potential. The full equation is rarely used since the $P\bar{V}$ term is usually negligible as well as τ. The Nernst equation is derived from this and is treated in most electrochemistry texts.

$$\psi^i - \psi^o = \frac{RT}{z_jF} \ln \left(\frac{C_j^o}{C_j^i} \right) \qquad (2)$$

Where ψ^i and ψ^o are the electric potentials inside and outside respectively, R is the gas constant, T the temperature in degrees Kelvin and C_j^i and C_j^o the concentration of the ion inside and outside respectively. (The experimental difficulties in obtaining activity coefficients for ions in protoplasma have forced workers to drop this activity term, although it obviously would be more correct to use them, especially for the high organic protoplasm.) This equation, of course, assumes passive flux, although it is quite clear from a large literature of results for both plant and animal cells that some ions are actively transported by energy obtained in some metabolic process. The assumption is made that the electric potentials are a linear function of the distance inside the membrane. It is also assumed that the potential steps at the two boundaries of each membrane with the bathing media can be ignored.

This bioelectrical potential is essentially a diffusion potential originating from the different role of diffusion of charged ions across the membrane. The usual assumption is that the principle ions are Na^+, K^+, Cl^-, which are expressed in the Goldman or constant field

equation
$$E = \frac{RT}{F} \ln \left(\frac{P_K[K^o] + P_{Na}[Na^o] + P_{Cl}[Cl^i]}{P_K[K^i] + P_{Na}[Na^i] + P_{Cl}[Cl^o]} \right) \quad (3)$$
where P_K, etc. are the respective ion permeability coefficients.

Let us now consider specifically the giant marine algal cell Valonia. The following data is chiefly that of Gutknecht (12, 13) redrawn as Figure 1. It is immediately seen that the concentrations of potassium, sodium and chloride change greatly between the external seawater (C_o) and the protoplasm (C_{cyt}). There is only slight change between the sodium concentrations in the protoplasm and the vacuole (C_{vac}) although chloride changes greatly. One must note there are far more problems in obtaining accurate concentrations in the cytoplasm than in the large volume of vacuolar sap or the external media. There are many compartments within the cytoplasm such as mitochondria, nucleii, chloroplasts which are all known to have their own membrane, which may accumulate ions. Protoplasmic values for Valonia reflect a homogenate of ionic composition all these organelles plus the protoplasm. Work on another giant algal cell Nitella translucens by MacRobbie and colleagues (see 5) using a very elegant technique indicates a pump system at the chloroplast membrane.

The electric potential profile shows the two membranes both having relatively high potentials (-71mV and -88mV). The summed potential is routinely reported as "the potential." The major problem in determining the plasmalemma potential is that the cell wall is very tough, so that impalement with a delicate microelectrode is nearly impossible. To overcome this Gutknecht (12) impaled the plasmalemma of a Valonia spore (0.3 mm diameter), which had virtually no cell wall and no vacuole for the first few hours after spore formation. From this plasmalemma potential of -71mV (cytoplasm to outside) he inferred the -88mV plasmalemma potential (cytoplasm to vacuole) from the summed potential of +17mV. A direct validation of these two potentials on a mature (0.6 cm diameter or larger) Valonia ventricosa cell is yet to be published. From this concentration and potential data, the electrochemical potential differences have been calculated.

$$\Delta \bar{\mu}_j = \bar{\mu}_j^i - \bar{\mu}_j^o = RT \ln C_j^i + z_j F \psi^i - RT \ln C_j^o - z_j F \psi^o \quad (4)$$
$$= z_j F (\psi^i - \psi^o) - RT \ln (C_j^o / C_j^i)$$

Since ($\psi^i - \psi^o$) is the observed electric potential difference between cytoplasm and the other phase (plasmalemma -71mV and tonoplast -88mV) and $(RT/z_j F) \ln (C_j^o/C_j^i)$ is the Nernst potential we then arrived at

$$\Delta \bar{\mu}_j = z_j F (E - E_j^N) \quad (5)$$
$$\frac{\Delta \bar{\mu}_j}{z_j F} = (E - E_j^N)$$

or $\overline{\Delta \mu_j}$

$\overline{z_j F}$ for sodium in <u>Valonia</u> cytoplasm is lower (-136mV) than in the outer media and the concentration is about 12.5 times lower. Potassium on the other hand is at a higher potential value (+21)(which must be multiplied by F) than the external seawater as is chloride. At the tonoplast the electrochemical potential differences are very low, especially for Na^+.

The second major technique we wish to discuss is the background of voltage clamp. The concept was introduced by Cole (14) and later used by Hodgkin and Huxley (15, 16, 17) in their prize winning series. An excellent review is by Moore and Cole (18).

Basically this is used to study the ionic permeability of biological membrane. The membrane potential is set at a certain value and made the independent variable rather than the dependent variable as just described. The conditions for voltage clamp are that the potential across the membrane system have a known value. Under ideal conditions the potential difference is constant over the area of the membrane through which current flow is measured. Since these conditions are difficult to achieve experimentally, Spyropoulos(19) has redefined the term to mean control of the potential difference between intracellular and extracellular potential electrodes. The methods specifically used are discussed below and have first been applied to <u>Valonia</u> by Gutknecht (13).

METHODS

Perfusion

The basic method for studying temperature effects will be to maintain two controlled temperatures, one inside and another outside the cell, while simultaneously perfusing the cell to measure the tracer fluxes in and out of the cell, the bioelectric potential and the resistance of the cell membrane system. This has been accomplished by internal perfusion with a temperature controlled fluid, while maintaining a second temperature-controlled fluid in the external media. The radiotracers can measure either as outflow or inflow.

The perfusion was accomplished in the following manner: a spherical cell 1 cm diameter was placed in a freshly filtered sea water media on a high glass stand with wide slits (Figure 2, E). The larger inflow pipet (0.2 mm tip diameter) was filled with perfusion fluid (J), inserted through a circular sheet of thin plastic and then impaled into the cell with a micromanipulator in the manner shown in Figure 2. The second fluid filled pipet (H) was inserted through the other side of the plastic and then impaled into the opposite side of the cell in a crossed position with respect to pipet to assure good mixing. The microthermistor (G) in a 25$_{gauge}$ needle was then inserted through the cell wall just to the plasmalemma by a third micromanipulator. The resistance of the thermistor was recorded on a General Radio conductance and capacitance bridge (Z).

The large inflow pipet was attached through a surgical tubing by a polypropylene coupling luer adapters to a 10 ml syringe. To maintain constant hydrostatic pressure, the level of the perfusate in the syringe was kept constant (35 cm above the cell) by dripping fluid (P) from a reservoir (L) fitted with a needle valve at the same rate as the fluid was leaving the outflow pipet.

The temperature control of the fluid in the syringe and tubing leading to the inflow pipet and of the bathing media is evident from Figure 2. An insulated glass jacket (K) fitted with rubber stoppers with one inflow (N) and two outflow arms (M) surrounded the syringe. The inflow was controlled by an impeller from a 20 liter insulated glass chromatography jar (T). The temperature was controlled by using a refrigeration unit maintained at -1.0 (accuracy $\pm$ 0.001°C) via a twelve turn copper coil (U) submerged in the bath opposed to two 250 watt heaters (S) controlled by a thermistor (Q) leading to a thermonitor (R). The control of temperature of the syringe was $\pm$ 0.01°C accuracy.

The temperature of the bathing media was controlled in a similar manner with the aid of an aluminum block which was bored through three sides for water to flow through the block. The entire block was insulated with layers of 1/4 inch styrofoam. The aluminum block (A) 8 by 8 by 4 inches was precision bored to fit a 250 ml absorption dish (ϕ). Between the bottom of the dish and the block a fitted circular mirror (B) was placed so that the bottom of the cell could be viewed. Full visibility of all areas of the cell exterior was very important since local plasmolysis opposite the inflow pipet occurred in many cells which then were discarded. The 250 ml dish was fitted with a magnetic stirer. The control in the *Valonia* dish was +0.01°C.

Experimental success required that the *Valonia* cells be in excellent condition. The cell was allowed to sit for several hours in order to form an electrically resistant seal around the pipets (Blinks (2). This was indicated by a gradual increase in potential. The perfusion was then begun until all the material inside the vacuole had been removed. During this time gentle suction was exerted on whichever pipet may have clogged due to the gelatinous organic material (Gutknecht (13)). A light-dark test previously fully explained was given to the cell at this time to insure health of the cell (Thorhaug (20)). If the cell appeared in excellent condition indicated by a potential over 8 mV, a light response of 1 mV or more, no aplanospore formation or plasmolysis, the temperature controls were turned on. The cell was allowed a time to come to equilibrium at this temperature and the tracer flow was begun. The light was natural Northern daylight since light is known to be a stimulant to active transport. The thermal coefficient of the system was 0.1 mV/°C. Drift was 0.05 mV/hour.

Electrical Measurements

Figure 2 shows the electrical circuits. Potential difference between vacuole and external sea water were measured to ± 0.05 mV through calomel electrodes by using a Keithly high impedance potentiometer 660Å (Y). The inflow syringe and the bathing media were connected to calomel electrodes by glass salt bridges (0.5 M KCl) fitted with vycor porous plugs. The resistance of the cell was measured by a General Radio conductance bridge.

The shunt circuit consisted of the outflow pipet, 0.5M KCl-agar bridge, Ag-AgCl electrode connected to a battery voltage divider and microammeter as prescribed by Ussing and Zerahn (21). The potential measuring and shunt circuits resistances were approximatly 400 and 200 k ohm respectively. The cell was bathed inside and outside with 75% artificial sap and 25% seawater. The assymmetry of the system was taken, the potentiometer offset by that amount and then sufficient current was applied to reduce the potential difference to zero, at 5 minute intervals.

ION FLUX

Radionuclide Measurements

The flow of tracer (Tho, Na^{22}, K^{42}, Cl^{36}) into or out of the cell could be measured over long periods. The outflow was seen by placing the tracer in the inflow syringe and taking samples from the unidirectional fluxes from the bathing sea water at discrete time periods. The inflow was measured by placing the tracer in the bathing solution and taking samples for unidirectional fluxes from the outflow pipet at discrete time periods. A discussion of the assumptions behind flux of water, sodium, potassium and chloride in Valonia is found in Aikman and Dainty (22): Gutknecht (13).

In order to ascertain the validity of the method we compared at 25°C (inside and outside fluids) for water diffusional permeability of the membrane system and found close comparison to those of Gutknecht (23): 1.2×10^{-4} cm. sec^{-1}.

RESULTS

Results of bioelectric measurements showed that between 15 and 30°C, the potential difference of two species, Valonia macrophysa and V. ventricosa, remained remarkably constant with very little standard deviation from the mean, Figure 3&4, Thorhaug (20). Both Species show a noted increase in potential difference above 30°C with an increase in standard deviation. The current models predict relatively simple continuous temperature dependencies for bioelectric potentials, yet abrupt changes were observed below 15 and 30°C. This indicates that the temperature coefficient of the resting potential of Valonia is not simply related to the RT/F term of the Nernst or Goldman equations.

There is a remarkable stability of the potential of both cells between 15 and 30°C. These results were reversible for short time periods.

When replotted into an Arrhenius plot of $\frac{1}{T}$ versus $\Delta\Psi/T$, (Figure 5) Thorhaug (24), the stability between 15 and 30°C is

$$\frac{\partial \Delta \Psi}{dT} = 0$$

in this middle range of temperature, whereas it becomes positive below and above this temperature range, which was postulated to be dependent on the active flow of materials in a theoretical analysis (Thorhaug(24).

Preliminary results from the short circuit measurements which many consider to be a relative measure of the active transport process, are seen in Figure 6. These cells were heated rapidly to the given temperature and then held at that temperature for long time periods. At 33°C there was a marked increase in short circuit current I_{sc} followed by a decrease to a plateau held for more than an hour and then a gradual decline to 0 short circuit and 0 potential at about 3 1/2 hours, indicating cell death. At 35°C a large increase in short-circuit current was seen in the first half hour followed by an abrupt decline for about 20 minutes and then a slower decline with cell death at about 3 1/4 hours. At 36°C there is an extremely marked peak with a rapid decline in the first hour. Then a more gentle decline for about 40 minutes followed by another abrupt decline to death after about 1 3/4 hours. At 38°C the peak in little less than one hour is far lower than the 36°C peak and also lower than the 33° or 35°C peaks. A rapid decline resulted in cell death within about 1 1/2 hours. There was no plateau present. This result was reversible for short time periods such as 1/2 hour.

These are preliminary results now being confirmed by radiotracer experiments.

DISCUSSION

Let us review the effect of high lethal temperature on <u>Valonia</u>. Bioelectric potential measurements made at continuous intervals abruptly increase in two species of <u>Valonia</u> above 30°C. These are reversible during short time periods. Such results are not described by current theoretical models, which would indicate a smooth, continuour temperature dependency for the bioelectric potential. A theoretical treatment based on non-equilibrium thermodynamic framework came to the conclusion that within the range of beneficial temperatures, reactions at the tonoplast and vacuolar membranes balance to give a

$$\frac{\partial \Delta \psi}{dT}$$

near 0. Above the boundary of heat death the reaction-linked flows of materials increase. Supporting this notion are short circuit current results, presented here, showing abruptly rising peaks at temperatures above the lethal limit of 31°C, preliminarily show increase in active transport above the boundary of heat death. These must be substantiated with radiotracer studies.

The large change in electrical properties (see Ginzberg and Hogg (26) for a discussion of short-circuit current) of the Valonia cell correlate closely with the temperature lethal limits of Valonia. Investigations of these cells using a polythermostat (multitemperature aluminum bar precision bored to hold 30 replicate sets of cuvetts) showed lethal limits of 5 Valonia species to be very near 15 and 31°C .(Figure 7). The thousands of individual cells used for these experiments were taken from various parts of the tropics from Bermuda (mean annual temperature 22.5°C) to Puerto Rico (mean annual temperature 28.5°C) Figure . No temperature differences in mortality were seen despite acclimation at different temperatures. Temperature limits were very abrupt, falling from almost no mortality to 100% mortality with 1 to 2°C. Other species of tropical macroalgae such as Penicillus, Udotea, Rhipocephalus and Laurencia showed approximately the same results.

These results were borne out through field studies. After three years of field studies of this area with two fossil fuel plants on line, we have found that the natural high summer mid-day temperature can reach 30.5°C falling to about 27.5°C at night. The artificially elevated effluent from the power plants had the following effect: In an area of +5°C above ambient, the normal grass and macroalgae died and instead a blue-green algal population became dominant, although it was not a suitable food base for the normal food web. In the +4°C area, all the macroalgae except two intertidal forms disappeared as well as the dominant grass, Thalassia testudinum. The +3°C area had algal death in the late spring when temperatures rose above 31°C; regrowth of short-lived forms was seen in the fall when temperatures fell below 30°C. When the heat was reduced during the past year by greater volume of cooling waters, regrowth occurred in some areas. In uneffected, control areas, the productivity (g dry weight produced/m^2/day) fell abruptly above 29°C (Thorhaug, Stearns and Pepper (27)). Thus, the tropics appear to be a place of stress during the summer months and may be more vulnerable to other pollutants than temperate marine waters, Figure 8.

The excellent fit between the membrane bioelectric potential short circuit current and the long-term metabolism of related algae in the field and laboratory studies would argue for the examination of bioelectric potential as a possible simple laboratory index for other pollutants on algal populations. As previously stated, there is a rather extensive literature on the effects of many chemicals at various concentrations and physical factors on marine algae, and giant algal cells in particular (reviewed in Blinks (2)). It appears that these organisms are quite sensitive to many pollutant conditions, which are reflected in the bioelectric phenomena: anoxia and high ammonia content are examples. However, a warning is issued by Todd (28) that "unnatural" pollutants (i.e. chemicals not usually found in the geological history of the group) may have quite a different effect on metabolism from pollutants which occur at some level in the usual geological environment of the group being examined. An example of the former would be DDT; the latter, extreme temperature. This would be an interesting path of investigation to follow for bioelectric properties. Both lethal and sublethal effects might be examined for a variety of physical and chemical pollutants in this rapid and relatively inexpensive manner.

ACKNOWLEDGEMENTS

The authors would like to express appreciation for the enthusiasm over these ideas and their guidance in the theoretical aspects of the late Dr. Aharon Katchalsky. Dr. John Gutknecht and Mr. David Hastings were kind to help us assemble the voltage clamp apparatus.

Support by the National Science Foundation, (GB033712), The U. S. Atomic Energy Commission (AT(40-1)-3801) and the National Aeronautic and Space Administration (NGL 10 007 010) are gratefully acknowledged.

REFERENCES

1. Osterhout, W.J.V. 1972. Injury, recovery and death in relation to conductivity and permeability. Lipencott, Philadelphia, Pa. 250 p.

2. Blinks, L. R. 1951. Physiology and biochemistry of algae. pp. 263-291. IN: Manual of Phycology - an Introduction to the Algae and their Biology. G. M. Smith (ed.). Chronica-Botanica, Co. Waltham, Mass.

3. Cole, K. S. 1968. Membranes, Ions and Impulses. Univ. of Calif. Press, Berkeley. 568 p.

4. Dainty, J. 1969. The ionic relations of plants. pp 453-486 IN: The Physiology of Plant Growth and Development. M. B. Wilkins (ed.) McGraw-Hill, London.

5. MacRobbie, E.A.C. 1970. Active transport of ions in plant cells. Quart. Rev. Biophysics. 3: 252-281.

6. Wolfe, D.A., G.W. Thayer, and R.B. Williams. 1972. Ecological effect of man's activities on temperate estuarine eelgrass communities. IN: Critical Problems of the Coastal Zone. B. Ketchum (ed.) M.I.T. Press, Cambridge, Mass. (in press).

7. Blinks, L. R. 1955. Photosynthesis and productivity of littoral marine algae. J. Mar. Res. 14: 363-373.

8. Ryther, J. A. 1963. Geographic variations in productivity. pp.347-380. IN: The Sea. Vol. 2, M.N. Hill (ed.) Interscience Bull. New York.

9. Mann, K. H. 1972. Ecological energetics of the sea-weed zone in a marine bay on the Atlantic coast of Canada. II. Productivity of seaweeds. Mar. Biol. 14(3): 199-209.

10. Thorhaug, A. and R. Stearns. In press. An ecological study of Thalassia testudinum in unstressed and thermally stressed estuaries. Ecology.

11. Dainty, J. 1962. Ion transport and electrical potentials in plant cells. Ann. Rev. Plant Physiol. 13: 379-402.

12. Gutknecht, J. 1966. Sodium, potassium, and chloride transport and membrane potentials in Valonia ventricosa. Biol. Bull. 130: 331-344.

13. Gutknecht, J. 1967. Ion fluxes and short-circuit current in internally perfused cells of Valonia ventricosa. J. Gen. Physiol. 50(7): 1821-1834.

14. Cole, K. S. 1949. Dynamic electrical characteristics of the squid axon membrane. Arch. Sci. Physiol. 3: 253-258.

16. Hodgkin, A. L., and Huxley, A. F. 1952c. The dual effect of membrane potential on sodium conductance in the giant axon of Loligo. J. Physiol. (London)116: 497-506.

15. Hodgkin, A.L. and A.F. Huxley. 1952a. Currents carried by sodium and potassium ions through the membrane of the giant axon of Loligo. J. Physiol. (London) 116: 449-472.

17. Hodgkin, A.L. and A.F. Huxley. 1952d. A quantitative description of membrane current and its application to conduction and excitation in nerve. J. Physiol. (London) 117: 500-544.

18. Moore, J. W. and K. S. Cole. 1963. Voltage clamp techniques. pp 263-321. IN: Physical Techniques in Biological Research. Vol. VI. W. L. Nastuk (ed.).Academic Press, N. Y.

19. Spyropoulos, C.S. 1959. Miniature responses under "voltage-clamp." Am. J. Physiol. 196: 783-790.

20. Thorhaug, A. 1971. Temperature effects on Valonia bioelectric potential. Biochem. Biophys. Acta. 225: 151-158.

21. Ussing, H. H. and K. Zerahn. 1951. Active transport of sodium as the source of electric current in the short-circuited isolated frog skin. Acta Physiol. Scand. 23: 110-122.

22. Aikman, P. P. and J. Dainty. 1966. Ionic relations of *Valonia ventricosa*. pp. 37-50. IN: Some Contemporary Studies in Marine Science. H. Barnes.(ed.). Allan and Unwin, London.

23. Gutknecht, J. 1967. Membranes of *Valonia ventricosa*: apparent absence of water filled pores. Sci. 158(3802): 787-788.

24. Thorhaug, A. 1971. Temperature controlled perfusion technique for the study of giant algal membrane systems. Proc. 1st Europ. Biophys. Cong. 8(E49): 419-428.

25. Thorhaug, A. 1972. Laboratory thermal studies: IN: An Ecological of South Biscayne Bay and Card Sound, Florida. R. G.Bader and M. A. Roessler (ed.). Prog. Rpt. to U.S. Atomic Comm.

26. Ginzberg, B. Z. and I. Hogg. 1967. What does a short-circuit current measure in biological systems. J. Theoret. Biol. 14: 316-322.

27. Thorhaug, A., R. Stearns and S. Pepper. 1972. The effect of heat on *Thalassia testudinum* in Biscayne Bay, Florida. Acad. Sci. (In press).

28. Todd, J.H., D. Engstrom, S. Jacobson and W.O. McLarne. 1972. An introduction to environmental ethology: a preliminary comparison of sublethal thermal and oil stress on the social behavior of lobsters and fishes from a freshwater and a marine ecosystem. Report to U.S.Atomic Energy Commission.

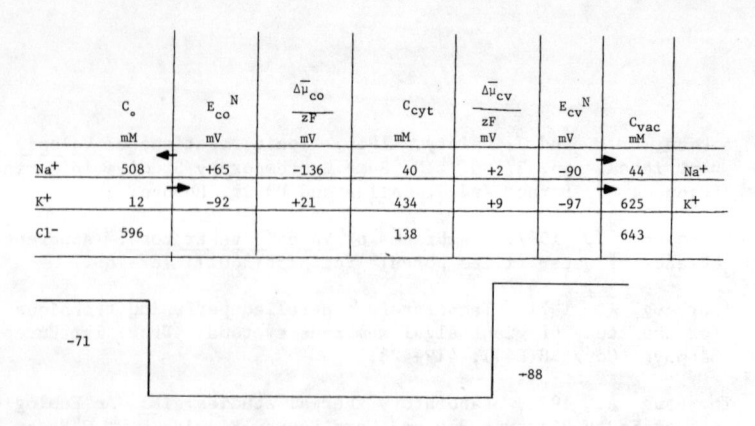

	C_o	E_{co}^N	$\dfrac{\Delta\bar{\mu}_{co}}{zF}$	C_{cyt}	$\dfrac{\Delta\bar{\mu}_{cv}}{zF}$	E_{cv}^N	C_{vac}	
	mM	mV	mV	mM	mV	mV	mM	
Na⁺	508	+65	−136	40	+2	−90	44	Na⁺
K⁺	12	−92	+21	434	+9	−97	625	K⁺
Cl⁻	596			138			643	

−71

−88

Figure 1. Ionic relations and potential differences for <u>Valonia ventricosa</u> redrawn from Gutknecht (12).

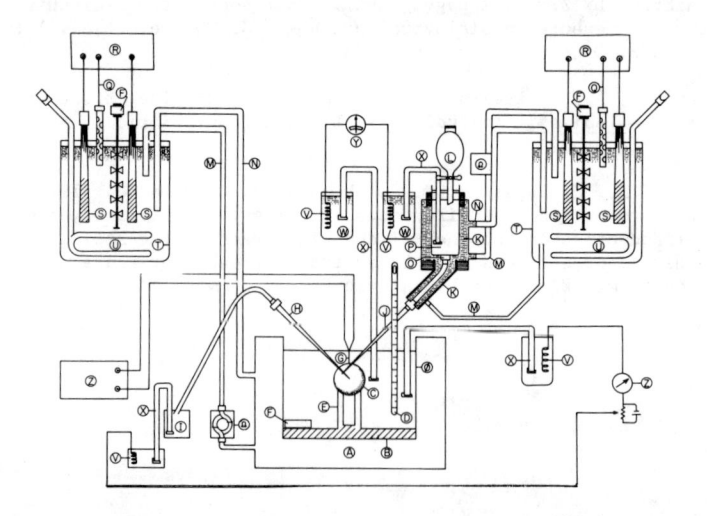

Figure 2. Assembly for <u>Valonia</u> perfusion with temperature control. A. Aluminum block. B. Mirror fitted to bottom of dish. C. <u>Valonia</u> cell. D. Thermometer. E. Glass stand for supporting cell. F. Stirring bar. G. Thermistor needle impaled into cell. H. Efflux pipet I. Influx pipet. J. 0.5 M KCl solution. K. Temperature jacket for influx fluid. L. Reservoir for inflow sap. M. Connecting inflow tubing from temperature control bath. N. Connecting outflow tubing to temperature controlled bath. O. Teflon adapter for syringe containing inflow fluid. P. Artificial sap. Q. Thermistor for temperature control apparatus. R. Temperature control apparatus. S. Knife blade heater. T. Temperature bath. U. Cooling coils. V. Electrode. W. KCL solution. X. Salt bridge with vycor tips. Y. Voltmeter. Z. Conductance bridge. Ω - pump. φ - Absorption dish.

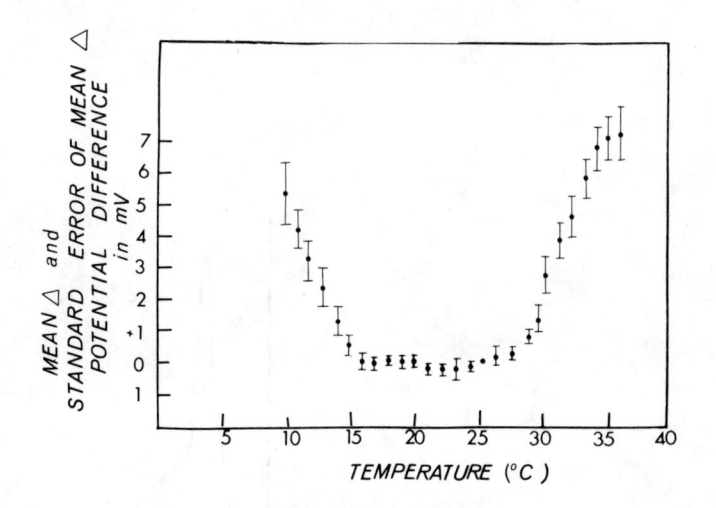

Figure 3. Change of mean potential difference from resting potential and standard error of mean change for 30 <u>Valonia</u> <u>ventricosa</u> cells versus temperature.

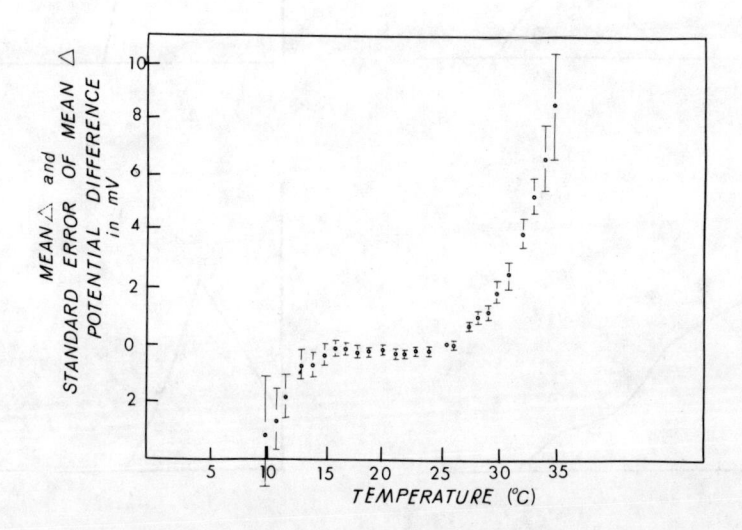

Figure 4. Change of mean potential difference from resting potential and standard error of mean change for 25 <u>Valonia</u> <u>macrophysa</u> cells versus temperature.

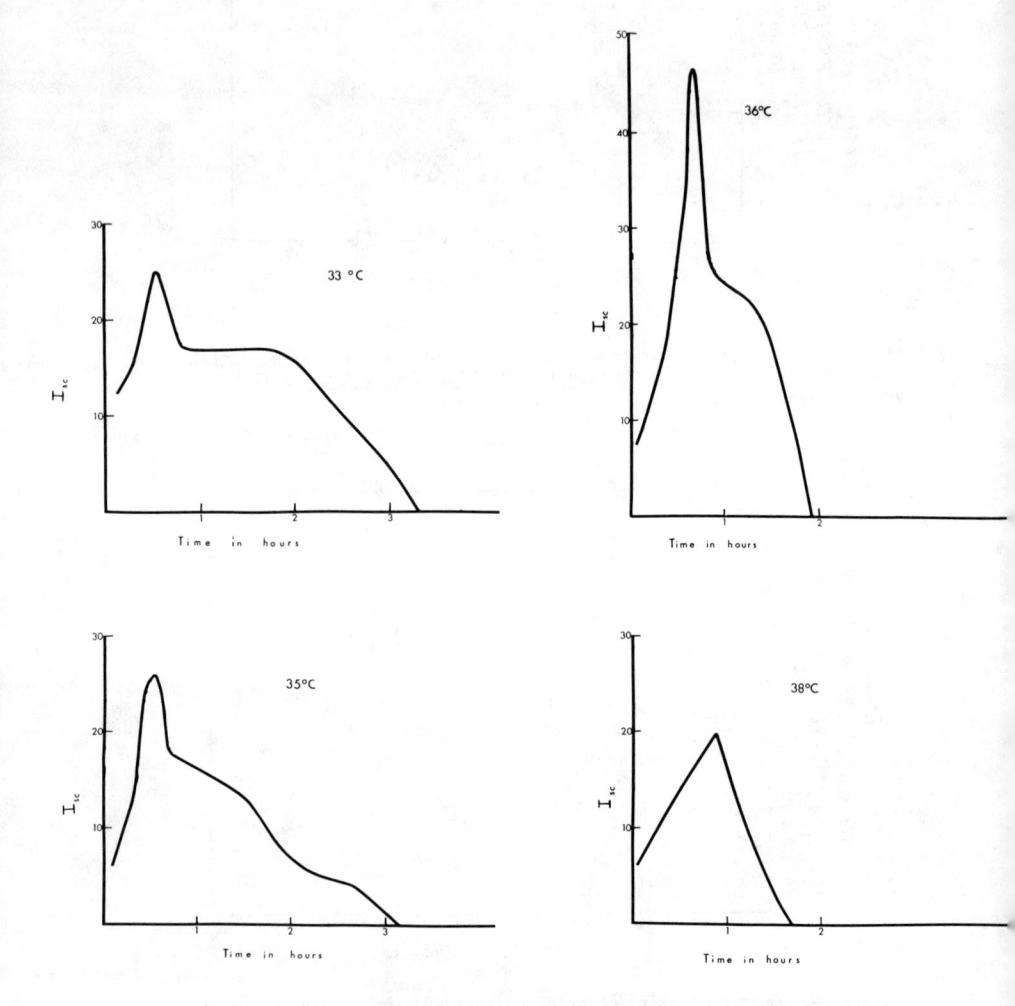

Figure 5. Short circuit current of <u>Valonia ventricosa</u> cells at various temperatures plotted as a function of time.

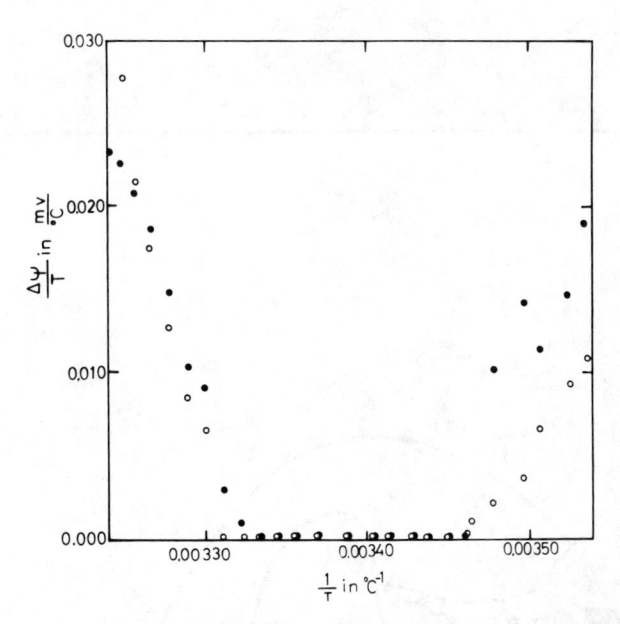

Figure 6. Plot of data from figure 3 (black dots) and 4 (open dots) for the potential difference across <u>Valonia</u> membrane system as a function of temperature. Change of potential per degree celcius is plotted versus reciprocal temperature.

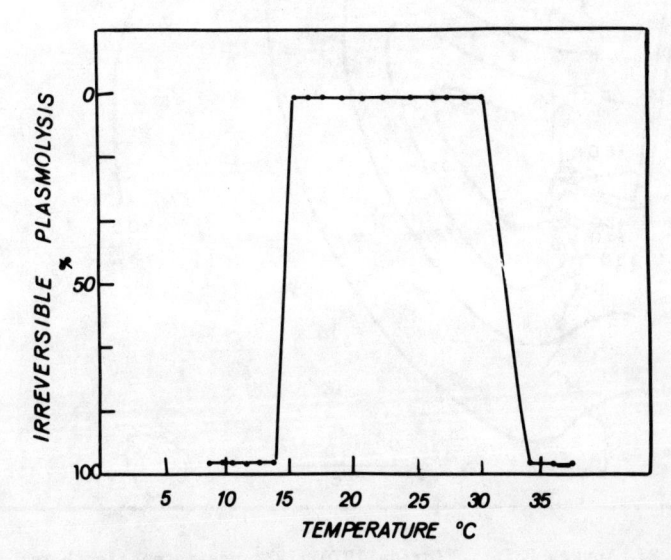

Figure 7. Irreversable plasmolysis versus temperature for <u>Valonia macrophysa</u> after 3 days exposure to the given temperature. Each point represents 40 cells.

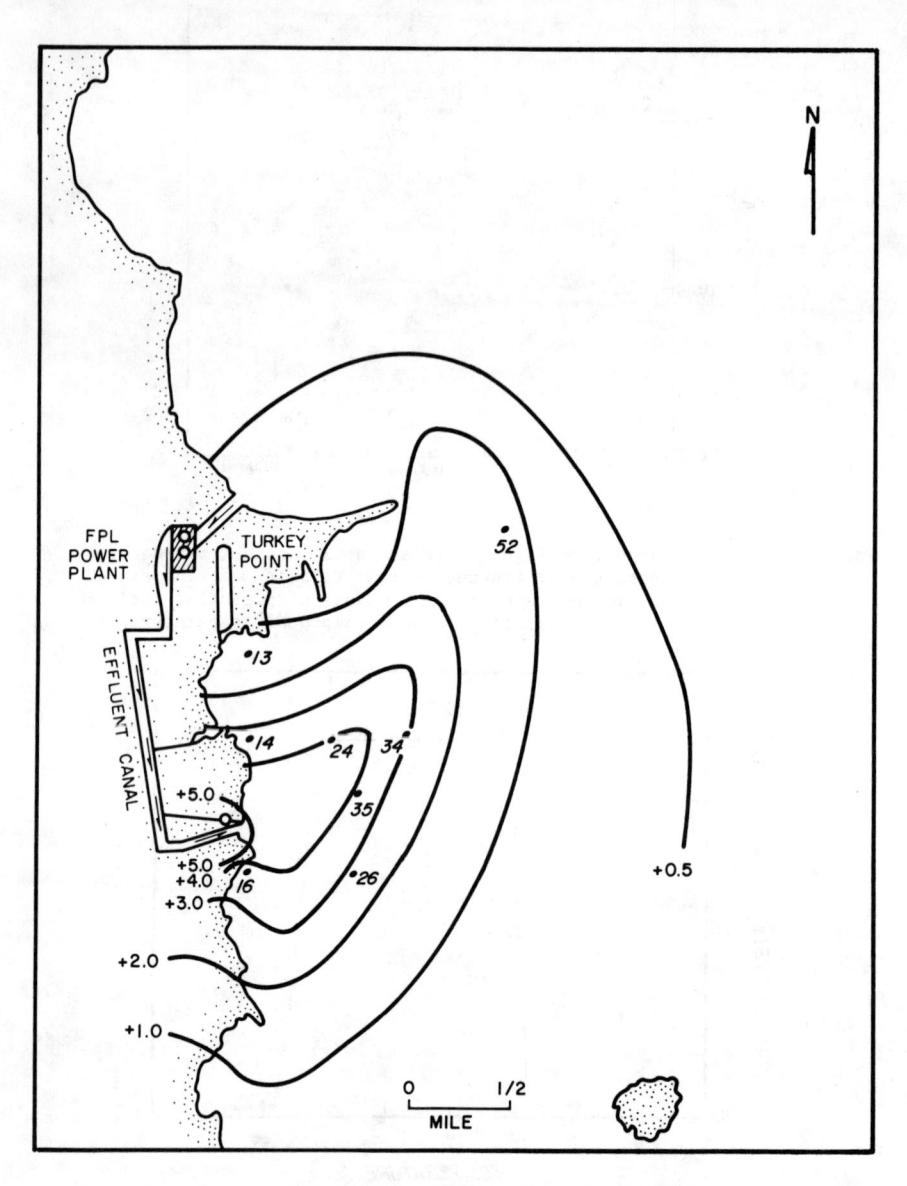

Figure 8. Turkey Point, Florida 1970, 1971 presistent isotherms in °C above ambient resulting from heated effluent of two fossil fuel generating plants. Stations observed byweekly are indicated by station number.

A REVIEW OF LEAD, SULFUR, SELENIUM AND MERCURY
IN PERMANENT SNOWFIELDS

H. V. Weiss
Naval Undersea Center
San Diego, California 92132

These investigations set out to determine whether the
flux of certain elements from the continents to the atmosphere has been influenced by man. Currently the lead
concentrations in Greenland are 500-fold greater than the
natural background. The primary source of atmospheric lead
is through lead aerosols generated in the combustion of
automotive fuel. Through the combustion of fossil fuels
a quantity of sulfur now is introduced into the atmosphere
which is about equal to the natural atmospheric component
derived from such processes as volcanism and biological
degradation. Selenium is mobilized in the atmosphere to a
much lesser extent than is sulfur in the combustion of
fossil fuels. This difference is ascribed to the chemical
behavior of their respective tetravalent oxides. Significantly higher rates of deposition of mercury in Greenland
snow has occurred in recent years. Part of this increase
may come about by an enhancement in degassing of the earth's
crust.

INTRODUCTION

Has the flux of elements from the continents to the atmosphere
been influenced measurably by the activities of man? Such processes
as the combustion of fuels, the roasting of ores, or the production
of chemicals with catalysts could, in principle, compete with other
natural phenomena to cause the injection of elements such as mercury,
lead, selenium and sulfur into the air. The purpose of this
investigation was to probe this possibility.

Permanent snowfields record the introduction of matter into the
atmosphere. These ices contain in essentially unchanged condition
both the water and the accumulated solids that precipitated from the
air as a function of time. Thus, it is possible to draw inferences
regarding the chemical composition of the atmosphere as it existed
centuries and even millennia ago through the analysis of this material,
particularly since the various strata are susceptible to reliable
dating by a variety of techniques including ^{210}Pb and fission product
geochronologies, the isotopic analysis of oxygen, and firn stratigraphy.

Dated glacial samples collected by C. C. Patterson (1) from Greenland and Antarctica were available for these studies. The Greenland samples were from the Camp Century area (77°10'N, 61°07'W) and from a virgin site 80 km east by southeast of this location. The older samples of the Greenland ice sheet were recovered from the walls of an inclined shaft at Camp Century, and the samples of more recent ages were taken from the walls of open trenches at the virgin site. The Anarctica samples were gathered from the Erebus Glacier Tongue (77°42'S, 166°40'E), the Meserve Glacier (77°32'S, 162°25'E), the New Byrd Station (80°01'S, 119°31'W) and a virgin site (78°52'S, 111°19'W). The samples were melted and stored in sealed polyethylene containers in 1966 immediately after removal from the glaciers.

LEAD (1)

The lead concentrations in permanent snowfields was sought since evidence appears to exist for the widespread atmospheric dissemination of this element through the activities of man. The upper layers of the ocean appear to be polluted with lead. Off-shore locations adjacent to industrial areas show greater lead concentrations than deep waters (2) and this effect is less pronounced in the open oceans (3).

In the lead determination ^{208}Pb tracer was mixed with a definite volume of the acidified sample. The ^{207}Pb in the melted ice which originated from the sea salts, terrestrial dusts and industrial pollutant aerosols mixed isotopically with the ^{208}tracer. The lead was isolated by a series of steps involving liquid extraction and ion exchange and the isotopic analysis was performed by mass spectrometry. The quantity of ^{207}Pb in the sample was determined from the ratio of ^{207}Pb/^{208}Pb observed.

The results of the analysis show that lead concentrations have increased markedly in Northwest Greenland snows over the last several centuries (Figure 1).

A measurement of the oldest ice from Camp Century (1753 A.D.), a time which coincides with the beginning of the European Industrial Revolution, shows that the lead value is already about 25-fold greater than natural levels. Lead concentrations tripled in Camp Century snows from 1753 to 1815. They appear to double again during the following century and over the three decades from 1933 - 1965, lead concentrations again rose by a factor of three. Currently lead concentrations in the Greenland snows are 500-fold greater than the natural background.

The levels of lead in the Antarctic snows are considerably lower than those of Camp Century (about a factor of ten); however measurements suggest that increases in lead concentration by a factor greater than 300 have occurred in this polar region during recent decades.

The primary source of lead put into the atmosphere today by man is through lead aerosols generated in the combustion of automotive fuel, while in the past the major contribution resulted from lead smelting. Estimates of lead aerosol production in the Northern Hemisphere compared with lead concentrations in Camp Century snow at different times are shown in Table I. The amount of lead deposited at Camp Century appears to be reasonably related to these aerosol production values.

SULFUR (4)

Currently over 3×10^7 tons of sulfur dioxide are probably being released into the atmosphere of the United States from fossil fuel burning. The United States consumes about 35 per cent of the fossil fuels; therefore, a world value of about 10^8 tons per year represents an estimate for the flux of sulfur dioxide into the atmosphere. This investigation sought to determine whether a flux of this magnitude could be identified in the global atmospheric budget.

An approach to this problem involves analyses of sulfate/chloride ratios in permanent snowfields. The chloride is presumed to be oceanic in origin, since the quantity of sodium, chloride, magnesium and calcium relate in these materials to each other in the same proportion as in seawater (1).

Sulfate on the other hand is produced both by nature and by man. The ratio of sulfate to chloride in seawater is 0.15. This ratio is an order of magnitude higher in rains and in pre-1900 polar snows and this difference of input into the atmosphere is attributable to sources which are other than marine. Significant sulfur sources seem to be volcanic activity and the release of hydrogen sulfide through the bacterial degradation of organic matter. The hydrogen sulfide is transformed first into sulfur dioxide and subsequently into sulfate. In the temperate areas, the sulfate is incorporated in atmospheric precipitation and returned to the earth in a period of about a week.

The analysis of chloride was determined spectrophotometrically by the mercuric thiocyanate method (5). Sulfate was also analyzed spectrophotometrically by a modified barium chloroanilate technique (4).

Sulfate values, corrected for the sea salt component for snows deposited in Greenland over the last six centuries are shown in Figure 2. Changes in this value are not apparent until 1964 - 1965. This Figure also shows the world production of thermal energy from fossil fuels. A sharp increase in power production occurred in the 1940's. The introduction of sulfur dioxide into the atmosphere through this process appears to be of the same level as that evolved from natural processes during the last decade.

Expression of the results as a ratio of sulfate to chloride indicates that in the Northern Hemisphere the sea-salt contribution and the combined continental volcanic and bacterial degradation inputs to Greenland snows had been relatively constant. The average of this ratio for 31 samples deposited prior to 1930 was 80 $\pm$ 20 ppb. The average chloride value is about 40 ppb and the sulfate/chloride ratio is 2 with a range which extends from 0.4 to 3.0. Since the seawater value for this ratio is only 0.15, the data suggest that the predominant sulfur input into the inland Greenland snows is by way of volcanic activity and bacterial organic degradation.

This thought is corroborated by the sulfate-chloride ratio of a coastal snow sample (dated 800 B.C.) in which the sulfate to chloride ratio was only 0.3. If the sea-salt contribution is subtracted, a a value of 80 ppb for sulfate remains. This value agrees with that obtained for the volcanic and organic derived sulfate observed in the older snows recovered from the interior of Greenland.

A co-variance in the concentrations of lead and sulfate during the 1964 - 1965 seasons is evident (Fig. 3) and this observation suggests that both of these pollutants could have comparable sites of injection into the atmosphere.

The quantity of lead entering the atmosphere of the Northern Hemisphere is estimated to be 10^5 tons/year (1). About half of this amount originates in the United States. In the burning of coal, oil and lignite, in the U. S. 4.6×10^7 tons of sulfate/year are discharged into the atmosphere. Thus a ratio of sulfate to lead of 920 is derived for the artificial injection of these species into the atmosphere in the U. S. This ratio for snows in Greenland is 600 (determined from average values of 0.2 and 120 ppb for lead and sulfate, respectively). The reasonable agreement between the lead and sulfate production and glacier concentration ratios indicates that the sources of lead and sulfate discharged into the atmosphere by man are similar.

SELENIUM (9)

The investigation of glacial material suggested that the amount of sulfur that enters the atmosphere from fossil fuel combustion is of the same order of magnitude as that which has resulted over historic time from the oxidation of organic matter and from volcanic activity (4). To determine whether the same is true of selenium was the purpose of this investigation. Selenium is a member of the group VIA elements in the periodic table, to which sulfur also belongs. Such an investigation might, therefore, also contribute to an understanding of the atmospheric chemistry of selenium as compared with sulfur. The value of measuring the concentration of closely related elements in geochemical studies is well recognized and, in relation to the current study, could extend our insight on the origin of sulfur in permanent snowfields.

Glacial samples were analyzed by neutron activation analysis (6) incorporating the technique of isotopic dilution. A known level of ^{75}Se activity was added to 1 - 4 liters of glacial waters. The water was evaporated to a small viscous drop and this material was irradiated with neutrons for 0.5 hours. Following the irradiation and a rapid radiochemical separation the beta-rays of 18 minute ^{81}Se were measured. The gamma rays of ^{75}Se were then counted to correct for losses of selenium that occurred in the initial evaporation and the radiochemical separation steps. The sensitivity of the procedure is about 1 nanogram and the error of replicate analyses, 9.0 per cent.

The Se content in glacial material of different ages extending in time from 800 B.C. to A.D. 1965 appears in Table 2. Included in the table are the S content and Se/S ratios for these samples. The pre-1960 samples have Se/S ratios varying between 2.9 and 5.9 x 10^{-4}; the post-1960 samples range in value from 1.1 to 1.6 x 10^{-4}. The 1960 sample appears to be transitional in value (2.2 x 10^{-4}).

The Se/S ratios in petroleums have been recently determined to range between 0.31 and 1.22 x 10^{-4} (8), and coal samples had an average value of 2.9 x 10^{-4} (9). Most crustal rocks have a Se/S ratio of the order of 1 x 10^{-4}. Similar values occur in snow, dust, and rain in the environs of the Massachusetts Institute of Technology. In general, the Se/S ratio in material of terrestrial origin is of the order of 10^{-4} (10), in sharp contrast with concentrations in seawater, for which this value is 10^{-7} (11). The relationship of Se and S in the glacial samples analyzed points to a terrestrial origin in both Greenland and Antarctica.

Several inductions concerning the atmospheric chemistries of Se and S can be derived from the glacial results and from the geochemical data given above. Fossil fuel combustion has not yet resulted in the long-range transport of Se as it has for S. The Se content has been relatively constant, but the S concentration in these glacial ices increased markedly in most of the post-1960 samples. This observation suggests a dilution of the Se by the S that was introduced from fossil fuel burning. As was previously noted (1, 4), the incorporation of atmospheric constituents into glacial matter is seasonally related; the lowest values occur in the fall. This study also reflects this effect (see Table 2). Therefore, while the absolute value of S in the 1964 fall sample is low, the Se/S ratio is of a magnitude that indicates S enrichment relative to Se.

The increase in S in glacial snows since 1960 has been uniquely attributed to fossil fuel combustion (4). The combustion of coal has been recognized as the primary source of S pollution. In 1960 an estimated 21 million tons of SO_2 pollution could be attributed to the combustion of coal and fuel oil, of which coal contributed 85 percent (12). Since the analytical values available for this fuel have a Se/S ratio similar to that of the pre-1960 glacial values, it appears that Se simply is not moved about the atmosphere after its input from fossil fuel burning.

Where and how do the differences in the geochemical behaviors of S and Se come about? It may be rewarding to examine a little more closely the two processes most probably responsible for the mobilization of these elements about the earth's surface: High temperature fossil fuel burning and low-temperature biological reactions. In high-temperature oxidation Se burns in air to form the white solid SeO_2. The SeO_2 is so readily reduced that even specks of oxidizable dust in the air react with it to form red elemental Se (13). On the other hand, S is oxidized to the relatively stable SO_2. Whether the high-temperature process is man-made or is natural, as in volcanic activity, this difference in chemistry between the elements would be manifest.

On the other hand, the low-temperature biological processes produce compounds of S and Se that are volatile. As organic compounds (whose identity is not yet established), Se is volatilized from land plants (14). Volatile organic Se compounds are also released in the process of plant degradation (14) (possibly as H_2Se). In either case, the very low concentrations in the atmosphere may survive against oxidation long enough to be transported in a similar manner as the gaseous S compounds H_2S and SO_2.

In summary, man initiates the translocation of compounds of S over great distances through fuel combustion, whereas Se is mobilized to a much lesser extent, perhaps owing to the differences in the chemical and physical properties of the tetravalent oxides.

MERCURY (15)

Environmental contamination by mercury is a current concern and it was of interest to determine whether its dissemination was so widespread to be manifest within recent deposits in permanent snowfields.

The analyses were performed by neutron activation. Samples were acidified, irradiated for an hour in a flux of 1.8×10^{12} neutrons $cm^{-2}\ sec^{-1}$ and submitted to a radiochemical purification specific for mercury.

Four days after the irradiation, the 77-kev gamma ray of ^{197}Hg was measured. The sensitivity of the measurement, which is defined by a count rate in excess of three standard deviations above the normal background count 4 days after the irradiation, is about 2 nanograms.

The mercury contents in glacial materials of different ages from 800 B.C. to A.D. 1965 (Table 3) indicate significantly higher rates of deposition of this element in recent years. Samples from waters deposited prior to 1952 had mercury contents ranging from 30 to 75 ng per kilogram of water with an average value of 60 $\pm$ 17. The mercury concentration in waters deposited between 1952 and 1965 averaged 125 $\pm$ 52 ng per kilogram of water with a range between 87 and 230.

This increased content of mercury in glacial snows in recent times is probably a reflection of man's impact upon the environment. What activities of human society could be responsible for the added inputs to the atmosphere?

Our approach to this problem is based upon oversimplified models of interactions between the earth's geospheres in which often fragmentary data on the geochemical parameters of mercury are used. Nonetheless, as a consequence of the following argument, we reach the conclusion that the mercury in the atmosphere probably arises predominantly from the degassing of crustal materials and that, if there is a measurable global impact by man upon the atmospheric burden of mercury, it may be through an enhancement of this degassing process.

The mercury content of unpolluted air appears to span the range of less than 1 to 10 ng/m^3 (16). If we use a value for the mercury content of 1 ng/m^3, and take the density of the atmosphere to be 1.3 x 10^3 g/m^3 and the weight of the atmosphere to be 5.1 x 10^{21} g, the atmospheric burden of mercury is 4 x 10^9 g. Rain effectively washes mercury from the atmosphere (17). If we take an average time between rains of 10 days, the flux of mercury from the continents to the atmosphere is 1.5 x 10^{11} g/year.

This calculation can be checked in the following way. The annual precipitation of the world is 4.2 x 10^{17} liters. The average content of mercury in the rainwater of the 1930's is 0.2 part per billion (ppb) (17). The flux of mercury from the atmosphere to the continents is 8.4 x 10^{10} g/year. If a mercury content of 0.06 ppb for apparently uncontaminated glacial ice (Table 3) is used, the flux is about 2.5 x 10^{10} g/year. The rate of exchange of mercury between the air and the solid surface of the earth appears to be in the range from 2.5 x 10^{10} to 1.5 x 10^{11} g/year.

The flux of mercury from the continents to the oceans by way of the rivers is much less than that of the continents to the atmosphere. If we use a value of 38 x 10^{15} liters as the annual river and ice-cap runoff and an upper limit of 0.1 ppb of mercury in rivers (18), an upper limit of transfer of 3.8 x 10^9 g of mercury per year is obtained.

The amount of mercury produced throughout the world in 1968 was 8.8 x 10^9 g. A substantial loss of mercury to the atmosphere occurs in chloralkali production, and about one-third of the mercury consumed annually in the United States is used in that process (19). If we apply this factor of 1/3 to the world consumption, the resultant flux is 3×10^9 g/year, a fraction of the flux involved in the degassing process.

On the basis of recent data on fossil fuel combustion, it has been computed that an upper limit of 1.6×10^9 g of mercury per year can be released to the environment through the burning of coal, oil, and lignites (20). Other industrial activities yield potential atmospheric fluxes of this order of magnitude. The amount of cement produced throughout the world in 1960 was 33×10^{13} g. Since cement, during its production, is heated to temperatures of 1400° to 1500°C, the mercury release from the limestone and shale components of the raw materials could conceivably discharge significant quantities of mercury to the environment. Since the mercury content of shales clusters around several hundred parts per billion and since the mercury content of limestones is usually less than 100 ppb, an upper limit of 300 ppb is taken as a conservative estimate of the mercury content of the building materials for cement. Thus, cement manufacture might result in the release of 10^8 g of mercury per year. Perhaps the roasting of sulfide ores releases significant quantities of mercury to the environment. Such mining operations involve 2×10^{14} g of ore per year (21). But even with a mercury content of 10 parts per million, this flux is only 2×10^9 g/year. Hence, a survey of industrial activity has not revealed mercury releases to the atmosphere that can rival that of the natural degassing rate, estimated to range between 2.5×10^{10} and 1.5×10^{11} g/year.

The following explanation is proposed to account for the doubling of the mercury content in a Greenland ice sheet. The background concentration of mercury in the atmosphere arises from the degassing of the upper mantle and lower crust. This view is supported by the observations that the mercury is markedly enriched in sediments as compared to igneous rocks (22). Moreover, the atmospheric mercury concentration is a function of barometric pressure (17). Increased exposure of such crustal materials can result in increased fluxes of mercury to the atmosphere. The variety of activities of man that result in greater exposure of the earth's crust through alteration of terrestrial surfaces allows more mercury vapor and more gaseous compounds to enter the atmosphere.

In conclusion, the recently measured mercury concentrations in such pelagic fish as tuna and swordfish are probably not far removed from the norm. The input of twice as much mercury to surface waters of the ocean in recent times can only increase the amount of mercury in the lower trophic levels by, at best, a factor of 2.

This argument can be approached in another way. There are 10^{12} g of mercury in the mixed layer of the ocean, if we assume a depth of 100 m and an average mercury content of 30 ng/liter. The increased mercury flux, 1.5×10^{11} g, presumed to result from an enhancement of earth degassing, would add annually only 15 percent to the mercury burden of this water body. With a residence time of mercury in this layer of as much as 5 years, the mercury content would be augmented by a factor of only 0.75. The increased mercury content in surface waters, if transmitted through the food web to such upper trophic levels as those of the swordfish and tuna, would, at best, double the mercury contents in these organisms.

Table 1. Lead aerosol production in the Northern Hemisphere compared with lead concentrations in Camp Century snow at different times *

Date	10^5 tons of lead smelted/ yr	Fraction converted to aerosols (%)	Tons of lead aerosols produced/ yr from smelteries (x 10^3)	10^5 tons of lead burned as alkyls/yr	Fraction converted to aerosols (%)	Tons of lead aerosols produced/ yr from alkyls (x 10^3)	Total tons of lead aerosols produced/yr (x 10^3)	μg lead per kg snow at Camp Century
1753	1	2	2	---	---	---	2	0.01
1815	2	2	4	---	---	---	4	0.03
1933	16	0.5	8	0.1	40	4	10	0.07
1966	31	0.06	2	3	40	100	100	0.2

* Taken from (1).

Table 2. Selenium and sulfur concentration in glacial samples. A single analysis was performed on the 1960 sample; the 1859 and 1965 (winter) samples were analyzed in triplicate, and all others in duplicate. The 1724 sample was recovered from Antarctica; the others, from a Greenland ice sheet. All dates in column 1 except the first are A.D.

Deposition of sample year	Se (ng/kg)	S (µg/kg)	Se/S x 10^4
800 B.C.	25.5 ± 1.6	56.0	4.5
1724	14.1 ± 1.6	26.0	5.5
1815	13.4 ± 2.6	26.7	5.0
1859	10.1 ± 1.1	34.7	2.9
1881	7.6 ± 0.3	21.7	3.4
1892	8.0 ± 0.2	26.0	3.0
1946	22.0 ± 2.3	37.7	5.9
1952	11.0 ± 0.2	35.7	3.1
1960	8.9	41.0	2.2
1964 (fall)	5.1 ± 0.5	38.7	1.3
1964 (Dec.)	9.7 ± 1.1	72.7	1.3
1965 (winter)	14.2 ± 2.2	98.0	1.5
1965 (spring)	8.0 ± 0.2	71.0	1.1
1965 (summer)	8.7 ± 0.3	53.7	1.6

Table 3. Mercury concentrations in the glacial samples. The sample deposited in 1724 was recovered from Antarctica, the others from Greenland. The sample deposited in the spring of 1965 was analyzed in triplicate, and the error shown gives the average deviation from the mean.

Time of deposition	Mercury content (nanograms per kilogram of water)
800 B.C.	62
1724	75
1815	75
1881	30
1892	66
1946	53
1952	153
1960	89
1964 (fall)	87
1964 (winter)	125
1965 (winter)	94
1965 (spring)	230 $\pm$ 18
1965 (summer)	98

REFERENCES

1. M. Murozumi, T. J. Chow, C. C. Patterson. Geochim. Cosmochim. Acta 33, 1247 (1969).

2. M. Tatsumoto, C. C. Patterson. Nature 199, 350 (1963).

3. T. J. Chow, C. C. Patterson. Earth Planet. Sci. Lett. 1, 397 (1966).

4. M. Koide, E. D. Goldberg. J. Geophys. Res. 76, 6589 (1971).

5. D. M. Zall, D. Fisher, M. O. Garner. Anal. Chem. 28, 1655 (1956).

6. H. V. Weiss, M. Koide, E. D. Goldberg. Science 172, 261 (1971).

7. H. V. Weiss. Anal. Chim. Acta 56, 136 (1971).

8. Y. Hashimoto, J. Y. Hwang, S. Yanagisawa. Environ. Sci. Technol. 4, 157 (1970).

9. K. K. S. Pillay, C. C. Thomas, Jr., J. W. Kaminski. Nucl. Appl. Technol. 7, 478 (1969).

10. Y. Hashimoto and J. W. Winchester. Environ. Sci. Technol. 1, 338 (1967).

11. D. F. Schutz and K. K. Turekian. Geochim. Cosmochim. Acta 29, 259 (1965).

12. D. Beinstock and F. J. Field. Air Pollut. Contr. Ass. 10, 121 (1960).

13. I. Rosenfeld and O. A. Beath. Selenium (Academic Press, New York, 1964), pp. 41 and 301.

14. B. G. Lewis, C. M. Johnson, C. C. Delwiche. J. Agr. Food Chem. 14, 638 (1966).

15. H. V. Weiss, M. Koide, E. D. Goldberg. Science 174, 692 (1971).

16. M. Fleischer. U.S. Geol. Surv. Prof. Pap. 213 p. 6, (1970).

17. J. H. McCarthy, Jr., J. L. Meuschke, W. H. Ficklin, R. E. Learned. Ibid., p. 37.

18. R. L. Wershaw. Ibid., p. 29.

19. *Mineral Industry Surveys* (U.S. Bureau of Mines, Washington, D. C., 1970).

20. K. K. Bertine and E. D. Goldberg. *Science*, *173*, 233 (1971).

21. L. J. Shannon, A. E. Vandegrift, P. G. Gorman. *Midwest Res. Inst. Rep. MRI-1023*, (1971).

22. K. H. Wedepohl, abstract of the International Geochemical Congress, Moscow, 20-25 July 1971, p. 936.

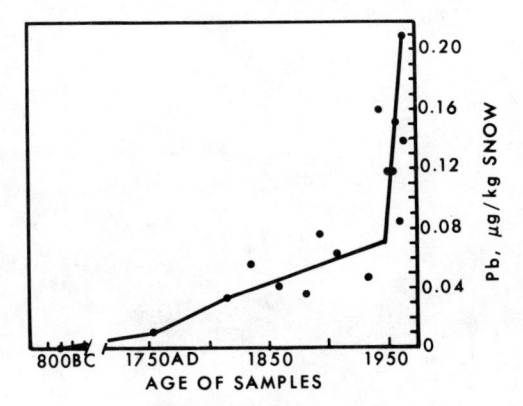

Figure 1. Lead concentrations in snows deposited in Camp Century at different times.

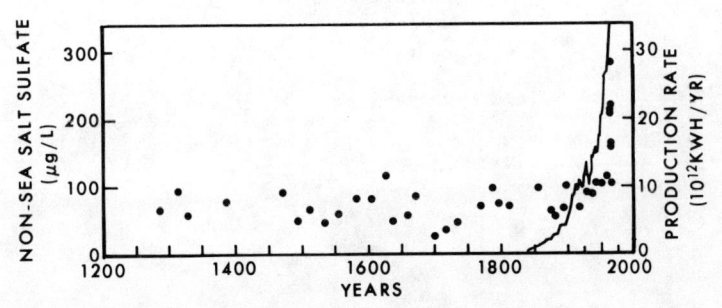

Figure 2. The salt-free concentration in northwestern glacier samples as a function of time. The solid curve represents production of thermal energy from coal, lignite and crude oil.

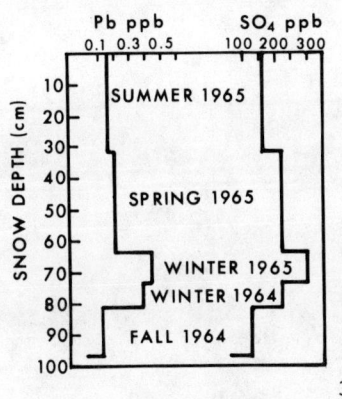

Figure 3. The sulfate and lead contents northwestern Greenland glacier a function of season during 1964-1965.

INDEX

A

Activation, 77
Activation volumes, 76
Active surface sites, 270
Activity coefficients, 115, 125, 127, 128, 129
Ag/AgCl electrodes, 38, 40, 99, 271, 335
Algae, 369
Aluminum, 246, 249
Ammonium ion, 249
Anionic speciation, 137
Anodes, 149
Anodic stripping voltametry, 14, 319
Aquaculture, 6
Arsenic, 344, 346
Artificial seawater, 116
Association constants, 135
Asymmetry potential, 100
Atomic absorption, 342, 346
Autec, 209

B

Baroclinic flow, 39
Barotropic flow, 39
Batteries,
  Activation, 204
  Cause of poor performance, 192
  Heat transfer, 181
  High charge rate, 181
  Power characteristics, 180
  Pressure compensation, 179
  Salt water, 141, 149
  Silver-zinc, 178
  Types of installation, 178
Bicarbonate, 12, 58, 116, 117, 333 ff
Bioelectric potential, 370, 372, 375
Biological productivity, 7
Boltzmann distribution, 40
Borate, 129
Bottom conductance, 36
Brackish water, desalination of, 278
Buffer, 336

C

Cables, submarine, 36, 37
Cadmium, 325
Calcareous deposit, 273
Calcium, 12
Calcium carbonate, 102, 130, 333
California, 37
Cantilever beam technique, 269
Cape Hatteras, 37
Carbon anhydrase, 334
Carbon cycle, 333
Carbon dioxide, 27, 58, 102, 199, 333, 334, 336
  , hydration, 333 ff
Carbonate saturometer, 102, 103
Carbonic acid, 333 ff
Carbonic anhydrase, 334, 335, 338
Cathode potential, 198
Cathodic polarization, 273
Cathodic protection, 224, 270
Cation hydration, 58
Cationic speciation, 137
Cavitation, 224
Cell pair,
  Area, 282
  Resistance, 281
  Voltage, 280
Cellular metabolism, 368
Charge-discharge performance, 169
Chelators, organic, 326
Chlorination, 247
Chlorine residual, 249
Chromium, 344
Clean-up, 356
Coastal currents, 11
Coastal waters, 5, 6, 24
Colligative properties, 9, 25
Compressibility, 55, 257, 258, 260
Computer modeling of seawater, 124
Concentration cell, 111, 114
  , incremental, 111, 116
Concentration polarization, 145, 199
Conductance, 77
Conductivity, 9, 25, 43, 59, 299
Conwed netting, 286
Copper, 130, 321, 324, 345
  Alloys, 220, 221, 247, 250
  Corrosion Index, 221

Corrosion, 15
　Crevice, tests, 223
　, rates in chlorinated seawater, 248
　"Accelerated" tests, 219
　, test program, 244
　Potential, 272
Costs, 283, 287, 347
　, electrodialysis apparatus, 288
Cuba, 37
Cuprous chloride cathodes, 141, 148
　Porosity of, 142
Current density, 198
Current efficiency, 281
Current meter, free-fall, 39, 46

D

Deep sea, 98, 99
Demister, 298, 303
Densimeter, 257, 258
Density, 8, 34, 42
Desalination, 242, 278, 296
　of brackish water, 278
　of sea water, 278
Desalting, degree of, 280, 284
　Factor, 298, 303
　Power requires, 282
Dielectric saturation, 261
Diffusion, 10, 61, 111, 112, 114, 118, 119, 144
Diffusion coefficient of Cl⁻, effective, 144
Diffusion layer thickness, 283
Discharge capability, 169
Dissolved solids, 300

E

Earth currents, 43
Earthquakes, 42, 44
Eddy currents, 42
"Electrode-less cell", 9
Electrode potential, 198
Electrodes, dropping mercury, 343
　, towed, 38, 45
Electrodialysis, 278-9
　High temperature, 279, 285, 286
　Design equations, 280
　Economics, 287
　Membrane, 287
　Spacer, 286
　Pressure limitation, 284

Electrolyte scrubber, cell design, 168
Electrometer, 101, 116
EMF, induced, 32, 33, 41, 45
Emission spectrography, 362
Encapsulation, 168
Energy storage, 188
English Channel, 32, 36, 37
Entrainment, 296, 297
Entropy, 260
　, ionic, 83
Equilibrium calculations, 126
Equilibrium model for seawater, 22
Erosion, 223
Euphotic zone, 24, 102

F

Faraday, M., 32, 33
Faraday's Law of induction, 33, 45
F/Cl ratio, 14
Flash Evaporator, 298, 299
Florida Current, 32-35
Florida Straits, 33, 35, 37
Fluorescence spectroscopy, 360
Fluoride, 14
Flywheel, high-speed, 188
Force, magnetic, on moving sea water, 41
Fouling, 247
Free energy, 127
Free-fall electromagnetic current meter, 39, 46
Fuel cells, 189, 209

G

Garrels and Thompson model, 112, 117, 118, 125
Gas chromatography, 361
GEK (Geomagnetic ElectroKinetograph), 11, 38, 40, 41, 45
Geochemical Ocean Section Study (GEOSECS), 7
Glaciers, lead, 385, 386
　, mercury, 385, 390, 391
　, selenium, 385, 389
　, sulfur, 385, 387, 388
Glass electrode, 98, 99, 333, 335
Glass membrane, 99
Glass transition temperature, 81
Gulf of Naples, 37
Gulf Stream, 7, 32, 33, 35, 37, 38, 42, 46

H

Hardware, compatibility, reliability, 167
Henderson's equation, 113, 114
Henderson's integration, 117
High pressure, 257
Horizontal motion in the sea, 33
Hydration, 80
Hydration number, 260
Hydrogen, electrolytically generated, 274
Hydrogen bond, 259, 260
Hydrogen embrittlement, 269, 275
Hydrogen evolution, 273, 275
Hydrogen sulfide, 249, 250

I

Industry associations, 244
Infrared spectroscopy, 81, 359
Instrumentation, 100, 101
Internal combustion engine, 189
International Biological Program, (IBP), 7
Iodate, 345
Iodide, 345
Ion activity, 13
Ion complexing, 112
Ion exchange membrane, 279, 287
Ion mobility, 60, 112, 113, 117, 118
Ion pair(s), 12, 13, 23, 25, 112, 58, 117, 118, 126, 129, 131, 225, 259
Ionic charge, 117
Ionic radii, 79, 260
Ionic speciation, 117, 118, 119
Ionic species, 344, 347
Ionic strength, 25
Ions, 113, 116, 117
Ion selective electrodes, 13, 14, 103
Irish Sea, 37
Iron, 272

J

Jones and Dole B-coefficient, 83

K

k-factor, 38
Key West, 37
Kinetic effect, 273
Kinetics, 333 ff
Kuroshio Current, 7, 37

L

Lanthanum fluoride electrode, 13-14
Lead, cathodically reduced
 Porosity, 142, 143
 Tortuosity factor, 144
Lead chloride cathodes, 141, 149
 Model of, 143
 Performance of, 147
 Porosity of, 142, 143
 Ultimate capacity of, 145
Lead chloride, reduction of
 Main period, 146
 Incubation period, 147
Lead, 321, 322, 325
 in glaciers, 385, 387
 in oceans, 386
Lead-acid battery, 169, 190
Lignand activities, 327
Limiting current density, 284
Liquid chromatography, 362
Liquid junction, 40, 111-119, 335
Liquid junction potential(s), 111, 112, 113, 115, 117

M

Magnesium, 12
Magnesium fluoride ion-pair, 14
Magnesium-silver chloride batteries, 141
Magnesium sulphate, 13
 ion pairs, 25, 60
Magnetic anomalies, 43-44
Magnetic field,
 of earth, 32, 33, 39, 42, 43
 of Gulf Stream, 42
 tidal, 42
Magnetometer, 44
Magneto-telluric method, 43, 44
Marine sediment, 111
Mass spectrometry, 361
Mass transfer, enhanced, 284

Materials Test Center, 243
Membrane potential, 372
Membranes, 287
Mercury, 390
  in glaciers, 385, 390, 391
Miami, 32, 33
Microalgae, 369
Microbial oxidation, 356
Mid-Ocean Dynamic Experiment (MODE), 7
Mid-ocean ridges, 44
Monocell Design; Battery Layout, 201
Mother ship installation, 203
Multi-stage flash evaporator, 296

N

Nansen bottles, 343
National Oceanic and Atmospheric Administration (NOAA), 10
National Oceanographic Instrumentation Center, 10
Navy, 5
Neutral buoyancy, 208
Nitrogen, 27
North Atlantic, 32, 37
Nuclear Magnetic Resonance, 361
Nutrients, 26

O

Ocean,
  dimensions, 34
  stratification, 34
  temperature, 34
  volume, 32
  circulation, 7
  currents, 7, 8
Oceanography, 342
Office of Saline Water, 243, 288
Oil
  spills, 359, 356
  pollution, 355, 357
  trace elements in, 362
  electrochemical analysis of, 363
  constitution of, 357
  bulk properties of, 359
Overvoltage, 143
Oxidation potential, 248
Oxygen, 27
  dissolved, corrosivity of, 246
Oxygen-temperature interrelation, 247

P

Pacific coast, 37
Pacific Ocean, 101
Partial molal ionic volume, 83
pH, 24, 62, 98-102, 333 ff
Photon activation, 14
Photosynthesis, 27, 28, 102
Piezometer, 55
Plant cells, 370
Polarization, 207, 270, 273, 274, 283
Polarization curves, 206
Polarography, 14, 320, 342, 346, 343, 347
Potable water, 242
Potential gradient, 33-34, 35, 45
  vertical, 39-40
Potassium sulphate, 13
Power density, efficiency, 170
Power losses; zinc anode, utilization, 200
Pressure, 53, 56, 57, 58, 59, 62, 76, 77, 126, 129, 131
Pressure compensated batteries, 116, 180
Pressure compensating fluids, 167
Pressure relief valve, 168
Probe, 101
Proton, 43

R

Radioactivity of seawater, 24
Radionuclide measurements, 374
Raman spectroscopy, 56, 57, 81
Rate constant, 333, 337, 338
Reducing electrolyte carryover, 193
Reference electrode, 98
Refractive index, 63, 64
Research vessel, 347
Resonance magnetometer, 44
Reversible electrodes, 100

S

Saling Water, Office of, 243
Salinity,
  variation with depth, 40
Salt bridges, sea water, 11, 39, 41, 46
Salt spray tests, 222
Santa Barbara, 355
Sargasse Sea, 37
Sea-mounts, 44
Sea salt, 258

Sea space ® power systems, 171
Seawater, 257, 269, 343, 346
  chemical composition of, 7, 8
    11, 22, 126, 246
  structure, 257
  conductance, 36
  volume, 32
  desalination of 278
  for testing, 245
  ions, 111
  thermodynamics, 127
Selenium, geochemical behavior, 390
  determination, 389
  in glaciers, 385, 389
Separator, 199
Sephadex elutions, 14
Shipboard, 346
Silver-zinc battery, 169, 170, 179, 190, 207
Single ion activity coefficients, 136
Skin depth, 42-43
Step loading, 271
Stratification, 34, 40
Stress corrosion cracking, 269
Stress intensity factor, 269, 270
Structural properties, 77, 78
Structure making and breaking ions, 82, 259, 261
Submarine Energy System, 196
Submersibles, small manned, 196
Sodium sulphate, 13
Specific volume of sea water, 55
Spectroscopic measurements, 56
Static permittivity, 62
Steel, 250
Sulfate, 12, 116, 117
Sulfide films, 221
Sulfur, geochemical behavior, 390
  covariance with lead, 388
  analyses, 387
  in glaciers, 385-388
Surface tension, 64
Suspended material, 23, 24
Swimmer delivery vehicle (SDV), 180
Synthetic sea water, 220

T

Temperature, 101, 126, 129
Thames River, 32
Thermal pollution, 369
Titanium, 273
  alloys, 269, 270
  corrosion potential, 272
  hydride, 270
Torrey Canyon, 355
Tortuosity factor, 144
Trace elements, 25, 130, 342, 344
  in oils, 362
Trace metals, 319
Transference number, 113
Transition state, 81
Transport numbers, 61, 77-79, 281
Tsunami, 42
Turbulence promoter, 284
Turbulent diffision, 15
Turbulent mixing processes, 10

U

Ultraviolet (UV) spectroscopy, 361
Uncontrollable chemical variables, 246
U. S. Coast Guard, 9

V

Valonia, 371
Velocity,
  conductivity-weighted average, 35-36, 38-39, 41-42, 45, 46
  surface, 38-39, 45
  true average, 36
Vine-Matthews hypothesis, 44
Viscosity, 61
Volume, 77

W

Water structure, 257
Woods Hole, 37

X

X-Ray fluorescence, 362

Z

ZENO; Monocell; Tactical, 195
Zinc, 321, 322
Zinc/Oxygen Cell, 197, 198